PHYSICAL ACOUSTICS

Principles and Methods

VOLUME VII

CONTRIBUTORS TO VOLUME VII

R. W. Damon

K. Dransfeld

M. Garbuny

Carl W. Garland

M. Gottlieb

C. K. Jones

D. H. McMahon

W. T. Maloney

J. A. Rayne

E. Salzmann

PHYSICAL ACOUSTICS

Principles and Methods

Edited by WARREN P. MASON

DEPARTMENT OF CIVIL ENGINEERING
COLUMBIA UNIVERSITY
NEW YORK, NEW YORK

and

R. N. THURSTON

BELL TELEPHONE LABORATORIES
HOLMDEL, NEW JERSEY

VOLUME VII

1970

ACADEMIC PRESS

NEW YORK AND LONDON

ACADEMIC PRESS, INC.
111 Fifth Avenue, New York, New York 10003

United Kingdom Edition published by
ACADEMIC PRESS, INC. (LONDON) LTD.
Berkeley Square House, London W1X 6BA

LIBRARY OF CONGRESS CATALOG CARD NUMBER: 63-22327

PRINTED IN THE UNITED STATES OF AMERICA

CONTENTS

1

Ultrasonic Attenuation in Superconductors: Magnetic Field Effects

M. GOTTLIEB, M. GARBUNY, and C. K. JONES

2

Ultrasonic Investigation of Phase Transitions and Critical Points

CARL W. GARLAND

3

Ultrasonic Attenuation in Normal Metals and Superconductors:
Fermi-Surface Effects

J. A. RAYNE and C. K. JONES

4

Excitation, Detection, and Attenuation of High-Frequency
Elastic Surface Waves

K. DRANSFELD and E. SALZMANN

5

Interaction of Light with Ultrasound: Phenomena and Applications

R. W. DAMON, W. T. MALONEY, and D. H. McMAHON

CONTRIBUTORS

R. W. DAMON
Sperry Rand Research Center, Sudbury, Massachusetts

K. DRANSFELD
Physik-Department der Technischen Hochschule, Munich, Germany

M. GARBUNY
Westinghouse Research Laboratories, Pittsburgh, Pennsylvania

CARL W. GARLAND
Department of Chemistry and Center for Materials Science and Engineering, Massachusetts Institute of Technology, Cambridge, Massachusetts

M. GOTTLIEB
Westinghouse Research Laboratories, Pittsburgh, Pennsylvania

C. K. JONES
Westinghouse Research Laboratories, Pittsburgh, Pennsylvania

D. H. McMAHON
Sperry Rand Research Center, Sudbury, Massachusetts

W. T. MALONEY
Sperry Rand Research Center, Sudbury, Massachusetts

J. A. RAYNE
Carnegie-Mellon University, Pittsburgh, Pennsylvania

E. SALZMANN†
Physik-Department der Technischen Hochschule, München, Germany

†Present address: Rohde and Schwarz, Munich, Germany

PREFACE

This volume treats four themes of current interest in physical acoustics: ultrasonic attenuation in metals and superconductors, ultrasonic investigations of phase transitions and critical points, interaction of light with ultrasound, and high frequency elastic surface waves.

The interaction of a sound wave with conduction electrons makes an important contribution to the ultrasonic attenuation in metals at low temperatures. If the metal becomes superconducting, this contribution decreases dramatically below the critical temperature T_c, and drops to zero at $T = 0$. The ultrasonic attenuation in the absence of a magnetic field, in both normal metals and superconductors, is treated in Chapter 3, where the current theory is reviewed and recent experimental results are discussed. Like the behavior as a function of magnetic field, the dependence of attenuation on frequency and propagation direction in the absence of a field is intimately related to details of the Fermi surface.

Chapter 1 is concerned with the effects of a magnetic field on ultrasonic attenuation in superconductors. The behavior in type I superconductors, which exhibit an intermediate state, is different from that in type II superconductors, which exhibit a mixed state. While there appears to be a good qualitative understanding of ultrasonic attenuation in the intermediate and mixed states, there are still some interesting discrepancies, and this chapter reviews the present experimental and theoretical situation.

Chapter 2 is devoted to another currently exciting topic, the ultrasonic investigation of phase transitions and critical points. As is explained, much can be learned from a study of velocity and attenuation changes in the neighborhood of a transition. A general discussion of the theory of ultrasonic properties in critical regions is followed by detailed sections giving the theory and available experimental results for liquid–vapor critical points, binary liquid phase separations, ferroelectric and antiferroelectric transitions and ferromagnetic and antiferromagnetic transitions. The cited experimental data include measurements both by conventional ultrasonic techniques and by light scattering. As remarked by P. A. Fleury in Volume VI, light scattering experiments are particularly useful in indicating phonon behavior near phase transitions.

Ultrasonic surface waves are a valuable research tool in thin-film physics and surface physics, and they appear to be promising in electronic devices.

Chapter 4 deals with the excitation, detection, and attenuation of high frequency elastic surface waves. The emphasis is on the experimental aspects and on attenuation mechanisms.

Chapter 5 outlines the progress made in applying the principle of light diffraction by elastic waves to the development of practical devices and to the measurement of material properties. The applications discussed include the imaging of elastic waves by Bragg diffraction, elastooptical light modulators, optical beam deflectors, and optical information-processing techniques using diffraction in both the Bragg and Raman–Nath limits.

The editors owe a debt of gratitude to the authors who made this volume possible, and to the publishers for their unfailing help and advice.

xi

CONTENTS OF VOLUME I—PART B

METHODS AND DEVICES

xii

CONTENTS OF VOLUME II—PART A

Properties of Gases, Liquids, and Solutions

CONTENTS OF VOLUME II—PART B

PROPERTIES OF POLYMERS AND NONLINEAR ACOUSTICS

xiv

CONTENTS OF VOLUME III—PART A

EFFECT OF IMPERFECTIONS

CONTENTS OF VOLUME III—PART B

LATTICE DYNAMICS

CONTENTS OF VOLUME IV—PART A

APPLICATIONS TO QUANTUM AND SOLID STATE PHYSICS

xvii

CONTENTS OF VOLUME V

CONTENTS OF VOLUME VI

$-1-$

Ultrasonic Attenuation in Superconductors: Magnetic Field Effects

M. GOTTLIEB, M. GARBUNY, and C. K. JONES

Westinghouse Research Laboratories, Pittsburgh, Pennsylvania

I. Introduction

An important contribution to the ultrasonic attenuation in metals at low temperatures arises from the interaction of the sound waves with the conduction electrons. If the metal becomes superconducting, this contribution to

1

the attenuation decreases dramatically below the critical temperature T_c, and drops to zero at $T = 0$. The electronic structure of the material is of considerable importance in determining this behavior, and its interpretation in terms of the topological and dynamic properties of the Fermi surface has been the subject of many investigations. It is the purpose of this present chapter to consider the ultrasonic properties of a superconductor in an applied magnetic field.

Very soon after the discovery of the electronic contribution to the ultrasonic attenuation α, it was observed that the magnetic-field-induced transition to the normal state was not an essentially discontinuous change like the onset of resistance, but occurred over a significant range in field. In general, α increases roughly monotonically from the superconducting value α_s to the normal-state value at the same temperature, α_n. It is observed that the field values at which this transition starts and terminates are functions of sample composition and geometry, field direction, and temperature. Two distinct modes of behavior can obtain which are determined by the electronic properties of the superconductor.

If the material is a type I superconductor, or superconductor of the first kind, it is possible for the sample to enter the intermediate state, whereby magnetic field penetration can occur at a field value $H_x(t)$ less than the thermodynamic critical field $H_c(t)$. The volume of the superconductor becomes divided up into discrete macroscopic domains of normal and superconducting material, with the relative proportions of these regions changing with increasing field up to $H_c(t)$, when the volume of the superconducting regions is reduced to zero and a return to the normal state occurs. The existence of the intermediate state is a direct consequence of the interfacial surface energy between the normal and superconducting domains being positive in sign.

If the interfacial surface energy is negative, the material is known as a type II superconductor, or superconductor of the second kind, and its behavior in a magnetic field is markedly different from that of a type I material. The most dramatic, and technologically important, difference lies in the ability of a type II material to remain in the superconducting state in applied fields greatly in excess of $H_c(t)$. Initial field penetration occurs, if there is no demagnetizing, at a field $H_{c1}(t)$ (the lower critical field) which is less than $H_c(t)$, but, because of the negative surface energy, there is no macroscopic domain structure created. Instead, a microscopically ordered array of individual flux vortices, or fluxons (known as the mixed state) comes into existence, whose packing density increases with increasing field up to $H_{c2}(t)$, the upper critical field, where the array becomes close packed, and a return to the normal state occurs. A schematic representation of the mixed-state fluxoid arrangement is shown in Fig. 1.

In both of the above situations complications are introduced when the

Fɪɢ. 1 Mixed-state structure (from Abrikosov, 1957).

behavior of the sample surface is considered. Surface superconductivity can occur in both type I and type II materials, whereby a very thin surface layer can remain superconducting at high fields even when the bulk of the sample has returned to the normal state. Since we are concerned with the interactions of bulk sound waves, rather than surface waves, with the magnetically induced structures in superconductors, this particular situation will not be the subject of further discussion here.

   In general, good qualitative agreement is obtained between experimental results and the theories of ultrasonic attenuation in the intermediate and mixed states, but some very interesting quantitative discrepancies still exist as the subjects of possible future investigations. It is the principal objective of this chapter to review the present experimental and theoretical situation in this area, in a manner attempting to relate the various aspects of the work to each other in a unifying way. The ultrasonic properties of superconductors are summarized, and the measurement techniques usually employed are discussed briefly, in Section II. In Section III the magnetic properties of superconductors of particular relevance to their ultrasonic properties are considered within the current theoretical framework. Ultrasonic attenuation in the intermediate state is then discussed in Section IV, and in the mixed state in Section V. In both cases the theoretical models are discussed, the experimental results are given, and the present situation reviewed and summarized.

II. Ultrasonic Attenuation in Superconductors

A. GENERAL PROPERTIES

The electronic contribution to the attenuation of sound in a superconductor varies rapidly with temperature below T_c, decreasing to zero at $T = 0$. For longitudinal waves it has been well established, by a large number of workers, that the temperature dependence is of the form

$$\frac{\alpha_s}{\alpha_n} = \frac{2}{e^{\Delta/kT} + 1}, \qquad \text{for} \quad \hbar\omega_{\text{phonon}} \ll \Delta \tag{1}$$

where α_s and α_n are the attenuation coefficients in the superconducting and normal states, respectively, at temperature T, and Δ is the effective value of the energy gap at that temperature. Under conditions of small electron mean free path, where $ql \ll 1$, or when the material has a spherically symmetrical Fermi surface, $\Delta(0)$ is single valued, and its value can be determined from Eq. (1) if the temperature dependence of $\Delta(t)/\Delta(0)$ is known. At a temperature t, the values of α in the superconducting and normal states are seen from Fig. 19 to be α_{st} and α_{nt}. The magnetically induced transition from α_{st} to α_{nt} has the general character shown in Fig. 2, where the fields H_1 and H_2 and the detailed properties of $\alpha = \alpha(H, T)$ vary significantly from type I to type II behavior.

The behavior of transverse waves is usually considerably more complicated, due to the complex nature of the interaction. Since there is no density modulation associated with a pure shear wave, there is no direct electrostatic coupling to the conduction electrons, such as occurs with longitudinal waves,

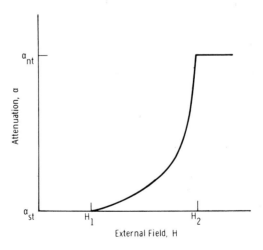

FIG. 2. General behavior of the field dependence of the ultrasonic attenuation of a type II superconductor.

due to the displacement of the ion cores in the propagation direction. Instead, the principal coupling occurs *via* the induced electromagnetic field
associated with the transverse ion-core motion whose electric field is normal
to the propagation direction. The effectiveness of this interaction diminishes
very rapidly with decreasing temperature below T_c due to the Meissner effect,
giving rise to an essentially discontinuous drop in α_t at this temperature. In
general, there is a residual contribution to the attenuation, α_c, which
decreases to zero in the same manner as for longitudinal waves. This component of α_t has two possible sources, the collision drag effect discussed
originally by Holstein (1956), and a contribution originating from the real
metal, or non-free-electronlike properties of the electronic structure. The
relative importance of these various mechanisms in determining α_t for
particular experimental situations has been considered by Leibowitz (1964).
The general behavior in a magnetic field at low temperatures away from T_c
is essentially the same as for longitudinal waves. The special case arising
near T_c is discussed later.

B. Measurement Techniques

The study of electronic contributions to the ultrasonic attenuation in
metals is almost invariably carried out at rather high frequencies, extending
into the microwave region. Several important improvements in technique
have been achieved recently, the most important of which include the development of the evaporated thin-film piezoelectric transducer (see deKlerk,
1966) and the use of thermocompression bonding. These refinements make
attenuation measurements in the gigahertz region an almost routine operation. Both pulse-echo and CW techniques can be employed, in essentially the
same form as discussed in previous contributions in this series by Bolef

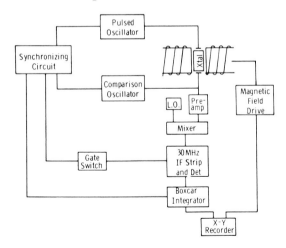

Fig. 3. Block diagram of apparatus for measuring ultrasonic attenuation.

(1966), Peverley (1966), and others. The use of integration techniques by means of either a boxcar integrator or lock-in amplifier is extremely convenient for continuous recording of the data, particularly under conditions where the experimental conditions are varying relatively rapidly. A typical system as shown in Fig. 3 was employed successfully in a number of the investigations to be discussed in this chapter, but possesses no very unusual features apart from convenience of operation and exceptionally good long- and short-term stability, particularly at very low pulse-repetition rates.

III. Magnetic Properties of Superconductors

A. Basic Characteristics

An ideal, unbounded superconductor possesses infinite electrical conductivity at temperatures below the critical temperature T_c. It also exhibits perfect diamagnetic properties in applied magnetic fields less than a critical value H_c. The phenomenological theory of superconductivity demonstrates that the superconducting state is in thermodynamic equilibrium, and as a consequence H_c can be related to T_c through the expressions

$$H_c^2(0) = \gamma T_c^2/0.18 \qquad (2)$$

and

$$H_c(t) = H_c(0)(1 - t^2) \qquad (3)$$

where γ is the electronic heat capacity and $t = T/T_c$.

The development of the microscopic quantum theory of superconductivity by Bardeen *et al.* (1957) has enabled most of the important results of the phenomenological theory to be derived from first principles. The superconducting ground state results from a net attractive interaction between conduction electrons of opposite spin and momenta. This interaction is the result of two opposing interactions, an attractive electron–electron interaction associated with virtual phonon exchange between the electrons, and the repulsive Coulomb interaction due to the like charges of the electrons. An attractive interaction between the conduction electrons can be considered as arising in the following way. A negatively charged electron moving with the Fermi velocity of about 10^8 cm/sec in the periodic potential of the positively charged lattice ions exerts an attractive force on the nearby ions, causing a small displacement in their positions. Relaxation to their equilibrium positions is relatively slow due to the large size of the ionic masses, and occurs at approximately the velocity of sound, about 10^5 cm/sec. It is immediately apparent that this distorted lattice field exerts an attractive force on any other electron in the vicinity within this time period, and hence is equivalent to a retarded, attractive electron–electron interaction which is strongly dependent upon the dynamic properties of the lattice. For this reason it is commonly referred to as the electron–phonon interaction. A small energy gap 2Δ is created between the pair ground state and the excited states of the conduction–electron energy spectrum. Superconductivity is destroyed when this pairing of electrons in the ground state is broken, and the energy gap vanishes, either by exceeding the critical temperature, or the critical field.

According to the theory, the superconducting transition temperature is given for a free-electron metal by the relationship

$$T_c \approx \theta_D \exp\left[-\frac{1}{N(0)V}\right], \qquad \text{for} \quad k\theta_D/E_F \ll 1 \tag{4}$$

where θ_D is the Debye temperature, E_F the Fermi energy, $N(0)$ the density of electron states at the Fermi surface, and V the net attractive interaction potential. The relationship between T_c and $\Delta(0)$, the energy gap at $T = 0$, is expressed by

$$2\Delta(0) = 3.56kT_c \tag{5}$$

where k is Boltzmann's constant. Later developments of the microscopic theory of considerably greater mathematical complexity and sophistication, by Bogolyubov *et al.* (1959), Eliashberg (1960), Gor'kov (1959, 1960), and many others have affected great increases both in its rigor and flexibility, so that the influence of strong coupling, retardation, and other real-metal effects can be taken into account.

As a consequence of the many-particle nature of the ground-state wave function, variation in the properties of the superconducting system can only occur over finite distances. Pippard (1953) originally introduced empirically the concept of the coherence length ξ_0 to take care of this non-local character of superconductivity in order to explain phenomenologically his microwave absorption experiments. The subsequent development of the microscopic quantum theory explained ξ_0 in a fundamental way, as a measure of the spatial extent of an electron pair wave function. The value of ξ_0 for a particular material is given approximately by the relationship

$$\xi_0 \approx (0.18/T_c)V_F\hbar \tag{6}$$

where V_F is the Fermi velocity. This expression becomes modified when the effects of finite electron mean free path are taken into account.

The perfect diamagnetism of a macroscopic superconductor results in the exclusion of an applied magnetic field from its interior until a critical value is reached, a phenomenon known after its discoverers as the Meissner–Ochsenfeld effect. This field exclusion is achieved by the presence of induced surface currents which generate an exactly equal and opposite magnetic field, which perfectly shields the interior of the superconductor. These currents exist only within a very thin surface layer characterized by the penetration depth λ, which is an intrinsic property of the material. Both ξ and λ are temperature dependent, but not in the same way, although obviously they decrease with increasing temperature, going to zero or the normal-state value, respectively, at T_c.

The behavior of a particular superconductor in an applied magnetic field can be characterized by means of the parameters $\kappa = \lambda/(\sqrt{2}\xi)$ and $\sigma_{\mathrm{ns}} = \xi - \lambda$. Two distinct modes of behavior are possible, type I or type II, a transition from one to the other occurring at the critical value of κ at which $\sigma_{\mathrm{ns}} \to 0$ and changes sign. The parameter σ_{ns} is actually a measure of the interfacial surface energy at the boundary between superconducting and normal regions and determines whether the material can enter either the intermediate or mixed state. The conditions for the formation of a boundary between a normal and superconducting phase are shown in Fig. 4.

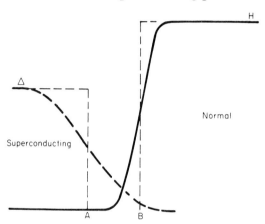

FIG 4. Boundary between normal and superconducting phases.

In the superconducting phase there are electron pairs with a binding energy of 2Δ. In the normal state $\Delta = 0$. However, the state of the electrons in the metal cannot change from superconducting to normal over distances less than ξ. Therefore, Δ varies approximately as shown in the figure. On the normal phase side of the boundary there is a magnetic field of magnitude H_{c} parallel to the interface; otherwise, there could not be equilibrium. The field in the interior of the superconducting region must be zero. This means that the field decreases from H_{c} to zero over a distance of the order of λ. If we replace the continuous variations in ξ and λ by defining effective boundaries at A and B without changing the areas under the curves, we are left with a region AB in which, on the one hand, the binding energy of the pairs is zero, so that it is like a normal metal, and, on the other hand, the field does not penetrate. This region has associated with it an equivalent surface energy $(AB)H_{\mathrm{c}}^2/8\pi$ per unit area, and it is evident that the net surface energy parameter corresponds to $\sigma_{\mathrm{ns}} = \xi - \lambda$. The condition for the Meissner effect is that $\xi > \lambda$, or that σ_{ns} is positive.

Magnetic flux penetration into the interior of a material at fields less than H_{c} is possible, even if σ_{ns} is positive, if the formation of a macroscopic domain structure of normal and superconducting regions can occur as a result of a particular sample geometry, value of σ_{ns}, and magnetic field direction. The material is then referred to as being in the intermediate state.

If σ_{ns} is negative, then the formation of macroscopic domains of normal and superconducting regions cannot occur. Instead, flux penetration into the interior is initiated by the creation of a microscopically ordered array of individual magnetic flux vortices. Each vortex is centered on a normal core, where $\Delta = 0$, and has associated with it a quantity of magnetic flux equal to a single flux quantum, 2.06×10^{-7} Oe/cm². Under these conditions the material is said to be in the mixed state.

An extremely important distinction between the intermediate and mixed states, particularly from a technological viewpoint, lies in the field at which a return to the normal state occurs. In both cases field penetration is initiated at a field less than H_c, but the return to the normal state occurs at H_c for a material in the intermediate state, whereas in the mixed state superconductivity can persist in fields considerably in excess of H_c.

B. The Intermediate State

In the vicinity of a macroscopic superconductor, that is, one large compared to the penetration depth λ, an applied field may be distorted due to the magnetization of the specimen, so that the surface field is not uniform. This behavior is strongly dependent upon the sample geometry and field orientation, and can result in situations where the local field exceeds H_c although the applied field is considerably below that value. Penetration into the interior of the sample then becomes possible, and, for type I

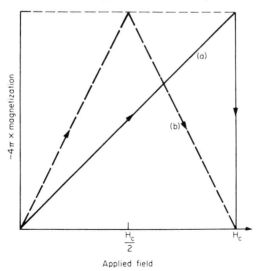

FIG. 5. Magnetization of a cylinder with axis perpendicular to field.

materials where σ_{ns} is positive, the creation of the macroscopic domain structure of normal and superconducting regions which constitutes the intermediate state can occur. In Fig. 5 the extreme situations possible for a long cylinder are shown. In a longitudinal field no penetration occurs until the applied field H_e reaches H_c, the magnetization M following the path a. For a transverse field, however, penetration begins at a field of $H_c/2$ with a linear decrease in M with H_e until total penetration is achieved at H_c. In the field range $H_c/2 < H_e < H_c$ the cylinder is in the intermediate state.

A model for the structure of the intermediate state has been proposed by Landau (1937, 1943) which appears to be generally applicable. The intermediate-state structure for the case of a slab of infinite extent and thickness d in a perpendicular applied field is shown in Fig. 6. It consists of a periodic

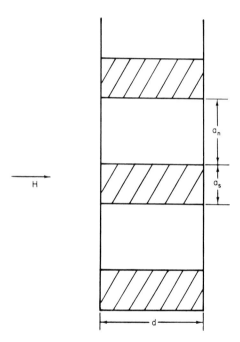

FIG. 6. Intermediate state of a slab in perpendicular field.

laminar structure of alternating normal and superconducting layers, the period depending upon the surface energy and field. For a periodic dimension of $a = a_s + a_n$, the superconducting lamellae have geometries defined by the relationships

$$\frac{a_s}{d} = \left[\frac{\pi}{\ln(1/2\eta)}\right]^{1/2} \frac{1-\eta}{\eta}\left(\frac{\delta}{d}\right), \qquad \text{for} \quad H \ll H_c \tag{7}$$

and

$$\frac{a_{\mathrm{s}}}{d} = 2.14 \left(\frac{\delta}{d}\right)^{1/2}, \qquad \text{for} \quad H \lesssim H_{\mathrm{c}} \tag{8}$$

where $\eta = H/H_{\mathrm{c}}$ and $\delta = 8\pi\sigma/H_{\mathrm{c}}^2$. Other sample geometries and field orientations change the direction of the domains, but this periodic structure is a fundamental characteristic of the intermediate state. Several other theories have been developed for the periodic structure of the intermediate state, but their results do not differ greatly from the above. The thin-plate geometry described above is an extreme case of the demagnetizing effects which produce the intermediate state; the magnetic field begins to penetrate the sample immediately as the external field is increased above zero. For other sample shapes flux penetration will commence at higher external fields, depending upon the extent to which the sample distorts the field distribution. It can be easily shown that an ellipsoid is the only sample shape that will support a uniform distribution of flux and of magnetization. For an applied magnetic field H_{e} which is parallel to a principal axis of the ellipsoid the field at the specimen is

$$H = H_{\mathrm{e}}/(1 - n) \tag{9}$$

where n is the demagnetizing factor. Correspondingly, the magnetization will be

$$-4\pi M = H_{\mathrm{e}}/(1 - n) \tag{10}$$

so that the slope of the magnetization curve will be steeper, the greater the demagnetizing coefficient. Special geometries of particular importance are the long cylinder with axis perpendicular to the field, for which $n = \frac{1}{2}$, and the sphere, for which $n = \frac{1}{3}$. Approximate demagnetizing factors for cylinders of varying length-to-diameter ratios have been computed by Bozorth and Chapin (1942).

In principle, knowledge of the interphase surface energy parameter permits a determination of the mixture of normal and superconducting domains by which superconductivity is destroyed for any configuration other than a long cylinder in a uniform magnetic field parallel to the axis. In practice, of course, only simple geometries are amenable to explicit solution, and then only in the absence of such disturbing factors as impurities or imperfections which may distort the flux distribution. Magnetization measurements have been the principal means of studying the gross features of the intermediate state, while detailed structure has been studied by magnetic microprobes in split samples by Meshkovsky and Shalnikov (1947), or by magnetic-powder-pattern measurements on the surface of samples in the intermediate state by Schawlow (1956). These techniques, besides being difficult, look only at the surface of the sample, where the intermediate state structure will almost certainly be distorted. The great advantage of ultrasonic techniques is that they probe the otherwise inaccessible interior of the specimen, and so are not subject to surface distortion. Another advantage in the ultrasonic methods over the microprobe methods lies in the range of sizes which may be resolved. Using ultrasonic methods, the resolution will be determined by the sound wavelength, which can be made much smaller than that achievable with probes.

C. The Mixed State

A superconductor in which σ_{sn} is negative enters the mixed state when an applied magnetic field exceeds its lower critical field H_{c1}, and returns to the normal state at the upper critical field H_{c2}. The magnetization curves of two materials with the same thermodynamic critical field, H_c, but surface energies of opposite sign are shown in Fig. 7. Demagnetizing effects

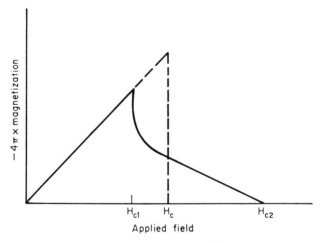

Fig. 7. Magnetization curve of a type II superconductor.

have been neglected. The areas under the magnetization curves are equal, since H_c, and hence the total energy difference between the normal and superconducting states, is the same in both cases. Pure materials, such as the elemental superconductors, are, in general, type I in behavior with σ_{ns} positive, and H_c is relatively small, typically $H_c(0) \lesssim 1000$ Oe. It is often possible to convert a pure type I material to type II by adding impurities or imperfections to the lattice, or by alloying. The electrons in such a superconductor are scattered by the defects or foreign atoms, and the effective range of correlation is thereby decreased. In a pure material the correlation length ξ_0 is typically $\sim 10^{-3} - 10^{-4}$ cm, the electron mean free path l being greater than this distance. In reducing l by impurity scattering the effective value of ξ can be decreased to a few hundred angstroms or less, resulting in a transition to type II behavior. It is possible for H_{c2} to become very much greater than H_c under these circumstances, and, in the case of those materials of technological importance in superconducting-magnet construction, can exceed 200 kOe.

The theory of the mixed state was developed originally by Abrikosov (1957) on the basis of the phenomenological Ginzburg–Landau (1950) equations for the particular case of negative σ_{ns}. The model of the mixed state as an ordered array of flux vortices, or fluxons, in the field region between H_{c1} and H_{c2} was an immediate outcome of this work. Subsequently it was shown by Gor'kov (1959, 1960) and other workers that Abrikosov's results could be obtained from the fundamental microscopic theory of superconductivity by using suitably sophisticated mathematical techniques. Only the results of this work will be presented here.

It is convenient to discuss the properties of type II superconductors in terms of the parameter κ. The transition from type I to type II behavior occurs at a critical value of κ corresponding to σ_{ns} passing through zero and changing sign. For $\kappa > 1/\sqrt{2}$ the material exhibits type II behavior, and the values of H_{c1} and H_{c2} can be expressed in terms of H_c and κ through the relationships

$$H_{c1} = (H_c/\sqrt{2}\kappa)[\ln(\kappa + 1.18) + 0.08] \qquad (11)$$

and

$$H_{c2} = \sqrt{2}\,\kappa H_c \qquad (12)$$

where H_c may be determined from the area under the magnetization curve, or from Eq. (2) if γ and T_c are known. The parameter κ is temperature dependent, reflecting, of course, the variation with temperature of λ and ξ, and may be written as

$$\kappa(T) = \kappa(T_c)A(T) \qquad (13)$$

where $A(T)$ varies from unity at T_c to 1.25 at $T = 0$. For a pure superconductor

$$\kappa_0(T_c) = \lambda(0)/\xi_0 \qquad (14)$$

and

$$\lambda(0) = [m^*c^2/4\pi Ne^2]^{1/2} \qquad (15)$$

where N is the number density of conduction electrons of effective mass m^* and charge e.

The order parameter κ is related to the magnetization near H_{c2} through the expression

$$-4\pi M = (H_{c2} - H)/[2\kappa^2(T) - 1]\beta, \qquad \beta = 1.16 \qquad (16)$$

and in practice it is obtained directly from the limiting value of the slope of the attenuation at H_{c2},

$$\kappa^2 = \frac{1}{8\pi}\left(\frac{dM}{dH}\right)^{-1}_{H_{c2}} + \frac{1}{2} \qquad (17)$$

The temperature dependence of κ is then determined directly from the temperature dependence of the magnetization slope. It will be shown in a later section that the slope of the attenuation near H_{c2} is also functionally dependent upon κ, but that the attenuation slope near H_{c2} is a much more rapidly varying quantity than is the magnetization slope. The relation between dM/dH and $d\alpha/dH$ is illustrated in Fig. 8. Thus, the attenuation near H_{c2} can be a much more sensitive measure of small changes in the properties of a type II superconductor near H_{c2} than is the magnetization.

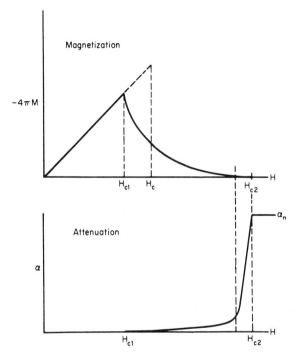

Fig. 8. Comparison between magnetization and attenuation of a type II super-conductor.

For impure superconductors or alloys in which the electron mean free path l is limited by impurity scattering, it was shown by Gor'kov (1959, 1960) that the effective value of $\kappa(T_c)$ is given by

$$\kappa(T_c) = \kappa_0(T_c) + 7.5 \times 10^3 \gamma^{1/2} \rho_n = \kappa_0 + \kappa_1 \qquad (18)$$

where γ is the electronic heat capacity coefficient in erg cm^{-3} deg^{-2} and ρ_n the normal-state residual resistivity in ohm-cm.

In general, pure materials are type I superconductors, and type II behavior is induced by increasing κ_1. It has been found, however, that niobium, due to the special nature of its electronic structure, is an intrinsic type II material even in the limit of infinite l, and it has therefore been the subject of a great deal of theoretical and experimental attention. For niobium $\kappa_0(T_c)$ is approximately unity.

The recent theoretical work of Maki (1964) has shown that the model description employed above where κ is a single parameter is actually an oversimplification, and, in fact, several "κ's" are rigorously required. However, it can be shown that for our present purposes it is adequate to equate the above κ with κ_2 in the more-sophisticated treatments.

It has been shown by Cape and Zimmerman (1967) that demagnetizing effects are of importance in type II materials, as well as in the type I case, where the intermediate state is an immediate outcome. They have shown that expressions for the magnetization M of the general form

$$-4\pi M = H_{\mathrm{e}} \frac{H_{\mathrm{c2}} - H_{\mathrm{e}}}{(\gamma + n)H_{\mathrm{e}}} \tag{19}$$

for type II materials are applicable, where n is the appropriate element of the configuration matrix N whose elements are the shape-dependent demagnetization coefficients, $\gamma = (2\kappa_2{}^2 - 1)\beta$, and β is a constant of order unity.

All of the above effects are particularly susceptible to investigation by ultrasonic techniques.

IV. Ultrasonic Attenuation in the Intermediate State

A. Introduction

Two types of intermediate-state effects may be studied using sound waves; the first is that in which the sound wave is used merely as a measure of the intermediate-state structure, and the other is that in which the sound itself interacts with the intermediate state to produce an observable effect. Examples of both types will be discussed here. Some gross features of the influence of sample geometry on demagnetization and flux-trapping effects in lead were studied with ultrasonic attenuation by Gottlieb and Jones (1966). Figure 9 gives attenuation curves for $\frac{5}{8}$-in.-long samples of square cross section ($\frac{1}{4}$ in. $\times$ $\frac{1}{4}$ in.) and of round cross section ($\frac{1}{4}$ in. diameter). From these curves the demagnetizing factors were determined from the penetration fields, and a measure made of the relative degree of flux trapped for different specimen geometries. For example, it is apparent from the four curves shown that a round-cross-section sample in a transverse field traps the smallest amount of flux of the four geometries. This has been explained by a model of Livingston (1968) in which that geometry requires the least magnetic force in decreasing field to drive flux from the sample interior, because of the diminishing length of the flux line as it leaves. Other details of flux trappings and penetration may also be seen on these curves. While this information could also be determined from magnetization, it is possible to look at geometries with ultrasonics that may be much more difficult for magnetization techniques. More important, magnetization can only measure that flux in a nonuniform specimen that crosses the outside surface. Ultrasonic methods are capable of studying internal flux rearrangements merely by changing transducer size and location.

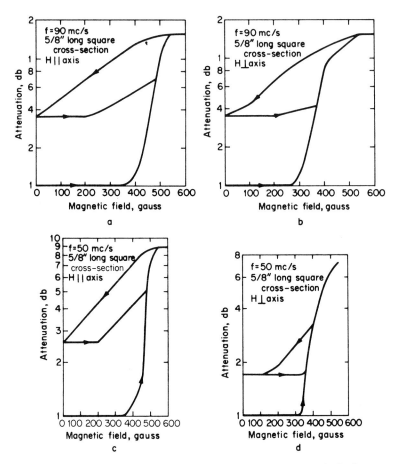

Fig. 9. Shape dependence of attenuation in intermediate state in lead.

B. Theories of Ultrasonic Attenuation in the Intermediate State

1. *Absorption due to Boundary Wall Motion*

The most important parameter of the intermediate state is the interphase surface energy per unit area, and the measurement of the period of the structure is virtually the only way of determining this. The structure has been made visible, as previously described, by means of finely ground diamagnetic or paramagnetic powder distributed on the surface of the specimen, and has been mapped by a small magnetic probe moved along the surface of a specimen. The laminar structure is, however, generally distorted near the surface, and it is highly desirable to examine the bulk of the material.

Ultrasonic waves offer a means to probe the interior of a superconducting specimen, and it has been shown theoretically by Andreev and Bruk (1966) that there exists an ultrasonic absorption mechanism that is specific to the intermediate state. A necessary condition for equilibrium between the normal and superconducting phases is that the magnetic field at the interface be equal to H_c. Since H_c in general depends upon temperature and pressure, the changes produced by the passage of a sound wave leads to a violation of this condition. As a result, the boundary between phases moves, and an ac magnetic field appears and induces eddy currents in the normal phase. There is thus an additional mechanism of absorption associated with the release of Joule heat.

The variation of the critical field at the interface is given by

$$H_c = H_c(T, P) = H_{c0}(1 + A u_{ii}) + (dH_c/dT)\,\Delta T \tag{20}$$

where H_{c0} is the critical field at equilibrium temperature and pressure, and u_{ii} is the divergence of the sound-wave displacement vector. The thermodynamic quantity A is given by

$$A = -\frac{\rho}{H_{c0}} \left(s_1{}^2 - \frac{4}{3} s_t{}^2 \right) \left[\left(\frac{\partial H_c}{\partial P} \right)_T - \left(\frac{\partial H_c}{\partial T} \right)_P \frac{T}{\rho c} \left(\frac{\partial \rho}{\partial T} \right)_P \right] \tag{21}$$

where ρ is the density, and s_1 and s_t are the longitudinal and transverse sound velocities, respectively. The magnetic field in the normal layers is computed from the usual equations

$$\frac{\partial \mathbf{H}}{\partial t} - \frac{c^2}{4\pi\sigma}\,\Delta \mathbf{H} = \nabla\nabla \times \mathbf{v} \times \mathbf{H}, \qquad \text{and} \qquad \text{div } \mathbf{H} = 0 \tag{22}$$

where $\mathbf{v}$ is the velocity of the medium and σ is the normal metal conductivity, and the current and Joule loss terms in the normal material are determined from

$$\mathbf{j} = (c/4\pi)\,d\mathbf{H}/dz \tag{23}$$

and

$$\dot{\mathscr{E}} = \int (|\mathbf{j}|^2/2\sigma)\,dz \tag{24}$$

The solutions will depend upon the relative magnitudes of the sound wavelength, the width of the normal layer, and the normal metal skin depth $\delta = c/(2\pi\sigma\omega)^{1/2}$. The attenuation obtained by Andreev and Bruk under conditions for which $\lambda_s \gg \delta$, a_n and $a_n \gg \delta$ is

$$\Gamma = \frac{4[A + 1 - (\mathbf{m}\cdot\mathbf{n})^2]^2}{\rho\sigma(a_n + a_s)s_1{}^3\delta} \left(\frac{cH_c}{8\pi} \right)^2 \tag{25}$$

where $\mathbf{m}$ and $\mathbf{n}$ are unit vectors in the direction of magnetic field and sound propagation, respectively.

The quantity A is typically not greatly different from unity, so that the attenuation due to this mechanism may be strongly anisotropic.

2. *Absorption due to Electron Orbiting in Magnetic Field*

The intermediate-state absorption mechanism associated with motion of the boundary walls is important for low sound frequencies, and not too high purities; i.e., for normal skin depths on the order of or larger than the period of the structure. Under conditions of higher frequencies, for which the sound wavelength is comparable to the layer thickness, and high purity, with electron mean free paths in the range 10^{-2}–10^{-1} cm, phenomena associated with cyclotron absorption in the normal layers should be observable. The theoretical development for this absorption was recently carried out by Andreev (1968), who pointed out that under the proper conditions measurement of the ultrasonic attenuation due to this mechanism could be used to determine the periodicity. Andreev carried out the solution for the two limits of the Larmor radius R of the electron in the critical magnetic field: R very large, and R very small compared to the electron mean free path l. The equation of motion of electrons in the normal domains is solved for a system in which sound is propagated in the x direction while the magnetic field is along the z direction; then the boundaries of the normal domains are taken at $x = 0$ and $x = a$. The energy of the electron in the presence of a sound wave is

$$\varepsilon = \varepsilon_0 + \lambda_{ik} u_{ik} \tag{26}$$

where u_{ik} is the deformation and λ_{ik} is the deformation potential. The equation of motion of the electrons can be expressed in the form (for $R \gg l$)

$$v_x(\partial\phi/\partial x) + \nu\phi = ev_i E_i - \Lambda_{ik}\dot{u}_{ik} \tag{27}$$

where the electron distribution function is of the form

$$f = f_0 + \phi\, \partial f_0/\partial\varepsilon \tag{28}$$

where v is the electron velocity, and ν the collision frequency. The departure of the deformation potential from its average over the Fermi surface is expressed by

$$\Lambda_{ik} = \lambda_{ik} - \bar{\lambda}_{ik} \tag{29}$$

ϕ is regarded as a quasistatic function, since the sound velocity is much less than the electron velocity. The sound absorption coefficient is calculated from the Maxwell equation

$$\boldsymbol{\nabla} \times \boldsymbol{\nabla} \times \mathbf{E} = (4\pi i\omega/c^2)\mathbf{j} \tag{30}$$

where $\mathbf{j}$ is the current density induced by the acoustic deformation, and

$$\mathbf{j} = \frac{e}{4\pi^3} \int \mathbf{v}\phi\, \frac{dS}{v} \tag{31}$$

where dS is an area element of the Fermi surface. The energy absorption per unit volume in the normal region is

$$Q(x) = \frac{1}{(2\pi)^3} \int \nu|\phi|^2\, \frac{dS}{v} \tag{32}$$

so that the average energy absorption per unit volume is

$$\bar{Q} = \frac{1}{a + a_{\mathrm{n}}} \int_0^{a_{\mathrm{n}}} Q(x)\ dx \tag{33}$$

where a_{n} and a_{s} are the thickness of the normal and superconducting regions, respectively. The results of Andreev's calculations for the various limiting cases are summarized as follows. For the sound wavelength much less than the normal layer thickness, $\lambda \ll a_{\mathrm{n}}$, the ratio of the intermediate-state attenuation to the attenuation for the fully normal state is

$$\Gamma/\Gamma^{\mathrm{N}} = \eta \tag{34}$$

where η is the fraction of sample that is normal, i.e., $\eta = a_{\mathrm{n}}/(a_{\mathrm{n}} + a_{\mathrm{s}})$. Both Γ and Γ^{N} are proportional to the first power of the sound frequency.

For the other limiting case, when $\lambda \gg a$, it is easily shown that

$$\Gamma/\Gamma^{\mathrm{N}} = \mathrm{const}\ \eta a_{\mathrm{n}}/\lambda \tag{35}$$

The intermediate-state absorption is here proportional to the square of the sound frequency, and will be of the order of magnitude of the normal metal absorption with $l \approx a$.

For the case that the Larmor radius is small compared to the electron mean free path, the equation of motion contains magnetic-field terms,

$$v_x \frac{\partial \phi}{\partial x} + \Omega \frac{\partial \phi}{\partial \tau} + \nu \phi = e v_i E_i - \Lambda_{ik} \dot{u}_{ik} \tag{36}$$

where Ω is the cyclotron frequency in the critical magnetic field and τ is the time of rotation of the electron in the magnetic field. The absorption coefficient calculated from this equation consists of a contribution which is monotonic with external magnetic field and one which is an oscillating function of external magnetic field,

$$\Gamma = \Gamma_0 + \Delta\Gamma \tag{37}$$

The monotonic part is

$$\Gamma_0 = \eta \Phi(2a_{\mathrm{n}}/\lambda) \Gamma_0^{(N)}(H) \tag{38}$$

where $\Gamma_0^{(N)}(H)$ is the field-dependent normal-state attenuation. The function $\Phi(2a_{\mathrm{n}}/\lambda)$ varies from 0 to unity, and is shown in Fig. 10. The oscillating part of Γ is smaller than Γ_0 by the factor $(R/a)^{1/2}$, and varies almost sinusoidally with the magnetic field.

Andreev suggested that the above relations may be used in making experimental determination of the periodic structure of the intermediate state. For $R \gg l$ the period can be determined from $\Gamma/\Gamma^{(N)}$ and the function $\Phi(2a_{\mathrm{n}}/\lambda)$, provided η is known. For a cylinder in a transverse magnetic field this is $\eta = (2H/H_{\mathrm{c}}) - 1$. For $R \ll l$ the procedure can be applied to the monotonic part of Γ, Γ_0.

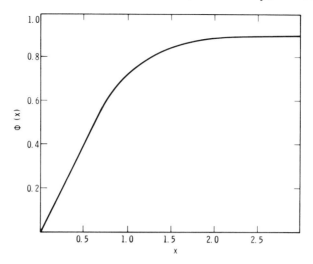

FIG. 10. Attenuation function $\Phi(2a_n/\lambda)$ $vs.$ $2a_n/\lambda$.

3. *Absorption due to Quasiparticle Relaxation*

An absorption phenomenon for superconductors in the intermediate state has been reported by Leibowitz and Fossheim (1968). The effect has been identified by Leibowitz (1968) as a relaxational absorption involving the excitation of quasiparticles in the region of the superconducting-normal (SN) state interfaces. The excitation may occur under the influence of transverse ultrasonic waves when the superconductor is in the intermediate state within several millidegrees of the critical temperature. It is necessary that the polarization vector, and hence the electric currents induced by the transverse wave, be normal to the intermediate state laminations. If the specimen temperature is sufficiently close to the critical temperature, electromagnetic screening of the quasiparticles dominates the interaction with transverse waves, and the induced fields modulate the energy gap in a volume centered on the equilibrium state SN boundaries. If the relaxation time with which the equilibrium energy gap is established is τ, the fraction of the Cooper pair binding energy dissipated per cycle depends on $\omega\tau$. The measurements of Leibowitz and Fossheim indicate a relaxation time of about 10^{-9} sec in indium for a temperature 5 millidegrees below the transition temperature. Note that even at this temperature $\hbar\omega \ll \Delta$, where Δ is the energy gap in a *uniform* superconductor. The experiments on indium will be described in a later section.

C. Attenuation in the Intermediate State—Experiments

Not all of the effects described in the theory of ultrasonic attenuation in the intermediate state have been observed, because of limitations on the availability of the necessary materials, or because of the inability to separate contributions to the attenuation due to different mechanisms. It is expected that techniques of measurement will be developed in the near future that will allow these theories to be tested. In this section we will describe those relevant experiments that have been done; they are, for the most part, incomplete, but possibly will suggest areas where further work is needed.

1. *Separation of Attenuation due to Phase-Boundary Motion*

The experimental determination of attenuation associated with the intermediate-state structure requires separating the usual normal state attenuation. A procedure has been developed by Gottlieb *et al.* (1967a) for performing this separation by comparing the measured attenuation with the magnetization of the specimen. The fraction of the sample that is normal, η_n, is obtained from the measured magnetization $4\pi M(H) = B - H$ by assuming that the magnetic field in the normal regions is always just H_c, as the specimen is in the intermediate state. Then the induction in the specimen is $B = \eta_n(H)H_c$, and the magnetization is $4\pi M(H) = \eta_n H_c - H$. Thus, measurement of $M(H)$ can be used to determine $\eta_n(H)$. From this value of η_n we proceed to compute the usual normal electron attenuation α_e due to electrons in the normal regions. If the laminar structure (i.e., external magnetic field) is transverse to the sound propagation direction, the ultrasonic signal in the intermediate state relative to the ultrasonic signal when the sample is fully superconducting will obviously be

$$I(H)/I_s = \exp(-0.23\eta_n\alpha_n) \tag{39}$$

where α_n is the attenuation coefficient of the fully normal state. In terms of the attenuation coefficients, the normal fraction is simply $\eta_n = \alpha_e/\alpha_n$, and the intermediate-state attenuation is the difference between the observed attenuation and the normal electron attenuation,

$$\alpha_i = \alpha_{obs} - \alpha_e = \alpha_{obs} - \eta_n\alpha_n \tag{40}$$

By substituting the value of η_n obtained from the magnetization, we obtain

$$\alpha_i = \alpha_{obs} - (\alpha_n/H_c)[4\pi M(H) + H] \tag{41}$$

If the magnetic field is parallel to the sound propagation direction, the intensity of the ultrasonic signal will be given by

$$I(H) = \eta_n I_n + (1 - \eta_n)I_s \tag{42}$$

where I_s has the same meaning as before, and I_n is the intensity of the ultrasonic signal for the fully normal state. From Eq. (42) the normal fraction is

$$\eta_n = [1 - \exp(-0.23\alpha_e)]/[1 - \exp(-0.23\alpha_n)] \tag{43}$$

and α_e can be expressed in terms of the magnetization,

$$\exp(-0.23\alpha_e) = 1 - \frac{1 - \exp(-0.23\alpha_n)}{H_c} [4\pi M(H) + H] \qquad (44)$$

The intermediate-state attenuation is obtained, as before, as the difference between α_{obs} and α_e.

Measurements of ultrasonic attenuation in the intermediate state of very pure lead single crystals (resistance ratio >1000) oriented with their cylinder axis along the (100) direction were made in the frequency range from 21 to 90 MHz with longitudinal sound waves, and in the temperature range from 1.33 to 7.2°K. A typical set of results for H parallel to the sound propagation direction at 50 MHz and $T = 4.2$°K is shown in Fig. 11. The

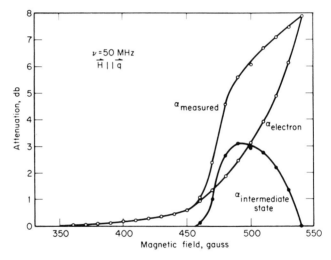

FIG. 11. Attenuation *vs.* field for intermediate state in lead.

attenuation is for the first sound pulse in a specimen $\frac{5}{8}$ in. long $\times \frac{1}{4}$ in. in diameter. The maximum value of α_i is 0.39 α_n and lies at a magnetic field value of about $0.9H_c$. The shape of the intermediate-state attenuation curve is the product of two factors: the volume of normal material, and the number of normal superconducting boundaries. The volume of normal material grows monatonically as H/H_c increases, so the peak in the curve is associated with a peak in the number of such interfaces. This may be compared with the curve of domain density *vs.* H/H_c for a flat-plate geometry shown in Fig. 12. This was determined visually using the magnetic-powder technique. The maximum value of the ratio α_i/α_n varied little with temperature, perhaps increased slightly with decreasing temperature, and increased from 0.34 at 21 MHz to 0.52 at 63 MHz. The reduced plots, α_i/α_n *vs.* H/H_c, shifted toward higher fields with decreasing temperature; a set of curves taken at 50 MHz at temperatures from 1.33 to 4.61°K is shown in Fig. 13.

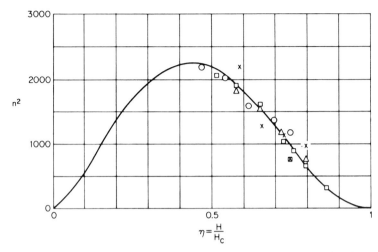

FIG. 12. Domain density *vs.* H/H_c for flat tin plates 0.32 cm thick. $T = 1.86°$K (from Schawlow, 1956).

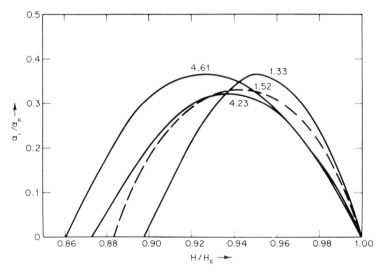

FIG. 13. Plots of α_i/α_n *vs.* H/H_c for the intermediate state in lead.

This variation of the reduced curves may be used to estimate the relative change with temperature between the volume free energy and the interphase-surface free energy.

A set of measurements similar to the ones described above was done for the magnetic field direction transverse to the direction of sound propagation. The maximum values of α_i estimated for this geometry were typically around 30 times smaller than for the parallel geometry; for example, at a

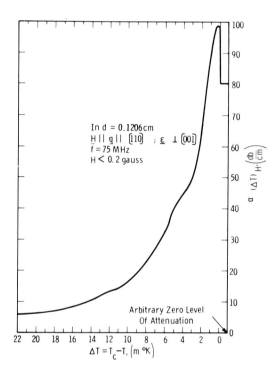

FIG. 14. Ultrasonic attenuation of indium on the intermediate state near T_c
[from Leibowitz and Fossheim (1968)].

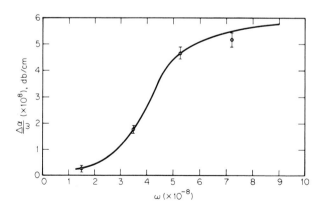

FIG. 15. Frequency dependence of absorption peak in indium [from Leibowitz (1968)].

frequency of 50 MHz and a temperature of 4.2°K the maximum value of α_i/α_n was 0.0126. Such a large difference is not consistent with the value of attenuation for the two orientations calculated using Eq. (25). The value of A [Eq. (2.1)] is near unity, using values of the parameters contained in A from the literature. Since $\mathbf{m} \cdot \mathbf{n}$ varies from 0 to 1 for the perpendicular to parallel orientations, there should be only about a factor of two change between the attenuation for the two orientations. Part of the difficulty may lie in the applicability of Eq. (25) to the frequency range in which these measurements were done. This formula is derived under the assumption that the sound wavelength is much larger than the domain size; using data of Schawlow (1956), we estimate that $a_n \approx 10^{-2}$ cm, which is about the same as the sound wavelength. Further work is necessary to clarify the discrepancy between theory and measurements.

2. *Ultrasonic Attenuation in Intermediate State near* T_c

Measurements of the ultrasonic attenuation in the intermediate state of indium very near T_c suggest the presence of the relaxation mechanism described in Section IV, B, 3. The usual behavior of the zero-field attenuation with temperature is an initial very rapid decrease just below the critical temperature, followed by a more gradual, monotonic decrease of attenuation at lower temperatures. It would be expected that the application of a small magnetic field would not make a qualitative change in this behavior, especially in the rapid fall region near T_c. It is just in this rapid fall region that the previously described absorption anomaly has been reported by Leibowitz and Fossheim (1968) and attributed by Leibowitz (1968) to the gap relaxation mechanism. Their data for pure indium are shown in Fig. 14. The sample geometry is that of a thin disk in a magnetic field (0.2 G for the measurements in Fig. 14) perpendicular to the plane of the disk. For such thin-disk geometry the demagnetizing factor is unity, and the intermediate state may begin for the smallest applied fields. Because of the sensitivity of these experiments to the smallest fields, the earth magnetic field had to be compensated with a set of Helmholtz coils. With the sound propagation direction also perpendicular to the plane of the disk, and polarization of the transverse wave perpendicular to the laminations, the very pronounced peak in the curve of attenuation *versus* temperature is observed for a sound frequency of 75 MHz. The additional absorption represented by the peak is about 25% of the total drop in attenuation below T_c. It was found that when the sound polarization was parallel in the lamellae, but under otherwise identical conditions, the absorption peak is totally absent. Furthermore, the absorption does not occur at lower temperatures for equivalent (i.e., higher magnetic field) intermediate-state conditons; this is consistent with the view that the acoustically induced magnetic field plays a central role. The frequency dependence of the absorption peak was measured in the range from 15 to 135 MHz, and this is shown in Fig. 15 as $\Delta\alpha/\omega$ *versus* ω. According to Leibowitz (1968) this suggests a relaxation time of 10^{-9} sec at a temperature 5 millidegrees below T_c.

V. Ultrasonic Attenuation in the Mixed State

A. INTRODUCTION

Because of the complexities of the interactions, theories of ultrasonic absorption in the mixed state have generally been very limited in their ranges of applicability. Caroli and Matricon (1965) treated the absorption due to a single vortex line in the field region very near to the lower critical field H_{c1}; Cooper *et al.* (1966) considered the absorption of longitudinal sound in a pure type II superconductor for temperature very near the critical temperature T_c and magnetic field near H_{c1}. In this work they suggested that there might be resonance absorption when the sound wavelength equals the spacing between the fluxoids of the mixed state. A phenomenological theory of ultrasonic wave propagation in the mixed state of high-field superconductors for low sound frequencies was developed by Shapira and Neuringer (1967); it is assumed in their work that the dominant effect on the attenuation is due to the magnetic field itself through magnetohydrodynamic interactions with the electrons.

The general behavior of the ultrasonic attenuation of a pure type II superconductor in the mixed state is well known, following more or less the magnetic field dependence illustrated in Fig. 16. The attenuation is constant as long as the sample is in the Meissner state. When the initial flux

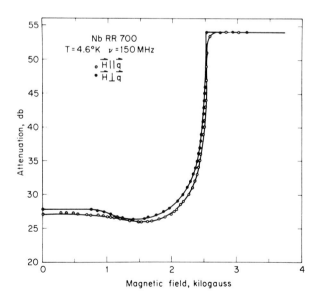

FIG. 16. Field dependence of ultrasonic attenuation in a type II superconductor (niobium).

penetration occurs the attenuation may increase or decrease, depending upon purity and temperature. In the figure the attenuation decreases, an effect which will be explained later. As the magnetic field is increased beyond H_{c1} the attenuation increases, at first slowly, then more rapidly. Near the upper critical field the ascent of the attenuation becomes extremely steep. The theory accounts, at least qualitatively, for these features.

B. THEORIES OF ULTRASONIC ATTENUATION IN THE MIXED STATE

1. *Simple Theory*

Some important features of the ultrasonic attenuation in the mixed state can be obtained simply from the BCS relation for attenuation

$$\frac{\alpha_s}{\alpha_n} = \frac{2}{1 + \exp(\Delta/kT)} \tag{45}$$

where Δ is the superconducting energy gap. Like most computations for the mixed state, the result is calculable only for magnetic fields very close to the upper critical field H_{c2}, for which the superconducting energy gap parameter Δ is much less than kT, the so-called gapless region of type II superconductivity. In this region the energy-gap parameter is (see Maki, 1964, 1967a, b)

$$\Delta^2 = A \frac{T^2}{H_{c2}(T)} \frac{H_{c2}(T) - H}{2\kappa^2(T) - 1} \tag{46}$$

for the limit in which $l \gg \xi$. The quantity A contains several material constants, and is independent of temperature, at least for the case in which the magnetic field direction is parallel to the sound propagation direction. If one assumes that the BCS relation for attenuation is still valid in its simple form for this gapless region, then the BCS relation simplifies to

$$\alpha_s/\alpha_n \approx 1 - [\bar{\Delta}(H)/2kT] \tag{47}$$

Since the slope of the energy-gap parameter with field is

$$\frac{d\bar{\Delta}(H)}{dH} = A^{1/2} \frac{T}{(H_{c2})^{1/2}} \frac{1}{(2\kappa^2 - 1)^{1/2}} \frac{1}{(H_{c2} - H)^{1/2}} \tag{48}$$

the slope of the attenuation with field near H_{c2} can be expressed as

$$\left(\frac{1}{\alpha_n} \frac{d\alpha_s}{dH}\right)_{H_{c2}} [H_{c2}(H_{c2} - H)]^{1/2} = \frac{B}{(2\kappa^2 - 1)^{1/2}} \tag{49}$$

This very simplified approach describes the most important qualitative feature of the attenuation in the mixed state in the pure limit; the steep rise which goes as $(H_{c2} - H)^{-1/2}$. Figure 8 shows the magnetization and the attenuation in the mixed state, indicating the much greater sensitivity of the attenuation in the critical region. It is this greater sensitivity which partly accounts for the usefulness of ultrasonic attenuation in studying the details of the mixed state.

2. *Microscopic Theory for Pure Limit, near* H_{c2}

Perhaps the most comprehensive theory of ultrasonic absorption in the mixed state is that due to Maki *et al.* (1964, 1967a, b) in a series of papers of considerable mathematical virtuosity. We will present here the results of the Maki theory for those limiting cases for which analytical expressions suitable for comparison with experiment are obtainable. The physically more interesting case is that for the pure superconductor, in which the electron mean free path l_e is much larger than the superconducting coherence length ξ.

Equation (49) can only be expected to give a rough description of the attenuation near H_{c2} because it has ignored any anisotropies imposed by the fluxoid structure, as well as any microscopic interactions. A strong anisotropy is to be expected, since the quasiparticles (normal electrons) bound to the fluxoids experience a weaker order parameter perpendicular to the fluxoids than parallel to them. Since the sound wavelength in the experimentally interesting region (say, 100 MHz to 2 GHz) might be typically 100 times larger than the fluxoid diameter, the structural variations of the order parameter cannot be simply averaged. In the analysis of Maki, which takes into account these details, the attenuation in the gapless region is generally of the form

$$\alpha_s/\alpha_n = 1 - [\bar{\Delta}(H)/kT]F(T, \theta, ql) \qquad (50)$$

where F is a function of temperature, angle between field and sound propagation direction, and ql; F is expressed in the form of an integral which can be evaluated for certain limiting cases of interest. Maki calculated the orientation dependence explicitly only for the case $ql \gg 1$, with the result

$$\frac{\alpha_s}{\alpha_n} = 1 - \frac{\Delta}{2kT} \int_{-\infty}^{+\infty} \phi_1(\alpha, \sin\theta) \cosh^{-2}\left(\frac{\alpha}{2kT}\right) d\alpha \qquad (51)$$

$$\phi_1(\alpha, \sin\theta) = \frac{2}{\pi^{3/2}\varepsilon} \int_0^1 \frac{dz}{(1-z^2)^{1/2}(1-z^2\sin^2\theta)^{1/2}}$$

$$\cdot \exp\left[-\left(\frac{\alpha}{\varepsilon}\right)^2(1-z^2\sin^2\theta)^{-1}\right] \qquad (52)$$

where z is in the direction of the magnetic field H, and

$$\varepsilon = v_F[\tfrac{1}{2}(e/c)\hbar H_{c2}]^{1/2} \qquad (53)$$

It is difficult to verify this formula experimentally, because of the high sample purity and/or high sound frequencies that are required. The anisotropy of the attenuation slope was extended from the Maki formulation by Kinder

(1968), and a set of curves of the reduced slope for longitudinal wave attenuation $C_L(\theta, ql, T)$, defined by

$$1 - (\alpha_s/\alpha_n) = C_L(\theta, ql, T)\,\Delta(H, T)/2kT \tag{54}$$

is given in Fig. 17 for ql values from 0.1 to 10. For niobium the curves shown correspond to a temperature around $4°K$, and indicate a surprisingly large anistropy for ql as low as 0.1.

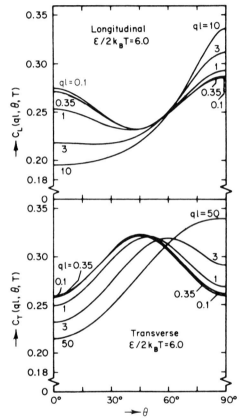

FIG. 17. Calculated anisotropy of normalized attenuation slope for Nb (from Kinder).

3. Microscopic Theory for Dirty Limit, near H_{c2}

A theory for ultrasonic attenuation in a very impure superconductor, i.e., one in which the electron mean free path is very much less than the coherence length, has been developed by McClean and Houghton (1967). As in the case of the pure superconductor, it is valid for H very near H_{c2}; in distinction to the pure superconductor, the attenuation has a quadratic

dependence upon the order parameter, rather than a linear one. Their result
for the impure superconductor is

$$\frac{\alpha_\text{s}}{\alpha_\text{n}} = 1 - \frac{1}{2}\left(\frac{\Delta}{2kT}\right)^2 F(p)$$

$$\approx 1 - \frac{3}{4\pi}\frac{mc^2}{Ne^2\tau^2 v_\text{F}^2}\frac{H_\text{c2} - H}{H_\text{c2}(2\kappa^2 - 1)\beta}\,[1 + L(p)] \qquad (55)$$

where m, N, e, τ, and v_F represent, respectively, electronic mass, density,
charge, collision time, and the Fermi velocity; $p = \tau v_\text{F}^2 eH_\text{c2}/6\pi kT$, and
$L(p)$ is defined in the previous reference.

4. Theories of Attenuation due to Isolated Fluxoids

A calculation by Cooper *et al.* (1966) explores the ultrasonic attenuation
in a type II superconductor ɪor the opposite limit of the previous theories,
i.e., magnetic field $H \ll H_\text{c2}$, so that the fluxoids are well separated in space.
Explicit solutions are possible near the transition temperature, $T \sim T_\text{c}$;
under these conditions the magnetic field is large inside the flux vortex and
decreases rapidly outside the vortex within a penetration depth δ. Theory
predicts that the fluxoids assume a periodic array in the mixed state, so that
the energy gap $\Delta(T)$ should also vary periodically, with a periodicity on the
order of δ.

The calculation of Cooper *et al.* proceeds by computing the response
of a superconductor with a periodically varying energy gap to a longitudinal
sound wave. The energy dissipation per unit volume is

$$Q(t) = \left\langle \frac{d}{dt}\langle H(t)\rangle \right\rangle \qquad (56)$$

the time average ratio of increase of the average energy of the system. The
Hamiltonian is

$$H(t) = H_0 + H'(t) \qquad (57)$$

where $H'(t)$ is the perturbation induced by the sound wave. The result of
the calculation for the energy dissipation after time averaging and Fourier
analyzing is

$$Q = \frac{1}{2\pi i}\int_{-\infty}^{+\infty}\frac{d\omega}{2\pi}\int\frac{d^3q}{(2\pi)^3}\,\omega\rho^\text{c}(\mathbf{q}, \omega)\Phi^*(q, \omega) \qquad (58)$$

where ρ^c is the charge response to the scalar field Φ induced by the sound
wave. A model is now assumed of a periodic fluxoid array which produces
a variation of the energy gap

$$\Delta(x, y) = \tfrac{1}{2}\Delta(T)[1 + (\cos ax)\cdot(\cos ay)] \qquad (59)$$

where $a = 2\pi/d$, and d is the fluxoid spacing. The one-dimensional representation of this is

$$\Delta(x) = \tfrac{1}{2}\Delta(T)[1 + \cos ax] \tag{60}$$

or

$$\Delta(\mathbf{k}) = \tfrac{1}{4}(2\pi)^3 \Delta(T)[2\delta(\mathbf{k}) + \delta(\mathbf{k} - a\mathbf{u}) + \delta(\mathbf{k} + a\mathbf{u})] \tag{61}$$

where $\mathbf{u}$ is a unit vector in the x direction. Cooper's results for the charge
response $\rho(\mathbf{q},\omega)$ contains terms that go as

$$\frac{1 - (\omega/2v_F|\mathbf{q} - a\mathbf{u}|)}{2v_F|\mathbf{q} - a\mathbf{u}|} \tag{62}$$

and

$$\frac{2}{\sqrt{L}}\left[\tan^{-1}\left(\frac{4v_F^2(\mathbf{q} \pm a\mathbf{u})\cdot\mathbf{q}}{\sqrt{L}}\right)\right] \tag{63}$$

$$L = 16v_F^4\left\{q^2|\mathbf{q} + a\mathbf{u}|^2\left(1 - \frac{\omega^2}{v_F^2|\mathbf{q} + a\mathbf{u}|}\right) - [(\mathbf{q} + a\mathbf{u})\cdot\mathbf{q}]^2\right\} \tag{64}$$

The attenuation that results from this will depend very strongly on the
direction between the magnetic field and the sound propagation. A logarithmic singularity occurs in the attenuation when

$$v_F|\mathbf{q} \pm a\mathbf{u}| = v_s\,q \tag{65}$$

The attenuation then becomes anomalous when $|\mathbf{q} \pm a\mathbf{u}| = (v_s/v_F)|\mathbf{q}|$; i.e.,
when the component of the sound vector $\mathbf{q}$ perpendicular to the magnetic
field is a, and when the component of $\mathbf{q}$ parallel to the magnetic field is
$(v_s/v_F)a$. If the orientation effects in the attenuation occur over wide ranges
of temperature and magnetic field, as suggested by Cooper that they may,
since they are so pronounced in the region of validity of the theory, then
they would provide a useful means of directly observing the fluxoid structure
of the mixed state.

An unexpected phenomenon that occurs in the attenuation of pure
niobium just above H_{c1} and at temperatures greater than about 4°K is an
initial decrease in the attenuation upon entering the mixed state; an illustration of this is shown by the attenuation *versus* field curve in Fig. 16. A
very plausible explanation of this effect was suggested by Forgan and Gough
(1966), who first reported observing it. They propose that the attenuation
decrease is caused by additional scattering of thermally excited unbound
quasiparticles (i.e., normal electrons in the superconducting matrix of the
mixed state). If this is the case, then there are two competing mechanisms
acting upon the attenuation at H_{c1}; first, as fluxoids enter the specimen
the attenuation increases because of absorption by the normal electron⌐

bound to these fluxoids (sometimes called bound excitations), and the density of these electrons will not be very temperature dependent. The mechanism described above competes with this, causing the attenuation to decrease as the mean free path of normal unbound electrons decreases. This latter mechanism will be highly temperature dependent, as the density of normal unbound electrons decreases with decreasing temperature. At low temperature the dominant effect near H_{c1} will be the absorption of ultrasound in the normal fluxoids, while at high temperatures the dominant effect is due to the decrease of normal electron mean free path. The crossover temperature for niobium is about $4°K$. The effect of the fluxoid scattering is estimated by assuming that the fluxoid can be characterized by a simple scattering diameter a. The density of fluxoids threading the specimen is $N = B/\Phi$, where B is the induction in the specimen and Φ is the flux quantum of the fluxoid. Then the mean free path of the electron in the presence of an induction B in the specimen is

$$1/l(B) = [1/l(0)] + (B/\Phi)a \tag{66}$$

In a pure superconductor, in which the attenuation is mean-free-path-limited, the BCS expression for attenuation will be modified as follows to take account of fluxoid scattering:

$$\frac{\alpha_s}{\alpha_n} = \frac{2}{\exp(\Delta/kT) + 1} \frac{l(B)}{l(0)}$$

$$= \frac{2}{\exp(\Delta/kT) + 1} \frac{1}{1 + [Bal(0)/\Phi]} \tag{67}$$

An estimate of the scattering diameter a may be obtained by differentiating the above expression,

$$\frac{1}{\alpha_n} \frac{d\alpha_s(T, B)}{dB} \approx \frac{2}{\exp(\Delta/kT) + 1} \frac{al(0)}{\Phi} \tag{68}$$

A value of 1.5×10^{-7} cm is arrived at for niobium.

An explicit microscopic calculation of the scattering cross section of quasiparticles by isolated flux vortices in a clean type II superconductor has recently been carried out by Cleary (1968). This is done for κ values near $1/\sqrt{2}$, applicable to niobium and vanadium. The excitation of quasiparticles is determined from the BCS equations in which the vortex modifies the order parameter $\Delta(T)$ from its value in the absence of a vortex. The new wave functions are determined in the presence of the additional potentials using the Ginzburg–Landau formulations. Solutions to this are obtainable for temperatures near T_c. The approximations used by Cleary for the variation of energy gap and magnetic field with r, the distance from the center of a vortex, are

$$\Delta(r) = \Delta(T)[1 - \exp(-r^2/4\lambda^2)]^{1/2} \tag{69}$$

$$H(r) = (c/4e\lambda^2)\exp(-r^2/4\lambda^2) \tag{70}$$

where λ is the penetration depth at temperature T. The above expressions were improved by numerical correction of the Ginzburg–Landau equations for κ values appropriate to niobium and vanadium. The scattering amplitude $f(\phi)$ is obtained by carrying out a partial wave expansion in the angular variable ϕ, the coordinate about the vortex axis. The differential scattering cross section is

$$d\sigma/d\phi = |f(\phi)|^2 \tag{71}$$

Cleary evaluated the scattering cross section,

$$\sigma = \int_{-\pi}^{\pi} d\phi \, (d\sigma/d\phi)(1 - \cos \phi) \tag{72}$$

as a function of particle energy E, and of θ, the polar angle of the trajectory with respect to the magnetic field direction. His results are shown in Fig. 18, in which the ordinate σ is plotted in units of $\lambda(T) \sin \theta$ against energy in

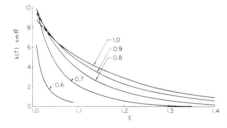

Fig. 18. Scattering cross section *vs.* energy (from Cleary, 1968).

units of $\Delta(T)$ for values of the parameter $\sin \theta$ from 0.2 to 1. The significant point here is that σ is very large for energies $E \sim \Delta(T)$, diverging logarithmically for $E = \Delta$. The thermal average of the cross section remains finite at all temperatures, increasing as temperature is lowered below T_c.

C. ULTRASONIC ATTENUATION IN THE MIXED STATE—EXPERIMENTS

1. *Temperature Dependence of Energy Gap from Attenuation*

The most important metal upon which experiments on attenuation in the mixed state have been done is niobium. This is because it is the only elemental type II superconductor that is available in relatively high purities, such that ql values substantially greater than unity are accessible to experiment, and in single-crystal form. More recently vanadium has become the object of mixed-state measurements, and results of these will be presented in this section.

Extensive investigations of the temperature dependence of the ultrasonic attenuation of niobium in the normal and superconducting states were carried out by Perz and Dobbs (1966). Their measurements were done over

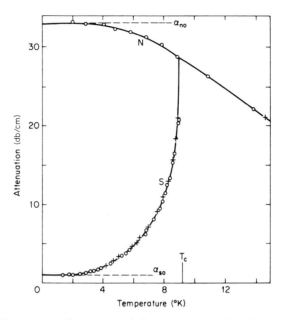

Fig. 19. Temperature dependence of ultrasonic attenuation of pure Nb (from Perz and Dobbs, 1966).

a range of ql values from 1.5 to 7, and for several different directions of sound propagation. The attenuation curve in Fig. 19 is typical of the variation observed in the temperature range from 1.4 to 14°K ($T_c = 9.2°$K). The branch of the curve marked N was taken with magnetic fields sufficiently large to make the sample completely normal. The normal-state attenuation increases with decreasing temperature down to $\sim 3°$K, reflecting the increase in the electron mean free path, which is dominated by phonon scattering. Below this temperature the mean free path is limited by impurity scattering, and so is independent of temperature.

Perz and Dobbs used the attenuation data to determine the anisotropy of the superconducting energy gap. At reduced temperatures less than about 0.45 the energy gap $\Delta(T)$ changes very little with temperature, and may be approximated by $\Delta(0)$. Then, by inverting the BCS equation, it is apparent that the slope of the straight line plot of $\ln^{-1}[2(\alpha_n/\alpha_s) - 1]$ versus T/T_c gives $1/\Delta(0)$. Assuming that the variation of the gap just below T_c goes as

$$\left[\frac{\Delta(T)}{\Delta(0)}\right]^2 = B\left(1 - \frac{T}{T_c}\right) \tag{73}$$

the attenuation measurements can be used to determine the constant B. Figure 20 shows a plot of the temperature dependence of the normalized energy gap versus reduced temperature from propagation measurements in the (111) direction. The discrepancy between the experimentally determined

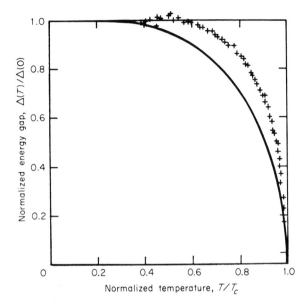

Fɪɢ. 20. Temperature dependence of the energy gap in Nb (from Perz and Dobbs, 1966).

energy gap and that predicted by the BCS theory was attributed by Perz and Dobbs to the high value of the coupling ratio T_c/θ_D. Their measured values of the anisotropy in the energy gap yielded extreme values in the parameter $A = 2\Delta(0)/kT_c$ of 3.52 in the (100) and (110) directions, and 3.61 in the (111) direction.

2. *Effective Energy-Gap Model for Entire Mixed State*

The agreement between temperature dependence of the attenuation in niobium with the BCS theory is relatively good, and attempts were made by Ikushima *et al.* (1966) and Tsuda *et al.* (1966) to explain the mixed-state attenuation in terms of the BCS theory. An assumption is made of an effective energy gap $\bar{\Delta}(H, T)$ over the entire mixed state, which is proportional to the root mean square of the order parameter ψ,

$$\bar{\Delta}(H, T) \propto \langle|\psi|^2\rangle^{1/2} \tag{74}$$

The Abrikosov theory relates the magnetization to the square of the order parameter,

$$M(H, T) \propto \langle|\psi|^2\rangle \tag{75}$$

for magnetic field very near H_{c2}, and Ikushima (1966) assumes its validity for all H in the mixed state. Then it is possible to relate the mixed-state attenuation to the magnetization through the BCS relation,

$$\frac{\alpha_S}{\alpha_n} = \frac{2}{\exp\{[\Delta(O, T)/kT][M(H, T)/M(H_{c1}, T)]^{1/2} + 1\}} \qquad (76)$$

where

$$\frac{\bar{\Delta}(H, T)}{kT} = \left[\frac{(\bar{\Delta}H, T)}{\Delta(O, T)}\right]\left[\frac{\Delta(O, T)}{kT}\right] = \left[\frac{M(H, T)}{M(H_{c1}, T)}\right]^{1/2}\left[\frac{\Delta(O, T)}{kT}\right] \qquad (77)$$

The results of measurements by Ikushima on niobium of resistance ratio $R(300°K)/R(4.2°K) = 1000$ at frequency of 90 MHz are shown in Fig. 21. The fit of the experimental curves to the BCS expression is reasonably good for temperatures near T_c and for magnetic fields near H_{c2}, but for temperatures below about 7°K the assumption of the validity of the Abrikosov relation between magnetization and order parameter is apparently an extremely

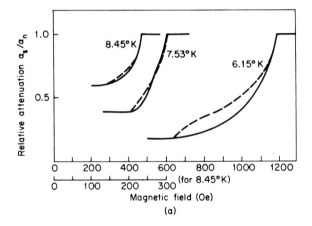

(a)

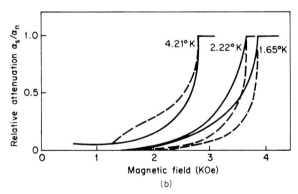

(b)

FIG. 21. Comparison of the measured attenuation in the mixed state with that calculated from BCS relation (from Ikushima, 1966).

poor one. Subsequent measurements by Tsuda *et al.* (1966) on niobium over a range of purities, from RR100 to RR600, yielded closer fit to BCS with decreasing impurity, but still showed considerable deviation. Tsuda attempted to explain the deviation from BCS on the basis of anisotropy of the energy gap, or possibly the existence of multiple gaps, but concluded that these could not account for the observed deviations.

3. *Measurements near H_{c2}*

The theory of Maki is the most comprehensive description of ultrasonic attenuation in the mixed state that is available. The most outstanding single feature of this theory is the prediction that for $l \gg \xi$ and for magnetic fields near to H_{c2} such that $(H_{c2} - H)/H_{c2} \ll 1$, $1 - [\alpha(H)/\alpha_n]$ is proportional to $(H_{c2} - H)^{1/2}$; but note that this result also follows simply from the BCS expression for attenuation, as indicated in Eqs. (45)–(49). For the very impure

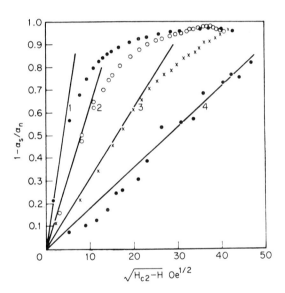

FIG. 22. Magnetic field dependence of attenuation near H_{c2} for $ql \ll 1$ (from Tsuda *et al.*, 1966).

limit, $l \ll \xi_0$, the result of McClean and Houghton (1967) is that $1 - [\alpha(H)/\alpha_n]$ is proportional to $H_{c2} - H$. A number of experimental groups have demonstrated the validity of these conclusions. The data of Tsuda *et al.* which were taken at a frequency of 30 MHz, and so correspond to $ql \ll 1$, are shown in Fig. 22 for RR100 to RR600; linearity with $(H_{c2} - H)^{1/2}$ is reasonable except for the least-pure sample. Measurements by Gottlieb *et al.* (1967b) were done on pure niobium of RR700 at a frequency of 210

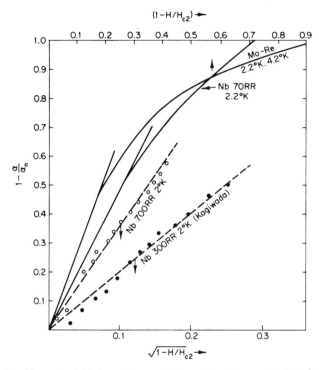

FIG. 23. Magnetic field dependence of attenuation for pure limit, $l/\xi \gg 1$ and for dirty limit, $l/\xi \ll 1$.

MHz, corresponding to a ql near unity. Figure 23 shows a plot of this data, linear with $(H_{c2} - H)^{1/2}$. Also shown are data of Kagiwada *et al.* (1967) on niobium of RR300 at a frequency of 104 MHz, and for the same temperature, 2°K. Note that the slope for the less-pure material is about one half that for the purer material.

The theory of McClean and Houghton, as outlined in Section IV, indicates that near H_{c2} the quantity $1 - [\alpha(H//\alpha_n]$ is proportional to $(H_{c2} - H)$ in very impure materials, for which $l/\xi \ll 1$. Experiments were done by Gottlieb *et al.* (1967b) in two systems to verify this behavior: niobium of resistivity ratio 70 with longitudinal sound waves propagated along the (110) direction, and the alloy Mo–25% Re of resistivity ratio 2.7. The sound propagation direction for the Mo–25% Re was along the (100) direction. Vacuum-deposited thin-film CdS transducers were used, and measurements made near 1 GHz. The attenuation curves for the impure niobium and the Mo–25% Re are also shown in Fig. 23, corresponding to the upper magnetic field scale. The linearity near H_{c2} is very good, in accordance with theory.

The normalized slope $(1/\alpha_n)(\partial\alpha_s/\partial H)_{H_{c2}}$, that is calculated for Mo–25% Re on the basis of the theory of McClean and Houghton is 1.2 G^{-1}, using the values $v_F = 5 \times 10^7$ cm/sec (from specific-heat measurements), electron mean free path 3.4×10^{-6} cm estimated from resistance measurements, $\kappa = 4$ from magnetization measurements, and the quantity $L(p)$ estimated to be 1. The slope measured from the experimental curve for Mo–25% Re is 2.7 G^{-1}, in reasonably good agreement with the theory.

An interesting set of attenuation measurements has recently been made by Tittman and Bömmel (1968) on an alloy system of apparently intermediate purity; i.e., the ratio ξ/l neither very small nor very large compared with unity. Their measurements on V–5% Ta, for which they estimate $\xi/l = 5$, are shown in Fig. 24. They have fitted their data to a field depen-

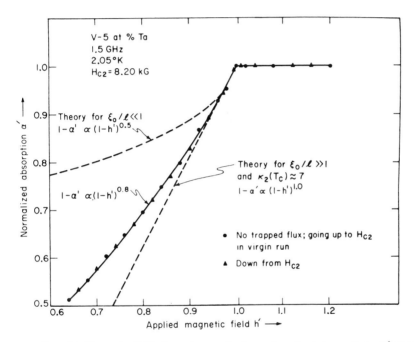

FIG. 24. Magnetic field dependence of attenuation for intermediate l/ξ (from Tittman and Bömmel, 1968).

dence $(H_{c2} - H)^{0.8}$; on the same figure are shown the clean and dirty limits, with the square root and linear dependences, respectively.

A detailed comparison of the experimental values of the slopes of the attenuation curves and those calculated on the basis of the Maki theory for the clean limit has been done by several groups. Kagiwada et al. (1967) fitted their data for the RR300 niobium, which corresponded to a ql of 0.19 at 104 MHz sound frequency, to the Maki theory by assuming a value of density of states $N(0) = 1.5 \times 10^{34}$ cm^{-3} erg^{-1}. The density of states estimated from specific-heat measurements of McConville and Serin (1965) and Leupold and Boorse (1964) is 5.6×10^{34} cm^{-3} erg^{-1}. The slope measured by Gottlieb *et al.* on niobium of RR700 for about the same ql value and temperature. as Kagiwada and Levy's measurements on niobium of RR300 is about twice as great as the latter, as can be seen on Fig. 23. Since the value of ql is the same, both samples should have yielded the same attenuation slope, according to the Maki theory for the pure limit. This suggests that slope values disagree with theory even worse with increasing purity. This effect with increasing purity was noted by Tsuda *et al.* (1966), who indicated the following values of $N(0)$ as adjustable parameter necessary to achieve agreement with theory for the attenuation slope: $N(0) = 5.8 \times 10^{34}$ cm^{-3} erg^{-1} for RR100, 2.5×10^{34} cm^{-3} erg^{-1} for RR200, 0.8×10^{34} cm^{-3}

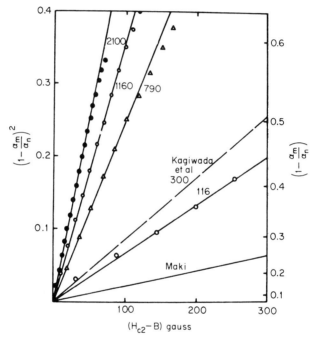

FIG. 25. Variation of attenuation slope near H_{c2} with purity (from Forgan and Gough, 1968).

erg^{-1} for RR400, and 0.2×10^{34} cm^{-3} erg^{-1} for RR600. The same trend at still higher purities was found by Forgan and Gough (1968), whose results are summarized in Fig. 25. For niobium with RR116–2100 they measure values of slope of attenuation for constant ql, which vary almost linearly with purity. It would thus appear that the theory is at fault with respect to purity dependence.

It is worth pointing out here a serious difficulty that has been encountered in measuring attenuation slopes in pure niobium near H_{c2} which may call into question the meaningfulness of comparisons with the Maki theory.

The Maki theory evaluates the slope of attenuation at H_{c2}, which in pure material becomes steeper, the closer the field is to H_{c2}. In an attempt to measure the true limiting field Gottlieb *et al.* (1968a) performed experiments to yield a field resolution of 1 G in the region of H_{c2}. It was found, however, that instead of a monotonic increase in the slope, the attenuation curve shows a marked rounding within about 10 G of H_{c2} with no discontinuous change at this point. The characteristic rounding observed is shown in Fig. 26 for a number of orientations between field direction and sound direction. We believe this rounding is intrinsic in origin for the following reasons: (1) it is independent of field-sweep rate, temperature, or sound power level, and (2)

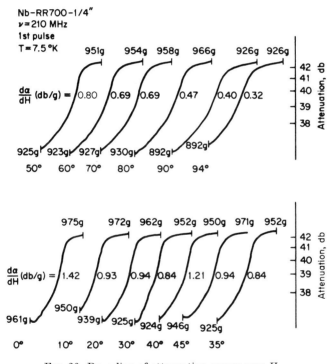

FIG. 26. Rounding of attenuation curves near H_{c2}.

it does not arise from magnetic-field nonuniformity, since one expects the field to be highly uniform near H_{c2} even for poor geometry, but as a check against this possibility, measurements were done in a thin slab of niobium $\frac{1}{4}$ in. in diameter and only 12 mils thick in a field parallel to the plane of the sample so that the field could not be significantly distorted. Under these conditions the same rounding was observed. Thus, there appears to be an attenuation mechanism active which is not accounted for by the theory, and which is most prominent just at that point where the theory evaluates the slope. As a matter of consistency, the slope measurements of Gottlieb *et al.*

were taken at the point of steepest ascent, typically about 10 G below H_{c2}. Measurements of dB/dH *versus* H done by Ikushima *et al.* (1966) show a discontinuity at H_{c2}, indicating a first-order phase transition, while the attenuation seems to indicate a second-order transition. This is further evidence of the inadequacy of the theory just in the critical region.

Other aspects of the Maki theory that may be explored experimentally are the temperature dependence of the attenuation slope, and the orientation dependence, as extended by Kinder. In order to compare experimental results with the Maki–Kinder theory, measurements were made by Gottlieb *et al.* (1968b) on the dependence of $(d\alpha/dH)_{H_{c2}}$ on magnetic field direction. These were carried out on electron-beam zone-refined niobium of resistivity ratio $R_{300}/R_{4.2} = 700$. Cylindrical samples $\frac{1}{4}$ in. in diameter, $\frac{1}{4}$–1 in. long, with the axis in the (100) direction were used, with longitudinal sound waves at a frequency of 210 MHz. This corresponds to a ql value of nearly unity. The results of these measurements are shown in Fig. 27. These are

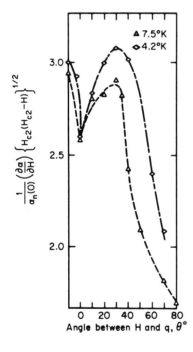

Fig. 27. Measured anisotropy of normalized attenuation slope in Nb.

in qualitative disagreement with the theoretical curves, which indicate a minimum slope for angle between 0° and 90°, while the experiments indicate a maximum in the reduced slope. It must be pointed out, however, that there is an essential difficulty when making experimental determinations of the anisotropy associated with interaction between sound wave direction and

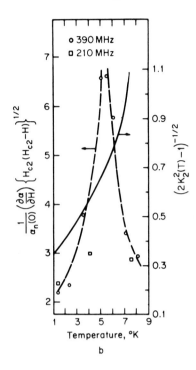

FIG. 29. Temperature dependence of normalized attenuation slope in Nb.

differs violently from that predicted by theory. At low temperatures the agreement between theory and experiment is not bad, taking into account the factor $T/H_{c2}^{1/2}$. These results cast serious doubt on the validity of the theory, which would be expected to yield closer agreement at higher temperatures.

4. *Measurements near* H_{c1}

The effects on the ultrasonic attenuation of isolated fluxoids near the lower critical field of a type II superconductor were first reported for niobium by Forgan and Gough (1966) and shortly later by Tsuda *et al.* (1966). The effect on the mixed-state attenuation is typically as shown in the curve in Fig. 16. There is a decrease in the attenuation just above H_{c1} that is much smaller than the change in attenuation experienced over the entire mixed state. No such decrease of attenuation is observed in impure niobium.

The same phenomenon described above was found by Sinclair and Leibowitz (1968) in pure vanadium, whose superconducting properties are very similar to niobium. A set of their curves for attenuation *versus* B over a range of reduced temperatures is shown in Fig. 30. From these curves they estimate a scattering diameter of 2.4×10^{-6} near T_c. The microscopic theory for this type of scattering proposed by Cleary (1968) describes the properties of isolated vortices for a type II superconductor with $\kappa = 0.707$

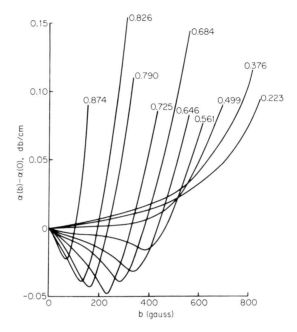

Fɪɢ. 30. Attenuation *vs.* induction near H_{c1} for V (from Sinclair and Leibowitz, 1968).

and for T near T_c. The scattering diameter so calculated is 2.1×10^{-6} cm, in good agreement with the result for vanadium, but considerably larger than for niobium.

   Some complex behavior in the hysteretic properties of the above phenomenon which cannot be accounted for in terms of the proposed scattering mechanism has recently been observed by Gottlieb *et al* (1968c). The data were obtained on a niobium rod of resistance ratio 700, $\frac{1}{4}$ in. long by $\frac{1}{4}$ in. in diameter. Longitudinal acoustic waves at a frequency of 510 MHz were used. Figure 31 shows the direct recorder tracing of attenuation *versus* magnetic field for both parallel and perpendicular orientations between field and sound propagation direction. The specimen was cooled in zero magnetic field from above T_c, then cycled in both senses of magnetic field direction to $H > H_{c2}$. The following observations can be made for the perpendicular orientation. On the virgin run (run 1) in increasing field the attenuation undergoes a small peak before the decrease at H_{c1}. This peak is not observed for H parallel to q, but disappears gradually as H is rotated away from the perpendicular orientation. This peak appears somewhat enhanced for the magnetic field increasing in the negative sense (run 2). The onset of the dip is around 4°K, while the onset of the peak is about 5°K, so these two phenomena may or may not be related. While the evidence is not very strong, it is interesting to note that the peak appears under conditions similar to the resonance absorption predicted by Cooper *et al.* (1966)

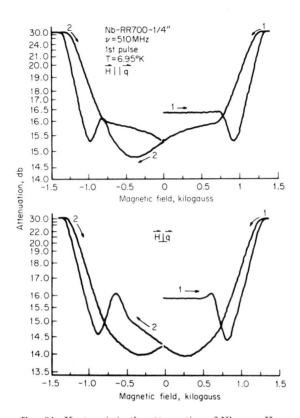

FIG. 31. Hysteresis in the attenuation of Nb near H_{c1}.

at magnetic fields near H_{c1}. It is further observed that the attenuation in decreasing field (run 1) goes to a significantly lower value below H_{c1} than the lowest value in increasing field. Assuming that the lowest value of attenuation corresponds to a certain fluxoid density, we would expect that the trapped flux would cause the attenuation minimum to be located at a lower external field in decreasing field, but that the minimum value would remain the same.

Figure 32 shows the hysteresis curves in the attenuation when the maximum field is kept below H_{c2}. In Fig. 32(a) the field is increased to 1 KG, a point at which the attenuation has gone through its minimum, and just returned to its zero-field value. When the field is now decreased, the attenuation increases, quite contrary to what is expected if flux is excluded. In Fig. 32(b) the procedure is carried out with the field increased further into the mixed state before stopping. Upon decreasing magnetic field, the attenuation again increases. At this point, however, the fluxoid density must be

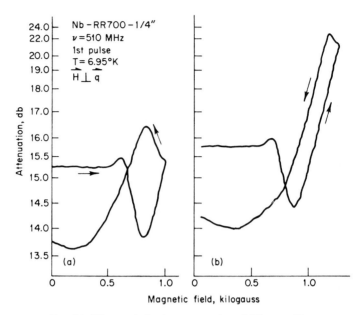

FIG. 32. Hysteresis in the attenuation of Nb near H_{c1}.

much greater than its value at the attenuation minimum, suggesting that the phenomenon is not associated with a mechanism involving isolated fluxoids.

REFERENCES

Abrikosov, A. A. (1957). *JETP USSR* **32**, 1442. [*Soviet Phys. JETP (English Transl.)* **5**, 1174].
Andreev, A. F. (1968). *Soviet Phys. JETP (English Transl.)* **26**, 428.
Andreev, A. F., and Bruk, Yu. M. (1966). *Soviet Phys. JETP (English Transl.)* **23**, 942.
Bardeen, J., Cooper, L. N., and Schrieffer, J. R. (1957). *Phys. Rev.* **108**, 1175.
Bogolyubov, N. M., Tolmachev, V. V., and Shirkov, D. V. (1959). "A New Method in the Theory of Superconductivity." Chapman and Hall, London.
Bolef, D. I. (1966). "Physical Acoustics" (W. P. Mason, ed.), Vol. IV, Part A, p. 113. Academic Press, New York.
Bozorth, R. M., and Chapin, D. M. (1942). *J. Appl. Phys.* **13**, 320.
Cape, J. A., and Zimmerman, J. M. (1967). *Phys. Rev.* **153**, 416.
Caroli, C., and Matricon, J. (1965). *Physik Kondensierten Materie* **3**, 380.
Cleary, R. M. (1968). *Phys. Rev.* **175**, 587.
Cooper, L. N., Houghton, A., and Lee, H. J. (1966). *Phys. Rev.* **148**, 198.
deKlerk, J. (1966). "Physical Acoustics" (W. P. Mason, ed.), Vol. IV, Part A, p. 195. Academic Press, New York.
Eliashberg, G. M. (1960). *JETP USSR* **38**, 996 [*Soviet Phys. JETP (English Transl.)* **11**, 696].
Finnemore, D. K., Stromberg, T. F., and Swenson, C. A. (1966). *Phys. Rev.* **149**, 231.

Forgan, E. M., and Gough, C. E. (1966). *Phys. Letters* **21**, 133.

Forgan, E. M., and Gough, C. E. (1968). *Phys. Letters* **26A**, 602.

Ginzburg, V. L., and Landau, L. D. (1950). *JETP USSR* **20**, 1064; Ginzburg, V. L. (1955). *Nuovo Cimento* **2**, 1234.

Gorkov, L. P. (1959). *JETP USSR* **36**, 1918; **37**, 883, 1407 [*Soviet Phys. JETP* (*English Transl.*) **9**, 1364; **10**, 593, 998].

Gottlieb, M., and Jones, C. K. (1966). *Phys. Letters* **21**, 270.

Gottlieb, M., Jones, C. K., and Garbuny, M. (1967a). *Phys. Letters* **24A**, 585.

Gottlieb, M., Jones, C. K., and Garbuny, M. (1967b). *Phys. Letters* **25A**, 107.

Gottlieb, M., Jones, C. K., and Garbuny, M. (1968a). *IEEE Symp. Ultrasonics.*

Gottlieb, M., Garbuny, M., and Jones, C. K. (1968b). *Phys. Letters* **27A**, 710.

Gottlieb, M., Garbuny, M., and Jones, C. K. (1968c). *Phys. Letters* **28A**, 148.

Gough, C. E. (1968). *Solid State Commun.* **6**, 215.

Hohenberg, P. C., and Werthamer, N. R. (1967). *Phys. Rev.* **153**, 493.

Holstein, T. (1956). Memo 60–94698–3–M17, Westinghouse Res. Lab.

Ikushima, A., Fujii, M., and Suzuki, T. (1966). *J. Phys. Chem. Solids* **27**, 327.

Kagiwada, R., Levy, M., and Rudnick, I. (1967). *Phys. Rev. Letters* **18**, 74.

Kinder, H. (1968). *Phys. Letters* **26A**, 319.

Landau, L. D. (1937). *JETP USSR* **7**, 371 [*JETP USSR* **13**, 377].

Leibowitz, J. R. (1964). *Phys. Rev.* **136**, A22.

Leibowitz, J. R, (1968). *Proc. XI Intern. Conf. Low Temperature Phys.* **2**, 807 (Allen, Finlayson, and McCall, Eds.).

Leibowitz, J. R., and Fossheim, K. (1968). *Phys. Rev. Letters* **21**, 1246.

Leupold, H. A., and Boorse, H. A. (1964). *Phys. Rev.* **134**, A1322.

Livingston, J. D. (1968). *In* "Superconductivity" (R. D. Parks, ed.), Chapter 21. Dekker, New York.

Maki, K. (1964). *Physics* **1**, 127.

Maki, K. (1967b). *Phys. Rev.* **156**, 437.

Maki, K., and Suzuki, T. (1966). *Phys. Rev.* **139**, 868.

Maki, K., and Cyrot, M. (1967a). *Phys. Rev.* **156**, 433.

McConville, T., and Serin, B. (1965). *Phys. Rev.* **140**, A1169.

McLean, F. B., and Houghton, A. (1967). *Phys. Rev.* **157**, 350.

Meshkovsky, A. G., and Shalnikov, A. I. (1947). *J. Phys. USSR* **11**, 1.

Perz, J. M., and Dobbs, E. R. (1966). *Proc. Roy. Soc.* **A296**, 113.

Peverley, J. R. (1966). "Physical Acoustics" (W. P. Mason, ed.), Vol. IV, Part A, p. 353. Academic Press, New York.

Pippard, A. B. (1953). *Proc. Roy. Soc.* **A216**, 547.

Schawlow, A. L. (1956). *Phys. Rev.* **101**, 573.

Shapira, Y., and Neuringer, L. J. (1967). *Phys. Rev.* **154**, 375.

Sinclair, A. C. E., and Leibowitz, J. R. (1968). *Phys. Rev.* **175**, 596.

Tittman, B. R., and Bömmel, H. E. (1968). *Phys. Letters* **28A**, 396.

Tsuda, N., Koike, S., and Suzuki, T. (1966). *Phys. Letters* **22**, 414.

—2—

Ultrasonic Investigation of
Phase Transitions and Critical Points

CARL W. GARLAND

Department of Chemistry and Center for Materials Science and Engineering,
Massachusetts Institute of Technology,
Cambridge, Massachusetts

I. Introduction

In recent years, ultrasonic studies have played an increasingly important role in characterizing the behavior of systems near cooperative phase transitions and critical points. One advantage of ultrasonic measurements is the fact that static and dynamic properties can be measured simultaneously. Low-frequency acoustic velocities provide precise information about the equilibrium adiabatic properties of the system, and the effects of temperature, pressure, and external fields can be readily studied. Ultrasonic attenuation data provide direct information about the dynamic behavior, and, from the dependence on frequency, as well as temperature, much can be learned about the mechanisms involved. Theoretically, new ways of describing critical phenomena in terms of fluctuation correlations have been of great importance. Emphasis on the dynamic aspects of the theory has increased markedly, and this has naturally focused more attention on ultrasonic work.

The ultrasonic investigation of phase transitions is still a rapidly developing field, with vigorous interplay between theory and experiment. This makes it an exciting field, but also one which is difficult to review definitively. Indeed, this chapter should be viewed more in the nature of a progress report. An attempt has been made to summarize in rather extensive tables all the pertinent experimental literature. Detailed comments are made on only a few of the better-studied systems. It is hoped that these systems will be sufficiently typical to characterize each of the types of transition described here. An overall discussion of a variety of theoretical ideas is given in Section II; more explicit applications of these theories are developed in subsequent sections.

A number of the investigations described in this chapter had not appeared as journal articles as of June 1969, when the chapter was written. Such work is cited with a 1969 date, but many of these papers may not appear until 1970. Although our principal concern is with ultrasonic measurements, some results at sonic ($f \lesssim 20$ kHz) and hypersonic ($f \gtrsim 1$ GHz) frequencies have also been included. In connection with equilibrium behavior, it is important to measure the velocity at low enough frequencies that the observed values are equal to the zero-frequency limit. In connection with dynamic behavior, it is of interest to know the magnitude and frequency range of the dispersion. The large variations in velocity u and amplitude attenuation α near a transition are usually referred to as "anomalous." We shall continue this usage, although such special variations are now expected and might better be described as "critical." In contrast to fluids, which are isotropic and do not transmit shear waves, the situation in solids is more complex. For single crystals, one must specify the direction and polarization of the sound wave or specify the values of the appropriate elastic stiffness constants c_{ij}.

It is difficult enough to avoid notational confusion in one's own work, but such difficulties are compounded in discussing a variety of papers by chemists, physicists, and engineers. The symbols used in this chapter are internally consistent, but may differ from those used in the original papers. Three warnings: κ is used for the compressibility (usually with the subscript S or T) and for the inverse correlation length; η is used for the order parameter, for the shear viscosity, and for the Ornstein–Zernicke critical index; α is used once or twice for the thermal expansion coefficient and the heat-capacity critical exponent, and it is used ubiquitously for the ultrasonic attenuation.

One obvious omission from this chapter is a discussion of transitions in superconductors. Such second-order transitions are closely related to the cooperative transitions discussed here; their exclusion is due only to the author's complete ignorance of the literature on superconductivity. Those interested should refer to Lynton (1969) or Ginsberg and Hebel (1969).

II. General Discussion of Theory

This section is intended to provide an overview of theoretical developments which relate to ultrasonic properties in critical regions. There are many papers concerned with the application of theory to individual systems or to special types of transitions; such papers will not be discussed here, but will be treated in later sections wherever they are appropriate. Fairly general ideas will be introduced in the present section, with a suppression of details and an emphasis on concepts. In Section II,E, in particular, an attempt is made to show the relationship among a large variety of recent fluctuation–correlation theories.

The first three parts (Sections II,A–C) are devoted to equilibrium or static properties—the behavior of the zero-frequency limit of the sound velocity. This subject has not been extensively developed, and more work is clearly needed. The last two parts (Sections II,D,E) are concerned with dynamic properties—velocity dispersion and attenuation of sound. Not much theoretical emphasis has been placed on the frequency dependence of the velocity, but a great deal of work has been done on ultrasonic attenuation. This field is developing rapidly, and the definitive formulation is by no means yet established. Thus, Section II,E should be viewed as an interim report.

A. GENERALIZED PIPPARD EQUATIONS

The Clausius–Clapeyron equation is a well-known relation between the slope of a first-order transition curve, dp/dT, and the discontinuities in entropy and volume associated with the transition. For second-order transitions (discontinuities in the second derivatives of the free energy), the Ehrenfest equations interrelate the slope of the transition line with the finite changes in C_p, α, and κ_T, the heat capacity at constant pressure, the isobaric volume coefficient of thermal expansion, and the isothermal compressibility, respectively. In the vicinity of a lambda-point transition, these thermodynamic quantities undergo exceedingly rapid variations, but no simple discontinuities; thus, one cannot apply the Ehrenfest equations. Pippard (1956) was the first to propose two new phenomenological equations relating such "anomalous" quantities near a λ-transition. In the exact form derived by Buckingham and Fairbank (1961), these Pippard equations are

$$C_p/T = (dp/dT)_\lambda \, V\alpha + (dS/dT)_t \tag{1}$$

$$\alpha = (dp/dT)_\lambda \, \kappa_T + (1/V)(dV/dT)_t \tag{2}$$

where $(dS/dT)_t$ and $(1/V)(dV/dT)_t$ are slowly varying quantities which can be treated as constants. The behavior of these quantities has been analyzed in terms of a compressible Ising model by Renard and Garland (1966b).

A generalization of these equations in terms of stress–strain variables was given by Garland (1964a); Janovec (1966) extended the theory to anisotropic dielectrics by including electric field variables. For our present purposes, the most important new result is the equation

$$s_{ij}^{T,E} = -(\partial T_\lambda/\partial X_j)\alpha_i + s_{ij}^{t,E} \tag{3}$$

where $s_{ij}^{T,E}$ is the isothermal elastic compliance at constant electric field, α_i is the linear thermal-expansion coefficient at constant stress and field, $(\partial T_\lambda/\partial X_j)$ is the variation of the lambda temperature with a given external stress, and $s_{ij}^{t,E}$ is effectively a constant. Thus, it is predicted that the anomalous increase in elastic compliance will be directly proportional to that in the thermal expansion whenever there is a coupling between thermal and mechanical variables. Equation (3) provides a convenient test of the consistency of data near lambda points, and has been used to analyze the elastic behavior near the α–β transition in quartz (Garland, 1964a) and the ferroelectric transition in triglycine sulfate (Janovec, 1966).

Since ultrasonic-velocity measurements yield adiabatic elastic constants, it is also convenient to give an expression for the adiabatic compressibility κ_S:

$$\kappa_S = \kappa_{S,\lambda} - \frac{T(dS/dT)_t^2}{VC_p(dp/dT)_\lambda^2} \tag{4}$$

where $\kappa_{S,\lambda}$ is the (finite) value of κ_S at the lambda point. For a fluid, this immediately gives an expression for the longitudinal sound velocity, since $\kappa_S = 1/\rho u^2$, whereas, for a solid,

$$\kappa_S = \sum_{i,j=1}^{3} s_{ij}^S .$$

The low-frequency ultrasonic velocity in helium near its λ-point has been discussed in terms of Eq. (4) by Chase (1959) and by Barmatz and Rudnick (1968). In the case of solids, the ultrasonic velocities in ammonium chloride and β-brass have been analyzed in these terms by Garland and Jones (1963) and by Garland (1964b), respectively.

B. COMPRESSIBLE ISING MODEL

Renard and Garland (1966a) have considered the static elastic behavior of a two-dimensional Ising ferromagnet in which the spins are located on mass particles which form a compressible lattice. Since the two-dimensional Ising problem has an analytic solution, it was possible to derive explicit expressions for the configurational contribution to the three independent elastic constants of a square lattice. The basic feature of the model is weak coupling between the lattice and spin systems. As a result of this assumption, the free energy can be written as the sum of two independent contributions: one due to a completely disordered lattice and the other to the ordering of a spin system (called the Ising contribution). The behavior of the disordered lattice is assumed to be similar to that of any normal crystal and should be adequately predicted by quasiharmonic theories. For the Ising contribution to the elastic constants, it was found that the constant-area temperature dependence of c_{11} (the compressional stiffness constant) is dominated by a term proportional to the configurational specific heat. In contrast, the angle shear constant (denoted by c_{44}) has a spin contribution which is directly proportional to the Ising internal energy. The other shear constant C' is a more complicated function of temperature, but it is similar to c_{44} in that both constant-area shear constants have an inflection point of infinite slope at T_c.

Although an analytic solution to the three-dimensional Ising problem is not yet available, the elastic constants of a cubic lattice will be very similar in form to those of a square lattice. Accordingly, the constant-volume elastic constants of a simple-cubic Ising lattice can be represented by

$$\frac{1}{\kappa_T} = c_{11}^T - \frac{4}{3}\,C' = \frac{1}{\kappa_{T,\,dl}} - \frac{vT}{J^2}\frac{C_I(0,\,H)}{N}\left(\frac{dJ}{dv}\right)^2 + \frac{v}{J}\frac{U_I(0,\,H)}{N}\left(\frac{d^2J}{dv^2}\right) \qquad (5)$$

$$C' = C'_{dl} - mG(0,\,H) - \frac{nU_I(0,\,H)}{NJ} \qquad (6)$$

$$c_{44} = c_{44,\,dl} - \frac{lU_I(0,\,H)}{NJ} \qquad (7)$$

where $C_I(0,H)/N$ and $U_I(0,H)/N$ are, respectively, the configurational heat capacity per spin and the Ising internal energy per spin as a function of $H = J/kT$; J is the interaction energy between nearest-neighbor spins; $G(O,H)$ is the three-dimensional analog of the function defined for a square lattice; and v is the unit cell volume. The isothermal character of the reciprocal compressibility $1/\kappa_T$ and the compressional stiffness c_{11}^T is denoted by the letter T; this is not necessary for the shear constants c_{44} and C', since the isothermal and adiabatic values are identical. The subscript "dl" indicates a disordered-lattice contribution, which corresponds to the essentially normal variations observed at temperatures far above T_λ. The coefficients m, n, and l are temperature-independent quantities, defined by Eqs. (39), (40), and (55) of the paper by Renard and Garland (1966a), except that σ must be changed to v. As shown explicitly in Eq. (5) and also by the detailed expressions for m, n, and l, it is necessary that the interaction energy J be a function of a given strain if the corresponding elastic stiffness is to exhibit an "anomalous" behavior.

This model has been used with considerable success to analyze the ultrasonic velocities in ammonium chloride (Garland and Renard, 1966b), and this analysis will be discussed in Section VII,B. It has also been applied to uranium dioxide by Brandt and Walker (1968). The principal weakness of the model lies in its disregard of fluctuations in the strain within the lattice. Every unit cell is treated as having an identical set of lattice parameters which are themselves stress-dependent, whereas, in fact, the lattice will be characterized by fluctuating parameters near a critical ordering point.

C. STATIC SCALING RESULT

There has been considerable recent progress on the general theory of equilibrium critical phenomena with special emphasis on the values of various critical-point exponents (Widom, 1965; Fisher, 1967; Kadanoff *et al.*, 1967). So far, these static scaling laws have involved "elastic" properties only in the case of the liquid–vapor transition. In that case, the isothermal compressibility κ_{T} has the same strong singularity as C_{p}, while the adiabatic compressibility κ_{S} diverges like C_{V}, which is only weakly singular. It follows that

$$
\begin{aligned}
u &\propto |T - T_{\mathrm{c}}|^{\alpha/2} \qquad \text{for} \quad \rho = \rho_{\mathrm{c}}, \quad T > T_{\mathrm{c}} \\
u &\propto |T - T_{\mathrm{c}}|^{\alpha'/2} \qquad \text{along coex. curve,} \quad T < T_{\mathrm{c}}
\end{aligned}
\tag{8}
$$

where α and α' are very close to zero (~ 0.1–0.2); see Section III,A. Thus, the low-frequency sound velocity should go to zero at a liquid–vapor critical point.

D. LANDAU THEORY

So far, we have been concerned with theories of the equilibrium properties near a cooperative phase transition. In terms of ultrasonics, such theories can only describe the zero-frequency limiting behavior of the sound velocity. If one wishes to discuss ultrasonic attenuation or dispersion, it is necessary to develop a dynamic theory. Landau and Khalatnikov (1954) were the first to develop such a theory in the vicinity of a lambda transition, and their treatment combines a mean-field (Bragg–Williams) approximation to the equilibrium statistical problem with the thermodynamics of irreversible processes. The result is expressed in terms of an anomalous relaxation time τ for the long-range order parameter.

Before describing the Landau approach, let us briefly review the results of relaxation theory as applied to a sound wave traveling through a condensed medium (Herzfeld and Litovitz, 1959). For a low-amplitude sound wave of angular frequency $\omega = 2\pi f$, the stress, the strain, and the characteristic order parameter (if it is coupled to the strain) will all vary as $\exp i(\omega t - \mathbf{q}^* \cdot \mathbf{r})$, where $\mathbf{q}^*$ is a complex wave vector. For a specified direction of propagation, q^* (or the equivalent complex sound velocity u^*) can be related to the real velocity u, and the attenuation α by

$$q^* = \omega/u^* = (\omega/u) - i\alpha \tag{9}$$

Thus, $u = \omega/\mathrm{Re}(q^*)$ and $\alpha = -\mathrm{Im}\ q^*$. One can also consider the problem in terms of a complex, frequency-dependent elastic constant $c^* = \rho(u^*)^2$. For a steady-state sinusoidal wave in a medium characterized by a single relaxation time τ involving the long-range ordering,

$$(u^*)^2 = u_\infty{}^2 - [(u_\infty{}^2 - u_0{}^2)/(1 + i\omega\tau)] \tag{10}$$

where u_∞ is the infinite-frequency ("frozen") velocity, which describes any variation of the stress which is so rapid that the order parameter cannot follow, and u_0 is the zero-frequency ("equilibrium") velocity, which pertains to the propagation of any sound wave which varies slowly enough that the system is in equilibrium at all times (i.e., the order parameter follows in phase). When $(\alpha u/\omega)^2 = (\alpha_\lambda/2\pi)^2 \ll 1$ (which is usually valid), one obtains $\rho u^2 = \mathrm{Re}(c^*)$ and $2\rho u^3 \alpha/\omega = \mathrm{Im}(c^*)$. The resulting expressions for u^2 and α are

$$u^2 = u_\infty{}^2 - \frac{u_\infty{}^2 - u_0{}^2}{1 + \omega^2\tau^2} \equiv u_0{}^2 + \frac{(u_\infty{}^2 - u_0{}^2)\omega^2\tau^2}{1 + \omega^2\tau^2} \tag{11}$$

$$\alpha = \frac{(u_\infty{}^2 - u_0{}^2)}{2u^3} \frac{\omega^2\tau}{1 + \omega^2\tau^2} \tag{12}$$

The relaxation time used above is $\tau_{S,x}$, the adiabatic relaxation time at constant strain. (Although the process is irreversible, one can consider it to be essentially isentropic for small sound amplitudes.) It is also possible to formulate these expressions in terms of $\tau_{S,X}$, the relaxation time at constant (zero) stress. For the low-frequency limit (i.e., $\omega^2\tau^2 \ll 1$), the attenuation expression then becomes

$$\alpha = [(u_\infty{}^2 - u_0{}^2)/2uu_\infty{}^2]\omega^2\tau_X \tag{13}$$

since the two times are simply related by $\tau_X/\tau_x = u_\infty{}^2/u^2$. Although it is occasionally convenient to use τ_X to discuss the ultrasonic behavior (see Section V,B), the use of τ_x is more common. We shall use τ without any subscript as the adiabatic constant-strain relaxation time. In the absence of piezocaloric effects, no distinction is necessary between isothermal and adiabatic conditions for shear waves. For a discussion of this difference in the case of longitudinal waves, see Garland and Jones (1965) and Garland and Yarnell (1966b).

In order to use Eqs. (11) and (12) to describe the absorption and dispersion of sound waves near a lambda-type transition, one needs an expression for the relaxation time τ. Landau and Khalatnikov (1954) derived such an expression by assuming that the free energy Φ could be expanded in terms of a long-range order parameter η:

$$\Phi(p,\,T,\,\eta) = \Phi_0(p,\,T) + \alpha(p,\,T)\eta^2 + \beta(p,\,T)\eta^4 + \cdots \qquad (14)$$

The equilibrium value η_0 is determined from the condition $\partial\Phi/\partial\eta = 0$, and one wants η_0 to vanish above a critical temperature T_c while being nonzero below T_c. Hence, one must have $\alpha > 0$ for $T > T_c$ and $\alpha < 0$ for $T < T_c$, where T_c at pressure p is determined from the condition $\alpha(p, T_c) = 0$. Landau chose the simplest form which would guarantee such behavior: $\alpha(p, T) = a(p) \cdot (T - T_c)$. The above assumptions lead to an equilibrium η_0 which is the same as that obtained in the mean-field theory of ferromagnets and the Bragg–Williams theory of order–disorder phenomena.

Furthermore, it is necessary that the equilibrium value of the order parameter be dependent on the type of strain associated with a given sound wave; i.e., there must be a coupling such that η changes during the passage of the sound wave. Then, the rate of approach of η to its equilibrium value η_0 was determined by the transport equation

$$d\eta/dt = -L\ \partial\Phi/\partial\eta \qquad (15)$$

where the kinetic coefficient L was assumed to have no singularities in the vicinity of T_c. Equation (15) represents the application of irreversible thermodynamics, in that the flux $\dot{\eta}$ is taken as proportional to a generalized force $\partial\Phi/\partial\eta$. With the kinetic coefficient introduced in this phenomenological way, its temperature dependence cannot be predicted. However, it is the hope of this Landau approach that the rapid variations in the equilibrium properties will explain the anomalous dynamic behavior even if L is essentially constant near T_c. On the basis of Eqs. (14) and (15), Landau and Khalatnikov predicted that the relaxation time associated with long-range ordering below T_c is given by

$$\tau = [La(T_c - T)]^{-1} \qquad (16)$$

Thus, τ should increase very rapidly as T approaches T_c from below, and Eq. (12) would predict a marked increase in the ultrasonic attenuation up to a maximum value at some temperature just below T_c, after which, α would drop to zero at T_c. Above T_c, Landau and Khalatnikov assumed that η was identically zero for all states, and thus no anomalies in α would occur. It should be stressed that this theory deals with isothermal relaxation processes. If the Helmholtz free energy A is used, $(\partial A/\partial\eta)_{T,x}$ is the generalized force, and the constant-strain relaxation time $\tau_{T,x}$ is obtained; if the Gibbs free energy G is used, $(\partial G/\partial\eta)_{T,X}$ will lead to the constant-stress value $\tau_{T,X}$.

Yakovlev and Velichkina (1957) improved on the Landau–Khalatnikov result in their discussion of ultrasonic attenuation in Rochelle salt. This theoretical development was prompted by the experimental fact that there is considerable excess attenuation above T_c, in contradiction to the above theory. These authors also used a mean-field-type model with a single long-range order parameter η to describe the equilibrium state of their system, but they considered relaxation in the *deviation* $\Delta\eta$ from the equilibrium value of this order parameter. Their results for the relaxation time τ^+ above T_c and τ^- below T_c can be given in the form

$$\tau^+ = A^+/(T - T_c), \qquad \tau^- = A^-/(T_c - T) \tag{17}$$

where A^+ and A^- are phenomenological constants which depend on the substance involved, but have a fixed ratio of $A^+/A^- = 2$.

Although both the Landau–Khalatnikov and the Yakovlev–Velichkina papers are concerned only with acoustic attenuation, an analysis of velocity dispersion based on Landau's model has been made by Kravtsov (1963).

Tanaka *et al.* (1962) have followed essentially the same theoretical approach, but have based their model on the Bethe approximation to the order–disorder transition in a ferromagnet. That is, they have combined the thermodynamics of irreversible processes with an equilibrium statistical-mechanical theory based on Bethe's approximation. Their model thus involves two order parameters (one for short-range order and one for long-range) and two relaxation times. The relaxation time τ_1 associated with the behavior of the short-range order is a slowly varying function of temperature which approaches the same finite value as T_c is approached from either above or below. In essence, τ_1 can be considered as a constant for a range of temperatures close to T_c. The other relaxation time τ_2 is a more complicated function of the parameters of the model. If the expressions for τ_2 above and below T_c are simplified by assuming that $|T - T_c|/T_c$ is very small and that no coupling exists between the relaxation of long- and short-range order, one finds that $\tau_2{}^+$ and $\tau_2{}^-$ are relaxation times for the long-range order which obey Eqs. (17) with $A^+ = 2A^-$. Again, it was predicted that the ultrasonic attenuation will vanish at T_c since $\tau_2 \to \infty$ there.

A more general statistical-mechanical treatment of the dynamic behavior near a lambda point has been made by Kikuchi (1960). He has investigated the β-brass order–disorder problem using a set of "path parameters," which can be related to long-range and short-range order parameters of the kind used in Bethe's approximation. Again, the irreversible approach to equilibrium in a cooperative system is characterized by two relaxation times. In the disordered state (above T_c), the relaxation of the long-range order is shown to be independent of that of the short-range order, while, in the ordered state (below T_c), they are coupled. For either state, the analytic expressions for τ_1 and τ_2 are complicated functions of temperature, such that the τ_1 values converge to a single finite value and the τ_2 values tend to infinity as the temperature approaches T_c from either side. In a temperature range close to the lambda point (i.e., when $|T - T_c|/T_c$ is small), one can show that Kikuchi's results for $\tau_2{}^+$ and $\tau_2{}^-$ are still of the form given in Eqs. (17). However, the constants A^+ and A^- now depend explicitly on microscopic transition probabilities, rather than on phenomenological kinetic coefficients, and the ratio A^+/A^- near T_c is ~ 1.6, rather than 2. For all practical purposes, τ_1 can be treated as a constant for a reasonable range of temperature close to T_c. In spite of considerable differences between

Kikuchi's treatment and that used by Tanaka *et al.*, their results are quite comparable (which perhaps justifies including Kikuchi's work in this section).

The above description of the original Landau approach (and of various calculations carried out in the spirit of Landau's theory) has been given in some detail because this was an early and influential dynamic theory. Many experimental observations have been analyzed in terms of Landau theory; this is especially true of numerous Russian investigations near ferroelectric Curie points. Thus, it is important to be familiar with this simple theory in order to read much of the literature in the field. However, there are several serious objections to this type of theory. First of all, it is known that the free-energy expansion given in Eq. (14) does not, in general, give a good description of static phenomena near transition points (Kadanoff *et al.*, 1967). This is due to the fluctuations in the order parameter. For the Landau expansion to yield valid results, static fluctuations in the order parameter must be small (in comparison to the order parameter itself) over distances comparable with the coherence length. It can be shown that the range of validity for all mean-field theories is $\varepsilon_c \ll |\varepsilon| \ll 1$, where ε is the reduced temperature $(T - T_c)/T_c$ and ε_c is a critical value characteristic of the substance. When $|\varepsilon| < \varepsilon_c$, fluctuations become very important and the Landau theory will fail. As the range of the forces increases, ε_c decreases, and the Landau theory becomes a better and better description. Thus, in ferroelectrics, where long-range Coulomb forces are important, Landau theory is expected to predict the static properties of all $|\varepsilon| \gg 10^{-4}$ (Kadanoff *et al*, 1967). Another objection to the Landau approach involves the phenomenological nature of Eq. (15) and the assumption that the kinetic coefficient L is well behaved near T_c.

E. Fluctuation Theories

In the past five years, most theories involving the propagation of sound waves in the vicinity of a critical point or cooperative phase transition have been based on a consideration of the large fluctuations which occur in such systems. The fluctuation-dissipation theorem (Kubo, 1966) states that the linear response of a given system to an external perturbation can be expressed in terms of the internal fluctuations of the system when it is in thermal equilibrium. In the case of ultrasonics, the dynamic property of greatest interest is the attenuation α, or the closely related sound-wave damping constant D_s ($\alpha = \omega^2 D_s/2u^3$). The fluctuations are characterized by a correlation function for the appropriate momentum density, although this may be replaced by the correlation function for the order parameter in some cases.

Theoretical work on a large variety of dynamic properties near phase transitions has become very active, and the situation is now in a state of considerable flux (Kadanoff, 1968). An attempt will be made here to sketch the most important current trends, although much progress (and, hopefully, clarification) should be expected in the next few years.

First of all, let us consider two Russian papers which are not closely related to the other papers which will be cited. Giterman and Kontorovich (1965) extended the theory of fluctuations of hydrodynamic quantities to the case of a medium with spatial dispersion (nonlocal connections between changes in pressure and density). They applied this approach to the velocity and scattering of sound waves in the one-phase region close to a liquid–vapor critical point, and found a dependence of u on wavelength, as well as a change in the character of the ultrasonic loss. Another paper, more in the spirit of the Landau–Khalatnikov theory, ascribes the ultrasonic attenuation near a lambda-type transition to the interaction between the sound wave and the thermal fluctuations of the order parameter (Levanyuk, 1966). Indeed, in the absence of a linear coupling between the acoustic strain and the order parameter, this mechanism would play the dominant role. It is assumed that the free-energy density can be expanded in terms of the deviation $\Delta\eta = \eta - \eta_0$ in the order parameter and the deviation $\Delta v = v - v_0$ in the volume strain, and a term $(\text{grad } \Delta\eta)^2$ is included to take into account approximately the correlation between values of $\Delta\eta$ at different points in space. The Levanyuk equations of motion do not include inertial terms (thus, η approaches its equilibrium value with a relaxation character), but they do include a random force which accounts for the fluctuations in η. For $T > T_c$, this theory reduces to the propagation of sound in a medium with random inhomogeneities of a known statistical character. Below T_c, the result can be represented as a sum of two terms—one due to the random thermal fluctuations, and one which corresponds directly to the Landau–Khalatnikov result.

Perhaps the earliest attempt to treat sound propagation near a critical point in terms of fluctuations was made by Fixman (1962, 1964) in his treatment of binary liquid mixtures.[1] The basic physical idea in his calculations was the recognition that long-range spatial correlations of the type predicted in Ornstein–Zernike theory would lead to enhanced fluctuations which could cause anomalous transport properties. Fixman considered the temperature variation associated with the sound wave and obtained a complex frequency-dependent heat capacity. Since the acoustic velocity depends on the specific heat ratio, he obtained an expression for the complex sound velocity as a function of frequency. Botch and Fixman (1965) later applied these same ideas to sound propagation near the liquid–vapor critical point.

A correlation-function analysis of the frequency-dependent bulk viscosity and the ultrasonic absorption in critical mixtures was made by Kawasaki and Tanaka (1967). These authors found that the bulk viscosity ζ shows a much stronger anomaly near the critical point than does the shear viscosity η, and they developed a theory of the anomalous attenuation in terms of ζ. Although their formulation is quite different from that used by Fixman and appears to give a physically different basis for the excess attenuation, their

---

[1] Pippard (1951) discussed ultrasonic propagation near the λ-point in liquid helium in terms of fluctuations arising from inclusions of the superfluid phase in the normal phase above T_λ (and vice versa below T_λ), but this represents a rather special kind of inhomogeneous medium.

result for $\alpha(\omega)$ is identical to that derived by Fixman (1962). Thus, the two descriptions are essentially equivalent. In both cases, the use of mean-field and Ornstein–Zernike theories may lead to an overestimate of the effect of critical fluctuations. However, there is reasonable agreement between these theories and experiment, as discussed in Section IV.

The early work of Fixman and Kawasaki indicated the importance of nonlinear couplings among hydrodynamic modes in determining the critical behavior of transport coefficients. In order to calculate the sound-wave damping constant near a critical point, Kadanoff (1968, 1969) made the bold assumption that one should consider decay processes in which a given sound mode breaks up into two heat modes or two sound waves (spin waves, in the case of a magnetic transition). In a system undergoing large density (magnetization) fluctuations, these decay processes will produce a divergent contribution to the damping. This mode–mode coupling theory was first developed by Kadanoff and Swift (1968) for the transport coefficients near a liquid–vapor critical point. It represents a semiphenomenological perturbation formalism which permits a direct estimation of the magnitude of the anomaly in the transport coefficients as well as the asymptotic functional form. An additional feature of the Kadanoff and Swift approach is the evaluation of correlation functions on the basis of static scaling laws rather than Ornstein–Zernike theory. This should result in a better prediction of the critical indices (in the case of ultrasonics, one is concerned with the power of $|T - T_c|$ which appears in the temperature dependence of α). In addition to the mode–mode treatment of the liquid–vapor critical point, this method has been recently applied to binary liquid mixtures (Swift, 1968), the λ-transition in helium (Swift and Kadanoff, 1968), and magnetic phase transitions (Laramore and Kadanoff, 1969). Specific predictions made in these papers will be discussed in the appropriate sections.

The mode–mode coupling results for the attenuation in magnetic systems are very similar to some recent results by Kawasaki (1968a, b, c) on critical relaxation in ferromagnets and antiferromagnets based on a random-phase approximation for decoupling the correlation functions. This work does, however, use Ornstein–Zernike theory (critical index $\eta = 0$) in deriving expressions for the dynamic correlation functions. Even more recently, Kawasaki (1968d, e, f, g) has reformulated his theory to include the possibility of finite η values. In this new approach, "dynamic-scaling" results are derived on the basis of static scaling assumptions.

Kadanoff's mode–mode coupling scheme and Kawasaki's critical-dynamic-variable approach can both be compared with the theory of dynamic scaling. The first development of dynamic scaling laws for systems near phase transitions was achieved by Ferrell et al. (1968) in their treatment of the lambda transition in liquid helium. Halperin and Hohenberg (1969) reformulated and generalized the theory and have applied it in some detail to ferromagnets, antiferromagnets, and liquid–vapor critical points, as well as the helium λ-transition. Only in the liquid–vapor case do they discuss the application of dynamic scaling to sound propagation, and we will describe the basic idea in that context. For a sound wave of angular frequency ω, one can write the dispersion relation in the form

$$\omega = \pm uq - (iD_s q^2/2), \tag{18}$$

where D_s is the damping constant. In *static* scaling theory, it is assumed that the coherence length ξ is a natural characteristic length near the critical point. From this, it follows that q always appears in the Fourier transforms of long-range correlation functions in the combination $(q\xi)$. For the time-dependent behavior of a system where there is a single dominant relaxation mechanism, dynamic scaling theory assumes that all relaxation times will be proportional to a single characteristic time multiplied by some function of $(q\xi)$. The result of all this is the prediction that $D_s \propto u\xi$, and thus the ultrasonic attenuation should diverge as the correlation length ξ (which varies along the critical isochore as $\varepsilon^{-\nu} \approx \varepsilon^{-2/3}$). Unfortunately, it may not be possible to handle sound propagation near liquid–vapor critical points by an approach based on scaling to a single characteristic relaxation rate. Further discussion of this point will be given in Section III.

In addition to the rather general theories which have been described above, there are a number of calculations based on well-defined models. Deutch and Zwanzig (1967) have treated a van der Waals model of binary liquid systems, while Mountain and Zwanzig (1968), and also deSobrino (1968), have developed dynamic theories for the anomalous transport properties of a single van der Waals gas near its critical point. Heims (1966) has discussed a time-dependent Ising model for binary liquid systems. These and other such calculations will be mentioned where appropriate in the following sections.

It should be obvious that the theoretical situation is developing rapidly, but has not yet stabilized. All of the current theories involve serious conceptual and mathematical difficulties, and it is not clear that different theoretical formalisms will always lead to the same results. In contrast to the rather universal character of all equilibrium behavior near a critical point, it appears that the detailed nature of the interactions will be important in determining the exact form of dynamic divergences in different phase transitions. As the remaining sections will show, the experimental situation is also developing vigorously. Recent and future experimental work should confirm or clarify the relative importance of different possible mechanisms. Indeed, the interaction between theory and experiment should be very productive in the next few years.

III. Liquid–Vapor Critical Point

A. THEORY

1. *Low-Frequency Sound Velocity*

One of the defining characteristics of the liquid–vapor critical point is the well-known fact that the critical isotherm has zero slope at the critical density, i.e., $(\partial p/\partial \rho)_{T_c} = 0$ at $\rho = \rho_c$. This is identical to saying that the isothermal compressibility κ_T has an infinite singularity at the critical point. If one considers the general thermodynamic expression

$$C_p - C_V = TV(\partial p/\partial T)_V \kappa_T \tag{19}$$

it is immediately clear that C_p is also divergent, since C_V is always positive. The question then arises as to the behavior of C_V and κ_S. Van der Waals' theory (or any other theory resulting in an equation of state having a finite number of analytical terms) predicts a finite maximum value for C_V at the critical point, and this was the accepted view for a long time. More recently, careful heat-capacity measurements all strongly suggest a divergence of C_V of the form $C_V \propto |T - T_c|^{-\alpha}$. The critical index α is different above and below T_c and has a value somewhere between 0 and 0.2, where $\alpha = 0$ indicates a logarithmic divergence.

Under conditions where the propagation of the pressure and temperature disturbance of an ultrasonic wave can be said to be adiabatic—that is, the variations are rapid enough that thermal equilibrium is not established on a local scale by heat conduction from the "crest" to the "trough" of the wave —the real part u of the complex sound velocity u^* is given by

$$u^2 = (\rho \kappa_S)^{-1} \tag{20}$$

where ρ is the mass density and κ_S is the adiabatic compressibility. From the thermodynamic relationship

$$\kappa_S^{-1} = \kappa_T^{-1} + T(\rho \tilde{C}_V)^{-1}(\partial p/\partial T)_V^2 \tag{21}$$

one obtains

$$\rho u^2 \approx T_c(\rho_c \tilde{C}_V)^{-1}(dp/dT)_{\text{sat}}^2, \qquad \text{as} \quad T \to T_c \tag{22}$$

since κ_T diverges very strongly as the critical point is approached and $(dp/dT)_{\text{sat}}$, the limiting value of the slope of the coexistence curve in the p–T plane, is essentially the same as $(\partial p/\partial T)_V$ near the critical point. Thus, the behavior of u^2 should be substantially the same as that of C_V^{-1} as the critical point is approached; see Eq. (8).

In general, u is a function of the ultrasonic frequency. The treatment given above is true only in the zero-frequency (static) limit. However, the measured velocity will become independent of f at sufficiently low frequencies. Therefore, velocities measured in the 0.1–10-MHz range can be taken as the zero-frequency velocity, *if the dispersion is shown to be negligible.*

2. Ultrasonic Attenuation

It is known that the ultrasonic attenuation increases very dramatically near the critical point, becoming larger than 400 dB cm^{-1}. This is due to two effects: the attenuation per wavelength (α_λ) increases sharply, and the wavelength of the sound diminishes along with the velocity. Before discussing such critical anomalies, let us consider the absorption of sound by normal fluids away from a critical point. A sound wave in a medium consists of a pressure wave and its associated temperature (or entropy) wave. The resultant gradients set up small-scale flows of matter and energy which involve dissipative processes—shear viscosity and thermal conductivity, respectively —giving rise to attenuation. The magnitude of these effects, known as the "classical" attenuation coefficient, can be calculated from the Navier–Stokes equation as

$$\alpha_{\text{class}} = 2\pi^2 f^2[(4\eta/3) + K(\bar{c}_V^{-1} - \bar{c}_p^{-1})]/\rho u^3 \tag{23}$$

where ρ is the mass density, η is the shear viscosity, $\bar{c}_p$ and $\bar{c}_V$ are the specific heats, and K is the thermal conductivity. In most fluids, even under normal conditions, the attenuation is greater than α_{class}. The discrepancy can be eliminated by the introduction of a volume ("bulk") viscosity ζ into Eq. (23) such that

$$\begin{aligned}\alpha &= \alpha_{\text{class}} + (2\pi^2 f^2 \zeta/\rho u^3) \\ &= 2\pi^2 f^2[(4\eta/3) + \zeta + K(\bar{c}_V^{-1} - \bar{c}_p^{-1})]/\rho u^3 \end{aligned} \tag{24}$$

As a rule, ζ is evaluated from measurements of the excess attenuation $(\alpha - \alpha_{\text{class}})$, so that its introduction might seem to be merely an *ad hoc* correction to the classical theory, but ζ does have a well-defined meaning (Herzfeld and Litovitz, 1959).

Since η was thought to have no divergence and K was thought to have only a weak divergence (like c_V), the cause of the anomalous attenuation was originally sought in scattering or "structural relaxation" due to large clusters with densities momentarily different from that of the surrounding medium. In terms of scattering, it was assumed that the density fluctuations known to be responsible for the critical opalescence of visible light could also act as scattering centers for sound waves. A formula originally proposed for the calculation of sound absorption in the ocean due to inhomogeneities in density (Chernov, 1960) has been applied to the critical region. If the mean diameter a of the inhomogenity is taken to be approximately 1000 molecular diameters ($\sim 10^{-4}$ cm), this formula would predict that the attenuation varies as the fourth power of the frequency, which has never been observed. However, it has been pointed out by Bhatia (1959) that, since the density fluctuations might be correlated over much larger distances, the resulting increase in the parameter a could give rise to a quadratic frequency dependence.

The possibility of structural relaxation was proposed on a phenomenological basis by Chynoweth and Schneider (1952) in their discussion of xenon. All fluids possess such relaxation—following the instantaneous compression of a volume of the fluid, it takes a finite time for the fluid to relax to its new equilibrium condition—but the relaxation time τ is normally on the order of 10^{-13} sec. In general, this relaxation involves both inter- and intramolecular processes, but only intermolecular relaxation is of interest here. (Note that rotational and vibrational relaxation cannot occur in a monatomic fluid.) At ultrasonic frequencies for which $\omega\tau \ll 1$, the relaxation expression for α given in Eq. (12) will become identical to the expression for the excess attenuation $(\alpha - \alpha_{\text{class}})$ given in Eq. (24). This equivalence is due to the fact that τ and ζ are related at low frequencies by $\tau = \zeta/\rho(u_\infty^2 - u_0^2)$. Mountain (1968) has discussed such relaxation times associated with volume viscosity in simple fluids away from their critical points. Near the critical point, however, structural correlations extend far beyond a few molecules, involving groups of 10^9 molecules, or perhaps far more. These clusters form and disappear (relax) with lifetimes much longer than 10^{-13} sec; and, since clusters of different sizes could have different lifetimes, one might expect a spectrum of relaxation times. As the critical point is approached, this spectrum would tend to become broader and to move toward longer times. As a result, the attenuation should increase dramatically as the critical point is approached. Velocity dispersion should also be observed [see Eq. (11)].

Botch and Fixman (1965) have provided a different theoretical foundation by postulating that a dynamic heat capacity associated with long-range density fluctuations in the fluid is responsible for the observed attenuation. In order to make a numerical calculation, the Ornstein–Zernike form of the correlation function was used along with the Debye form for the temperature dependence of the inverse correlation length. From the imaginary part of the complex sound velocity, the attenuation per wavelength, $\alpha_\lambda = \alpha\lambda = \alpha u/f$, was found to vary as

$$\alpha_\lambda \propto \omega^{-0.25} J_2(\varepsilon, \omega) \tag{25}$$

where the definite integral J_2 is a rapidly varying function of the reduced temperature ε and a very weak function of ω.

Let us now return to Eq. (24) and consider explicitly the anomalous behavior of the transport coefficients. The major factor contributing to the sound absorption is the term $[(4\eta/3) + \zeta]$, since the term $K(\bar{c}_V^{-1} - \bar{c}_p^{-1})$ appears to play a minor role in the frequency region of interest. Kadanoff and Martin (1963) give an expression for the low-frequency, long-wavelength limit of this quantity as

$$\lim_{\omega \to 0} \lim_{q \to 0} \omega^2 q^{-4} \, \mathrm{Im}[C(q, \omega)] = (4\eta/3) + \zeta \tag{26}$$

where $\mathrm{Im}[C(q,\omega)]$ is the imaginary part of the dual Fourier transform of the dynamic correlation function for density fluctuations $\langle[\rho(r,t) - \langle\rho(r,t)\rangle],$ $[\rho(0,0) - \langle\rho(0,0)\rangle]\rangle$. Kadanoff and Swift (1968) and Kawasaki (1968g) have calculated forms for $(4\eta/3) + \zeta$ using mode-coupling schemes involving heat-mode and sound-mode intermediate states.

The situation is complicated by the fact that there are at least three distinct frequency regions. In region I, the lowest-frequency region ($\omega \lesssim \omega_1 = K/\rho C_p \xi^2$), the sound wave breaks up into two heat modes and the resulting attenuation has a very strong divergence (roughly as ε^{-2}) as well as a quadratic frequency dependence. In regions II and III ($\omega_1 \ll \omega \ll u/\xi$), Kadanoff and Swift predict that $\alpha \propto \omega^2 \varepsilon^{-\nu-\alpha} \propto \omega^2 \varepsilon^{-2/3}$. In contrast, Kawasaki predicts two distinct behaviors: in region II ($\omega_1 \ll \omega \ll \omega_2$) the sound-wave damping is still due to heat-mode contributions to the volume viscosity, but $\alpha \propto \varepsilon^2$ (independent of ω); in region III ($\omega_2 \ll \omega \ll u/\xi$), sound-wave intermediate states dominate and $\alpha \propto \omega^2 \varepsilon^{-2/3}$, as in the Kadanoff and Swift result. The characteristic frequencies ω_1, ω_2, and $u\xi^{-1}$ are temperature dependent, and Kawasaki (1968g) has roughly estimated that they vary as $a_1 \varepsilon^2$, $a_2 \varepsilon^{4/3}$, and $a_3 \varepsilon^{2/3}$, respectively, where the a values are of the order of 10^6 MHz. Therefore, region I is usually below the ultrasonic range, and the upper limit u/ξ is clearly above ultrasonic frequencies for interesting and accessible values of ε. The above predictions can be summarized by

$$
\begin{aligned}
&\text{Region} \quad \text{I} \quad (\omega < \omega_1): && \alpha \propto \omega^2 \varepsilon^{-2} \\
&\text{Region} \quad \text{II} \quad (\omega_1 \ll \omega \ll \omega_2): && \alpha \propto \omega^0 \varepsilon^2 \\
&\text{Region} \quad \text{III} \quad (\omega_2 \ll \omega \ll u/\xi): && \alpha \propto \omega^2 \varepsilon^{-2/3}
\end{aligned}
\tag{27}
$$

It should be noted that, as the critical point is approached, ω_2 decreases rapidly, and the experimental frequency may shift from region II to region III. This would imply a complicated variation of α of the form $A\omega^2 \varepsilon^{-2/3} + B\varepsilon^2$, where both terms were of comparable magnitude.[2]

It should also be noted that Eq. (24) is valid only when $\omega\tau_{th} \gg 1$, where the thermal relaxation time is given by $\tau_{th} = \rho\bar{c}_v/Kq^2$. This condition is closely related to the condition for the adiabaticity of the sound wave.

Krasnyi and Fisher (1967) have also developed a theory of the volume viscosity near a critical point, but it is based on the assumption of an Ornstein–Zernike form: $[(4\eta/3) + \zeta] \propto (q^2 + \xi^{-2})^{-1}$. As a result, they predict that $\alpha \propto \omega^2\xi^2 \propto \omega^2\varepsilon^{-4/3}$ at low frequencies, in disagreement with Eq. (27).

As a final point, it should be mentioned that the result given in Eq. (27) for region III is consistent with the dynamic scaling laws of Halperin and Hohenberg (1969). The failure of dynamic scaling for the longitudinal momentum density as $q \to 0$ (regions I and II) suggests that the scaling approach must be modified to include several characteristic relaxation times, some of which scale differently than the order parameter.

3. *Velocity Dispersion*

The simplest theory for describing the frequency dependence of the velocity invokes relaxation processes by which the sound wave can be coupled to the internal or structural state of the medium. For a dissipative medium characterized by a single relaxation time, one has Eq. (11) for the velocity at frequency ω. The value of the frequency at which $\omega\tau = 1$ is called the dispersion frequency, and the dispersion region extends for about two decades on either side of this frequency; thus, dispersion is a rather broadband effect. Within this region, the velocity varies between u_0 and u_∞ and the attenuation per wavelength goes through a maximum. In order to find the values of τ, u_0, and u_∞ at some particular temperature, measurements of α_λ and u at two different frequencies, or of u at three different frequencies within the dispersion region, are necessary.

---

[2] Note added in proof: Very recently, Kawasaki (1970) has extended and improved the theory of ultrasonic attenuation near a liquid–vapor critical point. The new results differ appreciably from those described above.

If there are several relaxation times, caused by a number of independent modes of dissipation, then the terms involving $\omega\tau$ in Eqs. (11) and (12) must be replaced by a sum of such terms, and the calculations become greatly complicated. Indeed, the invocation of a spectrum of relaxation times renders the problem intractable unless the form of the spectrum is known. If spatial, as well as temporal, dispersion is present (i.e., if τ depends on the wave number q) then Eqs. (11) and (12) could be used, but with a different value of τ for each wavelength. In any case, relaxation theory only provides a formalism which may or may not be useful.

There does not yet seem to be any detailed statistical treatment of dispersion near a critical point.[3] A prediction based on the real part of the Botch–Fixman expression for the complex sound velocity does not seem to represent the ultrasonic velocity in Xe or HCl very well at all (Botch and Fixman, 1965). Kawasaki (1968g) predicts that the sound velocity in regions II and III should exceed the "adiabatic" (zero-frequency) value by a constant factor, but this aspect of the theory is not yet developed in enough detail to be useful in analyzing experimental data.

It should be clear that new theoretical work is still needed on all aspects of the liquid–vapor problem. Moreover, as we shall see, the experimental results are not yet in really good shape. A great deal of progress should be expected in the near future.

B. Helium

As indicated by Eq. (8) and discussed in Section II,A, the zero-frequency sound velocity is expected to go to zero as the critical point is approached. There have been many measurements of u in the critical region of various substances (see Table I) which appear to show only a finite velocity minimum. One possible explanation involves large velocity dispersion at ultrasonic frequencies. However, the temperature and pressure resolution of many of these investigations was sufficiently limited that they might not detect a very weak singularity at the critical point.

Up to the present time, helium is the only system in which the velocity has been studied with really high resolution, and Williamson and Chase (1968) present very strong evidence that, "along the critical isotherm, the quantity ρ/u^2 diverges approximately logarithmically as a function of $|p - p_c|$." An overall view of their results is shown in Fig. 1. [Additional measurements made along a near-critical isobar (Chase *et al.*, 1964) indicate that a plot of $\kappa_S = \rho u^2$ *versus* $\log|T - T_c|$ is linear at 1718 ± 1 Torr.] The locus of the sound-velocity minima in the immediate vicinity of the critical point falls along the extrapolated vapor-pressure curve (see Fig. 2), which is consistent with the available data on other materials. A detailed view of the velocity data along two isotherms ($T - T_c \approx 6$ mdeg and $T - T_c \approx 14$ mdeg)

---

[3] Note added in proof: Very recently, Kawasaki (1970) has extended his theory of the liquid–vapor critical point to include a detailed treatment of ultrasonic dispersion.

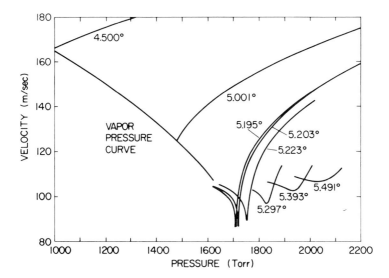

FIG. 1. Velocity of sound in helium for various subcritical and supercritical isotherms; $f = 1$ MHz (Williamson and Chase, 1968).

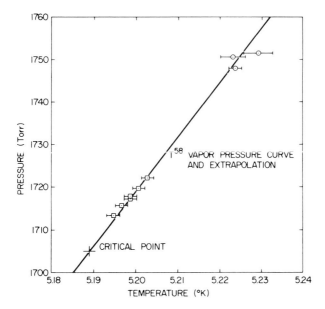

FIG. 2. Locus of sound-velocity minima near the critical point in helium (Williamson and Chase, 1968).

TABLE I

ULTRASONIC INVESTIGATIONS OF LIQUID-VAPOR CRITICAL SYSTEMS

Substance	T_c(°C)[a]	p_c(atm)[a]	Temperature range (°C)[a]	Pressure range (atm)[a]	Frequency (MHz)[a]	Quantity measured	Reference
Helium	5.189°K	1705 Torr	$T_c \pm 0.04$	$p_c \pm 40$ Torr	1	u	Chase et al. (1964)
			5.0–5.5°K	$p_c \pm 500$ Torr	1	u	Williamson and Chase (1968)
Argon	150.7°K	48.3	150.67°K	49–51	~1 kHz	u	van Dael et al. (1967)
Xenon	16.5C	57.6	15–19	$\rho \approx \rho_c$	0.25–2.25	u, α	Chynoweth and Schneider (1952)
			15.3–18.3	57–58.3	1–5	u, α	Mueller et al. (1969)
Hydrogen chloride	51.4	81.6	38–62	$\rho \approx \rho_c$	1–9	u $\alpha(T > T_c)$	Breazeale (1962, 1963)
Carbon dioxide	31.04	72.8	28–38	5–98	~0.27	u	Herget (1940)
			29–38	5–100	0.572	u, α	Anderson and Delsasso (1951)
			30–32	60–130	0.96	u	Noury (1951)
			26–38	40–110	0.5–2	u, α	Parbrook and Richardson (1952), Parbrook (1953)
			25–48	35–120	0.41	u	Tielsch and Tanneberger (1954)
			30.5–32	50–100	500 Hz	u	Trelin and Sheludyakov (1966)
			23–40	$\rho_c (T > T_c)$, coex. $(T < T_c)$	~440 near T_c	u	Gammon et al. (1967)
			31.4–47	ρ_c	~650 near T_c	u	Ford et al. (1968)
Nitrous oxide	36.5	71.7	31.0–32.5	$\rho \approx \rho_c$	Few kHz	u	Feke (1969)
			32–37	55–150	0.96	u	Noury (1951)

			Coex. curve				
Water	374	218			2	u	Nozdrev et al. (1962)
Ethane	32.3	48.2	29–40	100–200 Amagat	0.3–0.9	u	Tanneberger (1959)
Ethylene	9.9	50.5	9.7–23 7–18.7	35–75 0–90	0.27–0.6 0.5	u u, α	Herget (1940) Parbrook and Richardson (1952)
Suphur hexafluoride	45.55	37	42–49	$\rho_c(T > T_c)$, coex. $(T < T_c)$	0.6	u, α	Schneider (1951, 1952)
n-Hexane	234.8	29.9	b		2–3	u	Nozdrev (1955)
n-Heptane	266.8	27	b		2–3	u	Nozdrev (1955)
Methyl alcohol	240	78.5	b		2–3	u	Nozdrev (1955)
Ethyl alcohol	243.1	63.1	b		2–3	u	Nozdrev (1955)
Isobutyl alcohol	277	42.4	b		2–3	u	Nozdrev (1955)
Propyl acetate	276.2	32.9	b		2–3	u	Nozdrev (1955)
Ethyl acetate	250.4	37.8	b		2–3 5–9	u u, α	Nozdrev (1955) Nozdrev and Sobolev (1956)
Methyl acetate	233.7	46.3	b b		2–3 5–14	u α	Nozdrev (1955) Kalianov and Nozdrev (1958)
Cl$_2$(CH$_3$)$_2$Si	~247.5		b		3–4	u	Nozdrev and Stepanov (1968)
Cl(CH$_3$)$_3$Si	~224.5		b		3–4	u	Nozdrev and Stepanov (1968)

[a] Units as indicated unless otherwise noted.
[b] Coexistence curve, $T < T_c$; $\rho \approx \rho_c$, $T > T_c$.

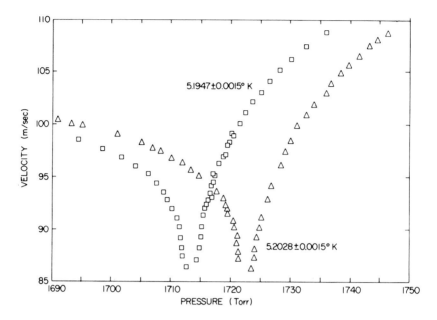

Fig. 3. Plot of 1-MHz velocity versus pressure at two slightly supercritical temperatures in helium (Williamson and Chase, 1968).

is shown in Fig. 3, where one can see the very rapid drop in velocity. Unfortunately, there is a gap of about 2 Torr where velocity measurements could not be made because of the high attenuation.

Williamson and Chase have fitted their data to the equation

$$\rho/u^2 = -C_\pm \ln|\Delta p| + D_\pm \tag{28}$$

where $\Delta p = (p - p_0)/p_c$ and p_0 is the pressure at which the minimum velocity is observed at a given temperature T. The quality of the fit for six isotherms is indicated by Fig. 4, and one can also see that $C_+ \approx C_-$ (i.e., the singularity is symmetric above and below p_0). Williamson and Chase cite scaling-law arguments which indicate that

$$u^{-2} \propto \Delta p^{-2\alpha/(2+\gamma-\alpha)} \qquad \text{for} \quad T = T_c \tag{29}$$

If the specific-heat index $\alpha = 0$, then Eq. (29) would predict a logarithmic singularity, in agreement with the empirical fit given in Eq. (28). If α is not zero, but is quite small, it might be difficult to distinguish experimentally between a logarithmic and a very weak power divergence. It is also interesting to note that, over the range 5.1947–5.2028°K, the observed sound velocities are essentially a function of density alone. This obviously cannot be true very close to ρ_c, since u presumably goes to zero at T_c, but not at any other temperature.

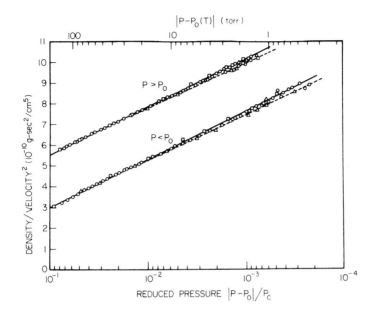

FIG. 4. Plot of ρ/u^2 as a function of ln $|\Delta p|$ in helium. The solid line represents a fit to the data for the 5.1947°K isotherm, while the dashed line is for the 5.2028°K isotherm. Data points represent four intermediate temperatures (Williamson and Chase, 1968).

All of the Williamson and Chase data were obtained at 1 MHz, so no indication of dispersion could be obtained. However, dispersion near the critical point would mean that the zero-frequency velocity would be lower than the 1-MHz values. This would require a stronger singularity in ρ/u^2 if significant dispersion does exist. Measurements of u at lower frequencies (1.5–50 kHz and ~ 0.25 MHz) are currently in progress (Barmatz, 1970; Williamson, 1970).

C. XENON

Chynoweth and Schneider (1952) determined the ultrasonic velocity and attenuation in xenon near its critical point, using a two-crystal interferometric technique. Velocity measurements were made as a function of frequency of 0.25 to 2.25 MHz at various temperatures between 15 and 19°C ($T_c = 16.59$°C). These experimental results indicated a velocity dispersion as large as 6% (see Fig. 5, which shows their data above T_c). As indicated in Fig. 6, the velocity as a function of temperature appears to go through a finite minimum (drawn as a sharp cusp) displaced somewhat from the "temperature of meniscus disappearance." No explanation was offered as to why

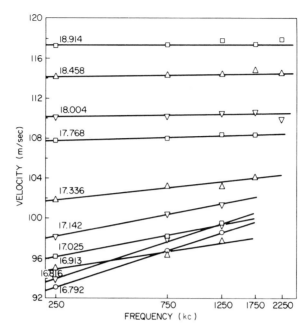

FIG. 5. Ultrasonic velocities in xenon at several supercritical temperatures (Chynoweth and Schneider, 1952).

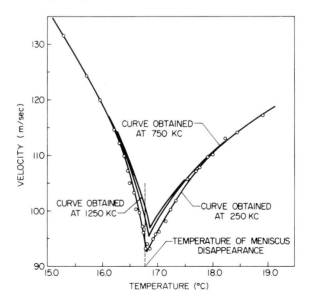

FIG. 6. Velocity versus temperature in xenon at 0.25, 0.75, and 1.25 MHz. Data obtained below T_c were all in the liquid phase (Chynoweth and Schneider, 1952).

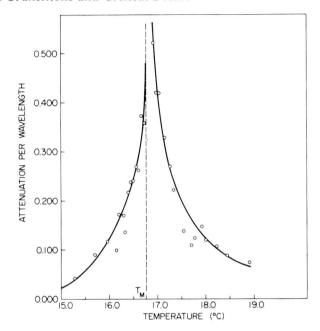

FIG. 7. Temperature variation of α_λ in xenon at 0.25 MHz. Measurements below T_M refer to the liquid phase (Chynoweth and Schneider, 1952).

the velocity minimum should be displaced by $\sim 0.2°$ from the critical temperature. Attenuation measurements were also reported at 0.25 MHz, and Fig. 7 shows the very sharp peak in α_λ which is observed near a critical point. At higher frequencies, the interferometer traces were not sufficiently accurate to permit a quantitative determination of α_λ, but it was reported that α_λ "did not increase with frequency to any large extent."

It now appears that Chynoweth and Schneider may not have come really close to the critical point in their measurements, although the *mean* filling density in their cell was close to the critical value. All of their data were obtained at a level several centimeters below the center of a rather tall cell. As T_c is approached, a fluid becomes very compressible and large density gradients are produced by the effect of gravity. These density gradients cause the local density below the center of the cell to exceed the critical value before the critical temperature is reached. Thus, the minimum value of their velocities would occur at a point where the decrease in u due to the approach to the critical temperature is offset by the increase in u due to the increasing deviations from the critical density. Although Chynoweth and Schneider were aware of density gradients caused by gravity, they did not discuss this particular effect.

The most interesting aspect of the Chynoweth and Schneider result is the observation of velocity dispersion at frequencies near 1 MHz. Since dispersion has not yet been observed at such frequencies in any other liquid–vapor system, it is of interest to confirm their result. Recently, Mueller *et al*

(1969) have essentially carried out the same measurements at a mean filling density of $0.95\rho_c$ and at frequencies of 1 and 3 MHz, using a pulse technique. The 1-MHz velocities were in good qualitative agreement with the Chynoweth and Schneider data (see Fig. 8), and dispersions $(u_3 - u_1)$ as large as 10 m sec^{-1} were observed. The attenuation had a temperature variation very similar to that shown in Fig. 7, but the α_λ values at 1 MHz were close to *one-half* the Chynoweth and Schneider α_λ values at 0.25 MHz. Somewhat limited measurements at 3 MHz indicated that α_λ at 3 MHz is very roughly *twice*

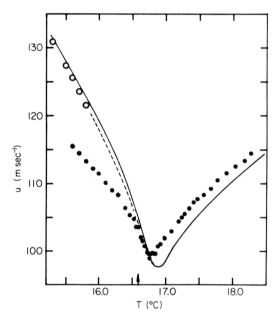

Fig. 8. Velocity at 1 MHz for xenon at a mean filling density of 0.95 ρ_c. Open circles are for measurements in the liquid, and filled circles are for the vapor and super-critical fluid. The dashed line represents an extrapolation of the liquid data to the critical temperature (marked by the arrow). The solid line is the 1-MHz velocity interpolated from the data of Chynoweth and Schneider (1952). The points would not be expected to conform to this line, due to differences in the filling density and placement of the transducers (Mueller *et al.*, 1969).

the 1-MHz α_λ values. As a result of these new data, one can conclude that the relaxation treatment given by Chynoweth and Schneider is not an adequate description of the critical behavor in xenon. More work at ultrasonic and hypersonic frequencies (see the discussion of Brillouin scattering in Section III,E) is currently in progress (Cummins, 1969; Benedek, 1969).

D. HYDROGEN CHLORIDE

Breazeale (1962, 1963) has measured the velocity and attenuation in HCl at frequencies between 1 and 9 MHz. As in the xenon work described above, the procedure involved a constant-volume cell filled to a mean density close to the critical value. All measurements were made as the temperature was lowered from an initial supercritical value, as suggested by Tanneberger (1959), and at least 4 hr were allowed for equilibration. The velocity as a function of temperature showed a sharp, finite minimum close to the critical temperature. However, no statement about the positioning of the transducers was given, and it is impossible to judge the effect of density gradients in changing the local density. Indeed, it is not clear whether data below T_c were obtained in the liquid or in the vapor phase. The most important observation concerns the question of velocity dispersion. Within the accuracy of the experimental velocities ($\sim 1\%$), no dispersion could be detected in the 1–9-MHz range.

The HCl attenuation results are shown in Fig. 9. Although the sound absorption was said to go "through the characteristic maximum as the critical point was approached," only supercritical measurements were plotted. It should be noted that the logarithm of the signal amplitude was

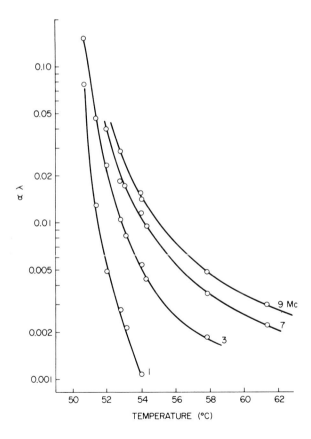

FIG. 9. Attenuation per wavelength in the supercritical region of HCl at several different frequencies (Breazeale, 1962).

fairly linear with distance, which indicates that finite-amplitude distortions were not excessive. The data in Fig. 9 could not be fit by the essentially f^4 dependence of a scattering model (Chernov, 1960), but Breazeale did show that α varied at f^2 at three supercritical temperatures (52, 53, 54°C).

Both the dispersion and attenuation results for HCl are in striking contrast to the results in xenon, which indicates that the characteristic relaxation frequencies near the critical point are quite different for these two substances.

E. CARBON DIOXIDE

Because of its ready availability, ease of handling, and convenient critical temperature, CO_2 has been the subject of many investigations (see Table I). Indeed, the existence of liquid–vapor critical points was discovered one hundred years ago when Andrews (1869) reported his pioneering work on the compressibility of carbon dioxide. In spite of extensive ultrasonic work, there is still some ambiguity as to the velocity variation near the critical point [see Tielsch and Tanneberger (1954) for a comparison of various data]. Perhaps the best and most extensive measurements are those of Tielsch and Tanneberger (1954), which are shown in Fig. 10. As the temperature is lowered toward T_c, the velocity minima become sharper, and, at 31°C, there is an exceedingly rapid variation near the minimum value of 140 m sec^{-1} (compare Fig. 10 with Fig. 1).

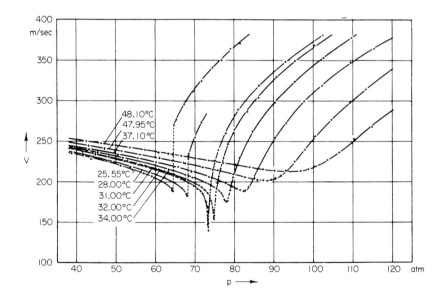

FIG. 10. Velocity in CO_2 as a function of pressure along various isotherms; $f = 0.41$ MHz (Tielsch and Tanneberger, 1954).

Although no velocity dispersion has been observed in the range 0.27–2 MHz, there are two low-frequency measurements which yield a lower minimum velocity than that shown in Fig. 10. Trelin and Sheludyakov (1966) report a minimum value of 132 m sec^{-1} at 500 Hz and 31.04°C. Very recently, Feke (1969) utilized a standing-wave technique at frequencies of a few kilohertz to obtain velocity minima of about 110 m sec^{-1} at temperatures between 31 and 31.1°C. The adiabatic compressibilities calculated from the data obtained at a given filling density are linear when plotted versus $\log(T - T_c)$. This linearity is observed over about two decades ($0.01 \leq \Delta T \leq 10$) and would strongly suggest a logarithmic singularity in κ_S. One problem with this method is the fact that a given velocity represents the average over a rather large vertical path length. Density gradients due to gravity will then play an important role when T gets sufficiently close to T_c.

Perhaps the most interesting feature about CO_2 is the fact that hypersonic velocities have been obtained in the critical region from the shifts in the Brillouin components of the light-scattering spectrum (Gammon *et al.*, 1967). These Brillouin shifts were measured below T_c in the liquid and in the vapor, at heights of -4.5 and $+4.5$ mm with respect to the meniscus, and along the critical isochore above T_c. No data were obtained in the range from 0.1°C below T_c to 0.2°C above T_c; thus, density gradients were not a

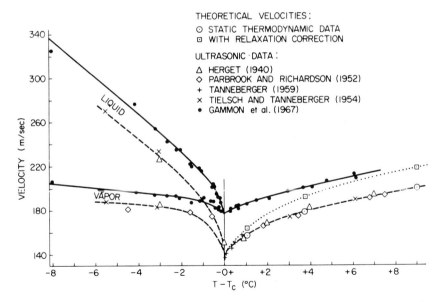

FIG. 11. Velocity of sound in CO_2 along both the liquid and vapor branches of the coexistence curve below T_c and along the critical isochore above T_c. The solid line and associated points represent the Brillouin data; the dashed line and its associated points represent a variety of ultrasonic data. The dotted line represents a "theoretical velocity" at hypersonic frequencies based on the ultrasonic values corrected for the effect of vibrational relaxation (Gammon *et al.*, 1967).

problem. The velocities of sound calculated from the Brillouin shifts correspond to frequencies ranging from 840 MHz away from T_c to 425 MHz near T_c. The hypersonic velocities are compared with the ultrasonic values in Fig. 11, and it is clear that there is a considerable difference. Part of this difference is due to the presence of internal *vibrational* relaxation in CO_2. The vibrational relaxation frequency for densities close to ρ_c is about 10 MHz (Madigosky and Litovitz, 1961), which is well above the 0–2-MHz ultrasonic frequencies, but well below the hypersonic frequencies. A correction for this vibrational contribution to u has been made in Fig. 11.

At temperatures more than 6°C above T_c, the difference between the Brillouin and ultrasonic velocities can be entirely attributed to vibrational relaxation. The vibrational contribution to u becomes less and less important as $T \to T_c$, and the hypersonic data should merge with the ultrasonic data if the only relaxation mechanism involved internal vibrational states. The sizeable dispersion near T_c ($\Delta u \approx 40$ m sec^{-1}) is presumably due to "structural relaxation" or an equivalent anomalous behavior in the volume viscosity. It would appear that the hypersonic velocities are close to u_∞ values, since $u(650 \text{ MHz}) \approx 189$ m sec^{-1} and $u(440 \text{ MHz}) \approx 180$ m sec^{-1} at $T - T_c = 0.2$°C (Ford *et al.*, 1968). Brillouin measurements in other substances, especially monatomic ones, will be of great value for investigating velocity dispersion and for determining the behavior of u_∞ near a critical point.

F. Other Systems

The other substances listed in Table I have been less thoroughly studied than those discussed above. The Russian work by Nozdrev and co-workers is especially sketchy, although valuable as an indication of the overall behavior in more complex molecules. Ethylene and sulphur hexafluoride are definitely the best studied of this group. The investigation of SF_6, a highly symmetric, nonpolar molecule, is interesting because data are available at two heights in a fairly tall (~ 7 cm) cell (Schneider, 1951). Measurements designated as "liquid" were made about 1 cm above the bottom of the cell, and measurements labeled "vapor" were made within 1 cm of the top of the cell. The resulting velocities and attenuation are shown in Figs. 12 and 13, although both quantities must be divided by 2 to correct for an interpretive error (Schneider, 1952). As these figures show, the u and α_λ values in the liquid and in the vapor seem, within the limits of the experimental error, to reach a common extremum at a temperature very close to T_c. Thus, it would appear that density gradients caused by gravity did not have an important effect on these measurements in SF_6. However, the minimum in the velocity is suspiciously shallow in comparison with helium and CO_2, which would suggest that the local density at the transducers was not equal to ρ_c. It is likely that the absolute deviation $|\rho - \rho_c|$ was the same in both the "liquid" and "vapor" runs, since the two heights were almost symmetric with respect to the center of the cell. If so, the velocities and attenuations above T_c might agree without either one representing data along the critical isochore.

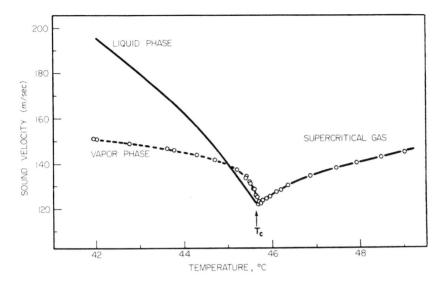

FIG. 12. Velocity data at 0.6 MHz in SF_6. The solid line represents a large number of observations made near the bottom of the cell; the open circles indicate measurements near the top of the cell (Schneider, 1951). (These values should be divided by 2; see Schneider, 1952.)

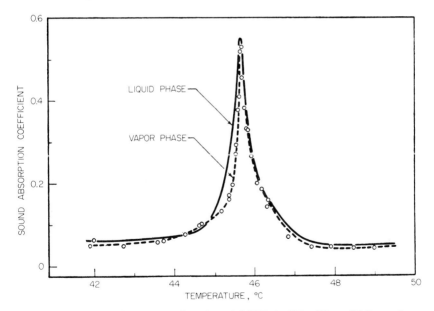

FIG. 13. Attenuation per wavelength at 0.6 MHz in SF_6. The solid line and open circles have the same meaning as in Fig. 12 (Schneider, 1951). (These values should be divided by 2; see Schneider, 1952.)

As a final comment on Fig. 12, it should be pointed out that, close to T_c, the velocity of sound in the saturated SF_6 vapor is higher than that in the liquid. This result was long considered dubious, but such an effect has also been observed by Tanneberger (1959) and by Nozdrev and Stepanov (1968). A thermodynamic analysis of the phenomenon has been made by Glinskii (1965).

IV. Binary–Liquid Phase Separation

A. THEORY

As discussed in Section II,E, two of the treatments of sound propagation in a critical binary mixture are very closely related. Fixman (1962) formulated a theory in terms of the effect that the temperature variation in the sound wave would have in perturbing the correlation of concentration fluctuations. This treatment seems to imply that the contribution to the attenuation due to the viscosity term is not important. However, Kawasaki and Tanaka (1967) showed that the bulk viscosity ζ increases enormously near the critical point. Their treatment is based on a correlation function expression for $(\frac{4}{3}\eta + \zeta)$:

$$\frac{4}{3}\eta + \zeta = (1/Vk_B T) \int_0^\infty dt \langle J^{zz}(t)J^{zz}(0) \rangle \tag{30}$$

where the current J^{zz} is the longitudinal component of an appropriate tensor. As long as the acoustic wavelength λ is much greater than the correlation length ξ, one obtains Eq. (24) as the low-frequency limiting form of the attenuation. Since the thermal conductivity exhibits no appreciable anomaly in binary liquids, Kawasaki and Tanaka drop the term involving K and ascribe the entire excess attenuation to $(\frac{4}{3}\eta + \zeta)$. They were able to carry out an approximate evaluation of Eq. (30) and show that $(\frac{4}{3}\eta + \zeta)$ diverges at the critical point. In spite of the obvious difference in the conceptual formulation of these two theories, Kawasaki and Tanaka have shown that both theories yield identical expressions for $\alpha(\omega)$.

The theories described above have two common features: a series expansion in powers of composition gradients, and the use of the Ornstein–Zernike form for the long-range part of the pair correlation function. By studying an explicit model system—a binary van der Waals mixture—Deutch and Zwanzig (1967) were able to avoid such approximations. In this treatment, the time dependence of the currents appearing in the time correlation function, Eq. (30), was determined from linearized hydrodynamic equations. Kawasaki and Tanaka (1967) were able to show that their result reduces to the Deutch–Zwanzig result for a van der Waals mixture. This does not, however, show that their assumptions are justified in the case of a more realistic model. In general, one should be cautious about applying the results of a long-range force model to a real system where the interactions have a short range. In addition to this warning, Deutch and Zwanzig also state that "the determination of bulk viscosity from ultrasonic absorption

measurements cannot be accomplished until the theory of thermal conductivity has been worked out."

This latter point may be cleared up by a recent mode–mode coupling
calculation by Swift (1968). A perturbation formalism is used to develop a
more detailed theory which allows heat-conduction modes (as well as sound
modes) to contribute to the anomalous part of the transport coefficients.
Static scaling laws are used rather than the Ornstein–Zernike formula, but
the result obtained for $(\frac{4}{3}\eta + \zeta)$ is essentially the same as Kawasaki and
Tanaka's result except that the critical index $\eta \neq 0$ in the new treatment.
Thus, all of the four theories described above seem to predict about the
same behavior.

Since the most recent experimental papers have been interpreted in
terms of Fixman's theory, we shall present a brief summary of his results.
According to Fixman (1962), the excess attenuation and velocity dispersion
associated with critical phenomena in binary liquids can be represented by

$$\alpha_\lambda = \pi A f^{-1/4} \mathscr{I}(d) \tag{31}$$

$$(u - u_0)/u_0 = -(A/2)f^{-1/4}\mathscr{R}(d) \tag{32}$$

where

$$d = Bf^{-1/2}|T - T_c| \tag{33}$$

and u_0 is the adiabatic sound velocity in the "absence of critical composition
fluctuations." The quantities A and B are independent of frequency and
are only weakly dependent on $|T - T_c|$; they are given explicitly in terms
of heat capacities and composition variables as well as the scale parameter a
and the inverse correlation length $\kappa \equiv \xi^{-1}$ of the long-range pair correlation
function. The quantities $\mathscr{I}(d)$ and $\mathscr{R}(d)$ are definite integrals which are
known analytically and have been evaluated numerically (Kendig *et al.*,
1964); they do not depend strongly on the frequency, but do show a very
rapid variation with $|T - T_c|$. In obtaining Eq. (33), the Debye form for κ^2
has been assumed:

$$\kappa^2 = 6l^{-2}|T - T_c|/T_c \tag{34}$$

where l is a short-range parameter characterizing the range of the intermolecular forces between two molecules. A more general form of Eq. (34),
based on scaling laws, would be $\kappa^2 \propto \varepsilon^{2\nu}$, where the critical index ν equals
2/3. However, one should not attempt to simply incorporate such a change
into Eqs. (31) and (32) as a correction, since the quantity A contains
$(\partial\kappa^2/\partial T)^2$ as a multiplicative factor. The use of any value of $2\nu > 1$ in Eq.
(31) will lead to a vanishing value of α at T_c, in conflict with experiment
(Anantaraman *et al.*, 1966).

Fixman (1962) attempted an experimental test of his theory on the
aniline + n-hexane system (Chynoweth and Schneider, 1951), with moderate
success. Since the expression for α_λ is quite sensitive to the value of l, one
way to judge the success of this theory in predicting the magnitude of the
excess attenuation is to consider the physical reasonableness of the l value.
In the aniline + n-hexane system, l has the fairly plausible value of 4.0 Å.

TABLE II

ULTRASONIC INVESTIGATIONS OF BINARY-LIQUID SYSTEMS AT 1 ATM

System	T_c (°C)	Composition studied	Temperature range (°C)	Frequency (MHz)	Quantity measured	Reference
Aniline + n-hexane	68.3	47.6 w% n-hex.	55–74	0.6	u, α	Chynoweth and Schneider (1951)
Aniline + cyclohexane	~30	51 w% an.	15–55	3	u, α	Cevolani and Petralia (1952)
	30.7	20–63 mole% an.	29–34	1.5–5	u, α	Brown and Richardson (1959)
	30.9	37.5 v% an.	31.2, 34.9	3.4–50	α	Kruus and Bak (1966)
Aniline + CCl_4	?	0–100%	−5, −20	9–27	α	Cevolani and Petralia (1958)
Benzene + methyl alcohol	239	10 w% benz.	220–270	?	u	Nozdrev and Tarantova (1962)
		0–100%	20–290	?	α	Makhanko and Nozdrev (1964)
Methyl alcohol + cyclohexane	49.1	29 w% alcohol	37–55	0.8–25	α	Singh and Verma (1968)
Nitrobenzene + n-hexane	23.2	53.2 w% nitrob.	23–28	1–9	α	Alfrey and Schneider (1953)

	21.02	0–100%	25	8	α	Sette (1955)
	21.0	0–100%	15–60	5–95	u, α	D'Arrigo and Sette (1968)
Nitrobenzene + n-heptane	19.4	61 v% hep.	18–32	3.4–50	α	Kruus and Bak (1966)
Nitrobenzene + iso-octane	30.2	0–100%	30.5–44.8	4.5–16.5	u, α	Anantaraman et al. (1966)
Perfluoro-cyclohexane + CCl$_4$	28.05	0–100% (mostly 56 v% CCl$_4$)	28–34	3.4–60	v, α	Kruus (1964)
Water + triethylamine	17.9	44.6 w% am.	10–28	0.6	u, α	Chynoweth and Schneider (1951)
	19.5	34 w% am.	10–20	1–9	α	Alfrey and Schneider (1953)
	17.9	44.6 w% am.	15	7–54	α	Sette (1955)
	18.2	25 w% am.	5–28	1.5, 2.5	u, α	Shimakawa (1961)
	17.8	0–100%	17–18	3.4–60	α	Kruus (1964)
Water + phenol	66	34 w% ph.	51–80	3	u, α	Cevolani and Petralia (1952)

Two of the systems listed in Table II will be discussed below in terms of Fixman's theory, and several other systems will be mentioned briefly at the end of this section.

The discussion above is based on the assumption that scattering of sound does not make an important contribution to the excess attenuation. This assumption seems to be valid for the systems described below, but Brown (1967) has presented evidence for the existence of scattering in the system aniline + cyclohexane. He considers the angular distribution of scattered ultrasonic energy in terms of a correlation function for concentration fluctuations. According to this view, scattering centers with radii as large as 0.3 mm exist and contribute significantly to the attenuation.[4]

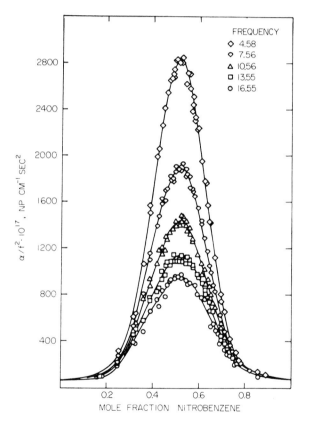

FIG. 14. Quantity α/f^2 as a function of composition in nitrobe zene + iso-octane at 34.8°C. Frequency values are in MHz (Anantaraman *et al.*, 1966).

---

[4] A very recent study of aniline + cyclohexane by D'Arrigo *et al.* (1970) does *not* indicate that scattering plays an important role in determining the absorption and dispersion of ultrasonic waves in this system.

B. NITROBENZENE + ISO-OCTANE

Anantaraman *et al.* (1966) used a pulse technique to measure the ultrasonic attenuation in this system over the full composition range, at temperatures of 30.5, 34.8, and 44.8°C ($T_c = 30.2$°C), for five frequencies between 4.5 and 16.5 MHz. Their experimental absorption data at 34.8°C, shown in Fig. 14, illustrate the very large excess attenuation at the critical composition (0.515 mole fraction nitrobenzene). Indeed, at the lowest temperature (30.5°C), the peak value of α is 47 times greater than the normal value observed in the pure liquids. No measurements were made below 30.5°C because of difficulties involving the determination of the composition. The variation of α/f^2 with temperature is shown in Fig. 15, and it is clear that anomalous attenuation persists to quite high temperatures. Figure 16 gives the frequency dependence of the attenuation for several compositions at 34.8°C.

These attenuation data could not be represented by a relaxation model involving only a single relaxation time; however, the Fixman theory was in reasonable agreement with the temperature and frequency dependence of the data. This is shown in Fig. 17, where the peak absorption at the critical composition is represented. The various parameters of the Fixman theory were empirically determined by fitting all the data at 16.5 MHz and the 30.5°C and 44.8°C points at 4.5 MHz. The dashed lines in Fig. 17 were then

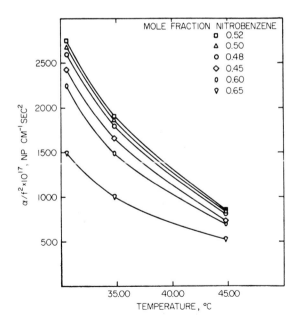

FIG. 15. Variation of α/f^2 in nitrobenzene + iso-octane as a function of temperature at 7.5 MHz (Anantaraman *et al.*, 1966).

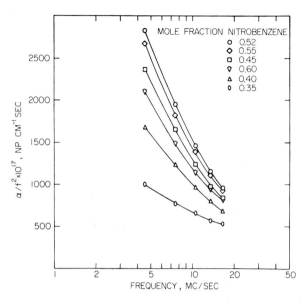

FIG. 16. Dependence of α/f^2 on frequency in nitrobenzene + iso-octane at 34.8°C (Anantaraman *et al.*, 1966).

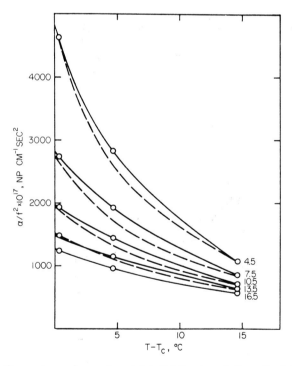

FIG. 17. Comparison of experimental (solid lines) and theoretical (dashed lines) sound absorption in a critical mixture of nitrobenzene and iso-octane. Frequency values are in megahertz (Anantaraman *et al.*, 1966).

calculated from Eqs. (31) and (32). The agreement with experiment is seen to be good, but far from exact. In particular, the predicted frequency dependence ($\alpha/f^2 \propto f^{-5/4}$) is verified within experimental error, and the parameter l has the reasonable value of 6.6 Å.

Since velocities were only measured at 4.5 MHz, one cannot discuss dispersion. However, the Fixman theory also predicts a monotonic increase in the value of $(u - u_0)/u_0$ at a fixed frequency as $(T - T_c)$ increases. Anantaraman *et al.* observed a 3.3% *decrease* in u on going from 30.5 to 44.8°C, which is in agreement with the direction of the change observed in aniline + cyclohexane by Brown and Richardson (1959) and in water + triethylamine by Chynoweth and Schneider (1951).

C. Nitrobenzene + n-Hexane

It so happens that the other binary system which has been recently investigated over a wide range of composition, temperature, and frequency is very similar to the system described in Section IV,B. D'Arrigo and Sette (1968) measured both the attenuation and the velocity as functions of f, T, and the mole fraction X of nitrobenzene. The critical constants for their system were $T_c = 21.0$°C and $X_c \approx 0.37$, whereas the critical composition commonly reported in the literature is around $X = 0.43$. In general, the results for this system conform very well with the behavior shown in Figs. 14–16 for nitrobenzene + iso-octane. A typical curve for the α/f^2 variation

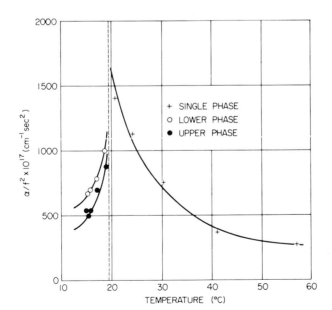

Fig. 18. Variation of α/f^2 with temperature at 15 MHz and $X = 0.33$ in nitrobenzene + n-hexane (D'Arrigo and Sette, 1968).

with temperature both above and below the phase separation temperature is
shown in Fig. 18 for $X = 0.33$ and $f = 15$ MHz, and the supercritical data
could be well represented by Fixman's theory. In fact, Singh *et al.* (1966)
have used the experimental attenuation data of D'Arrigo and Sette to test
Eq. (31) with respect to the temperature and frequency dependence at the
critical concentration.

D'Arrigo and Sette have attempted to deduce the composition dependence of α from Fixman's theory by using the Flory–Huggins approximation for the inverse correlation length κ. As shown in Fig. 19, the

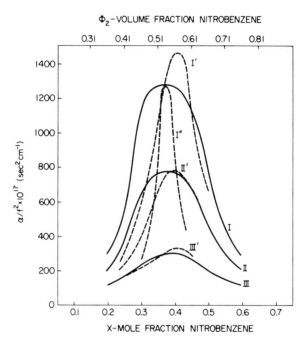

FIG. 19. Comparison of experimental (solid lines) and theoretical (dashed lines)
values of α/f^2 as a function of composition in nitrobenzene + n-hexane. I, II, and III,
designate temperatures of 22, 30, and 50°C, respectively. The two dashed lines labeled
I' and I'' represent different empirical choices of the theoretical parameters (D'Arrigo
and Sette, 1968).

agreement between their calculated and experimental values of α/f^2 is not
very good. This may be due to the use of the Flory–Huggins expression
for $(\partial\mu_1/\partial\phi_2)$, or it may indicate a failure of the Fixman theory. In a sense,
the very fact that the theory does predict the existence of a well-defined
maximum as a function of composition might be considered as a success.

The experimental observation that u *decreases* by about 1.5% when the
temperature is increased from T_c to $T_c + 5$ is in apparent contradiction with
predictions based on Eq. (32); see also Section IV,B.

D. OTHER SYSTEMS

Another system of interest is water + triethylamine, which has a *minimum* critical solution temperature. That is, there is a single-phase region for temperatures below T_c, and the system separates into two phases above T_c. Chynoweth and Schneider (1951) studied this system at a fixed composition of 44.6 wt.% amine, for which $T_c = 17.9°C$. Their velocity data, shown in Fig. 20, follow the same qualitative trend as that observed in systems with a maximum solution temperature [e.g., see D'Arrigo and Sette (1968)]. The attenuation has also been measured in this system by many investigators (see Table II). The most recent of these studies was by Kruus

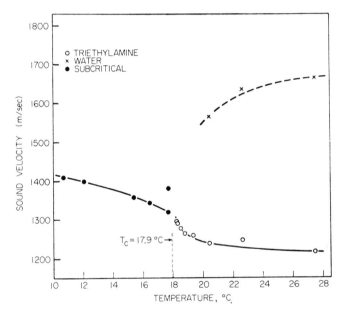

FIG. 20. Variation of u with temperature for water + triethylamine: $f = 0.6$ MHz (Chynoweth and Schneider, 1951). (These values should be divided by 2 to correct for an interpretive error; see Schneider, 1952.)

(1964), who analyzed his data in terms of the Fixman theory. He found that the frequency dependence of α was better described by $\alpha/f^2 \propto f^{-5/4}$ than by the relaxation form $\alpha/f^2 \propto [1 + (f^2/f_c^2)]^{-1}$. However, the l value required to fit Fixman's theory to his data was 2.0 ± 0.3 Å, which seems like a rather unattractively small value.

Kruus (1964) has also studied the attenuation in $C_7F_{14} + CCl_4$. In fitting those results to Fixman's theory, he finds that l values between 6.6 and 14.6 Å are consistent with his rather sparse data. Singh and Verma (1968) have made a more extensive investigation of attenuation in the methyl alcohol + cyclohexane system and have found only qualitative agreement between their data and Fixman's theory.

In addition to the binary systems listed in Table II, there have been two rather limited investigations of multicomponent liquid systems. Grechkin and Nozdrev (1964) studied u as a function of temperature and composition in the critical region of the ternary system benzene + methyl alcohol + toluene, while Nozdrev and Yashina (1966) have studied the quaternary system benzene + methyl alcohol + toluene + ethyl acetate. One wonders if they will soon add a fifth component to this mixture!

V. Ferroelectric and Antiferroelectric Transitions

A. THEORY

Due to piezoelectric and/or electrostrictive effects, there is a strong coupling between mechanical and dielectric behavior in all ferroelectric materials. In the static limit, the relationship between the anomalous dielectric and elastic properties can be predicted from thermodynamics (Janovec, 1966). We shall be concerned here with dynamic effects associated with the propagation of acoustic waves in either the ultrasonic or Brillouin (hypersonic) region.

The idea of coupling acoustic and polarization waves was treated by Sannikov (1962) from a phenomenological point of view, with the damping of the mixed polarization–sound wave explicitly included. The velocity and attenuation of polarized sound waves near a Curie point were considered for a crystal of arbitrary symmetry, but detailed expressions were developed for waves propagating along the crystallographic axes of rhombic and cubic ferroelectrics. In rhombic crystals, such as Rochelle salt in its paraelectric phase, the piezoelectric (linear) effect is dominant, and electrostriction can be neglected. In cubic crystals, such as paraelectric $BaTiO_3$, the piezoelectric constants vanish because of symmetry, and the coupling is due entirely to electrostrictive (quadratic) effects. In both cases, anomalous variations in u and α were predicted for waves of a specific polarization. Sannikov (1962) carried out numerical calculations, which were compared with experimental results in the case of Rochelle salt. A very closely related discussion of the dispersion relations for such mixed acoustic–polarization modes in $BaTiO_3$ has been presented by Dvorak (1968).

In another paper, Dvorak (1967) has treated the interaction of acoustic waves with polarization waves in ferroelectrics by using linear response theory to determine the "effective" elastic constants. For crystals which are piezoelectric in both the ferroelectric and paralectric phases, it was shown that the frequency and spatial dispersion of the effective elastic constants are directly related to the dispersion of the linear dielectric susceptibility of the clamped crystal. In tensor notation,

$$c_{\alpha\beta\gamma\delta}(q, \omega) = c^{P}_{\alpha\beta\gamma\delta} - a_{i\alpha\beta} a_{j\gamma\delta} \chi_{ij}(q, \omega) \tag{35}$$

where c^{P} is the elastic stiffness at constant polarization, a is the piezoelectric stress constant, and χ is the clamped susceptibility. A discussion of Eq. (35) as applied to KH_2PO_4 is given in Section V,B.

When the frequency ω_0 of the polarization oscillations (optical soft mode) is much greater than the acoustic frequency ω, Eq. (35) will give rise to relaxation formulas just like Eqs. (11) and (12) with a critical relaxation time of the same form as Eq. (16). In this case, the Landau kinetic coefficient L is proportional to the inverse of the damping constant Γ for the ferroelectric soft mode, and the temperature dependence of τ reflects the temperature variation of ω_0^{-2}. That is, for small ω,

$$\tau = 2\Gamma/\omega_0^2 \qquad \text{(damped soft mode)} \qquad (36)$$

$$\tau = \chi(0)/L \qquad \text{(Landau model)} \qquad (37)$$

where $\chi(0)$ is the static susceptibility, and both equations predict identical variations for τ.

There are three recent papers which discuss acoustic anomalies in ferroelectrics directly in terms of soft-mode models.[5] Lefkowitz and Hazony (1968) have used a lattice-dynamic theory in the diatomic (i.e., two atoms per unit cell) approximation to show that the "softening" of the transverse optic mode will cause an anomalous decrease in the shear elastic constants. The temperature dependence of the TO and TA dispersion curves in $SrTiO_3$ and $KTaO_3$ were discussed qualitatively in terms of this model. Tani and Tsuda (1969) calculated the attenuation and velocity variation in the paraelectric phase of displacive-type ferroelectrics. They used the one-dimensional Silverman model and assumed that the acoustic phonon was coupled to the soft optical phonon via third-order anharmonic interaction (which has the same form as an electrostrictive interaction). The anomalous increase in α_λ and decrease in u were predicted to vary like $T/(T - T_c)^{3/2}$, but these results are limited to rather high frequencies and temperatures not too close to T_c. Inoue (1969) has adopted a similar model for his treatment of the critical attenuation in KH_2PO_4-type ferroelectrics. He based his calculation on the Kobayashi proton-tunneling model, assumed that the acoustic phonon is coupled to the soft ferroelectric mode via third-order anharmonic interaction, and also considered the damping effect of the random force. Unfortunately, this is also a one-dimensional treatment, and the result can be applied only to the case of a sound wave propagating parallel to the polar axis. Thus, none of these three papers is of direct usefulness in interpreting current experimental data.

In a paper which is in the general spirit of Landau's theory as described in Section II,D, Geguzina and Krivoglaz (1968) have considered the influence of long-range forces in ferroelectric crystals which are piezoelectric. The original application of Landau's theory to ferroelectrics was made by Yakovlev and Velichkina (1957). In that treatment, it was assumed that the relaxation rate $\partial P/\partial t$ at a given point depended only on the value of the polarization P

---

[5] Note added in proof: Very recently, Barrett (1970) has discussed the acoustic properties of perovskite materials (especially $KTaO_3$ and $SrTiO_3$) in terms of their soft-mode behavior.

TABLE III

ULTRASONIC INVESTIGATIONS OF FERROELECTRIC AND ANTIFERROELECTRIC SYSTEMS

Substance	T_c (°K)	Temp. range (°K)	Electric field (kV/cm)	Frequency[a]	Quantity measured	Reference
Ferroelectrics						
BaTiO$_3$	393	300–425	0–20	10	α_L	Hueter and Neuhaus (1955)
		293–413	0–6	0.05–1	Q^{-1}	Ikeda (1957)
		298–423	10	~6	s_{11}, s_{12}, s_{66}	Berlincourt and Jaffe (1958)
		295–408	Yes	Low	$s_{11}, s_{11}^P, 2s_{12}+s_{66}$	Huibregtse et al. (1959)
		393–433	0	Low	$s_{11}, 2s_{12}+s_{44}$	Fushimi and Ikeda (1966)
SrTiO$_3$	108[b]	108–300	0–30	10 or 30	c_{11}, c_{12}, c_{44}	Bell and Rupprecht (1963)
KTaO$_3$	None[c]	2–300	0	180–900	$u_L, u_T, \alpha_L,$ $\alpha_T(100)$	Barrett (1968, 1969)
Triglycine sulfate (TGS)	~322	293–328	0.09 bias, 0.9 poling	Few kHz	$s_{11}, s_{22}, s_{33},$ $s_{11}^D, s_{22}^D, s_{33}^D$	Ikeda et al. (1962)
		303–333	0–8	~100 kHz	$s_{11}, s_{33}, \delta_{11}, \delta_{33}$	Shuvalov and Pluzhnikov (1962)
		321–323	—	15–125	u_L, α_L	O'Brien and Litovitz (1964)
		321–323	0, 0.5, 1	5–15	$\alpha_L(001)$	Minaeva and Levanyuk (1965)
		307–327	Value unspecified	~7.5 GHz, ~18GHz	u_T, u_L	Gammon and Cummins (1966)
		$T_c \pm 0.6$		10	$\alpha_L(x, y, z)$	Minaeva et al. (1969)
		$T_c \pm 0.25$	0–0.1	30	$\alpha_L(1\bar{0}1)$	Gammon and Verdieck (1969)
TGSe	~298	288–302	—	5, 10, 15	$s_{33}, \alpha_L(001)$	Minaeva et al. (1967)
Triglycine fluoroberyllate	343	333–347	—	5, 10, 15	$s_{33}, \alpha_L(001)$	Minaeva et al. (1967)
KH$_2$PO$_4$	122	122–363	—	~100 kHz	Six s_{ij}, s_{66}^P	Mason (1946)
		80–140	0–4	10–200 kHz	$s, \delta(Z45°)$	Shuvalov and Mnatsakanyan (1965)
		119–122	0	5, 15	$\alpha_T(x_y)$	Golubeva and Shustin (1968)

Material	T_c (K)	Range (K)		Frequency[a]	Quantities	Reference
KD₂PO₄	211[d]	118–300	0	0.5–5 GHz	c_{66}	Brody and Cummins (1968)
		122–145	0	1–90	$c_{66}, \alpha_T(x_y)$	Garland and Novotny (1969)
		110–200	0–3.9	0.5–5 GHz	c_{66}	Brody (1969)
		120–125	0–4	15, 25	$c_{66}, \alpha_T(x_y)$	Litov and Garland (1969)
	205.6[d]	150–300	0	10–200 kHz	s_{66}	} Shuvalov and Mnatsakanyan (1966)
		200–300	0	10–200 kHz	Other 5 s_{ij}	
		200–225	1.5	5–45	$c_{66}, \alpha_T(x_y)$	Litov and Uehling (1968)
RbH₂PO₄	~143	143–300	—	Low	s_{66}, s_{66}^D	Steinemann (1952)
	~148	100–300	0	10–200 kHz	Six s_{ij}	Mnatsakanyan et al. (1966)
Rochelle salt	$T_u \approx 297$,	280–313	0–0.8	10	$s_{44}, \alpha_T(Z)$	Price (1949)
	$T_l \approx 255$	288–308	—	5	$\alpha_T(z_y)$	Yakovlev et al. (1957)
		291–308	0, 0.6	5	$\alpha_T(z_y)$	Yakovlev et al. (1958)
		283–313	0–3	Low	log dec δ_L	Shuvalov and Likhacheva (1960)
		245–273	—	5	$\alpha_T(z_y)$	Shustin et al. (1961)
		253–270	Yes	6	$\alpha_L(x, y, z)$	Merkulov and Sokolova (1962)
		289–305	0	5, 15	$\alpha_T(z_y)$	Baranskii et al. (1963)
		243–308	0, 0.05	20–70 kHz	log dec δ_L	Shirokov and Shuvalov (1964)
NaNO₂	436	225–500	0, 3	~100 kHz	$s_{11}, s_{22}, s_{33}, s_{44}, s_{66}$	Hamano et al. (1963)
K₄Fe(CN)₆·3H₂O	247	213–293	—	10	u_L and u_L^D, (010), (101), (10$\bar{1}$)	Schacher (1967)
Antiferroelectrics						
NH₄H₂PO₄	148	153–373	—	~100 kHz	Six s_{ij}, s_{66}^P	Mason (1946)
ND₄D₂PO₄	242	243–343	—	~100 kHz	Six s_{ij}	Mason and Matthias (1952)
PbMg₀.₅W₀.₅O₃	~308	283–343	—	50–100 kHz	s_{11}, log dec δ	Shuvalov and Minaeva (1963)
		$T_c \pm 18$	—	5, 10, 15	α_L	Minaeva et al. (1966)

[a] In megahertz, except where otherwise noted.

[b] Transition due to instability at [111] zone boundary, not a ferroelectric Curie point: see Shirane and Yamada (1969).

[c] While T_c would be expected to be a few degrees Kelvin, paraelectric phase is stabilized by zero-point fluctuations, and the transition does not occur.

[d] Incomplete deuteration: $T_c = 222°K$ for 100% D.

at that point. However, if the acoustic wavelength is smaller than the Debye screening length, dipole–dipole interactions between the variations in P at different points in the crystal will become important. When this nonlocal coupling was included, Geguzina and Krivoglaz found that the attenuation and velocity variation of ultrasonic waves had a very strong directional dependence. In particular, anomalous ultrasonic behavior in a uniaxial ferroelectric can only occur when the acoustic wave vector is perpendicular to the polar axis. The frequency and wave-vector dependence of u and α were analyzed for the case of Rochelle salt, but the treatment is equally valid for KH_2PO_4 or for the ferroelectric phase of triglycine sulfate (TGS). However, the results of Geguzina and Krivoglaz are only valid when piezoelectric coupling is the dominant mechanism for the interaction between acoustic waves and the polarization.

In the case of a uniaxial ferroelectric which is *not* piezoelectric in its paralectric phase, Levanyuk *et al.* (1969) have shown that there is still anomalous attenuation caused by the interaction of the sound wave with thermal fluctuations in the polarization. Since such fluctuations create an electric field and become correlated over large distances, the very large fluctuations typical of a lambda-type transition are partially suppressed in ferroelectrics. For biaxial and triaxial ferroelectrics, the behavior of the complex elastic modulus is essentially that given by an earlier theory of Levanyuk (1966), which deals with fluctuations correlated only at short distances (see Section II,E for a brief description of this theory). Uniaxial ferroelectrics form a special class, since polarization fluctuations are very strongly suppressed. Indeed, only the fluctuations in the component of the polarization parallel to the polar axis, $P_{\parallel}(q)$, increase as $T \to T_c$ and then only for propagation vectors $\mathbf{q}$ which are perpendicular (or almost so) to the polar axis. Levanyuk *et al.* (1969) predict that the largest anomalies in u and α for uniaxial, nonpiezoelectric crystals are associated with acoustic waves generating strains which are coupled by electrostriction to $P_{\parallel}^2$. Such anomalies should be of observable magnitude within $\sim 0.2°C$ of T_c, and it is predicted that they should be *independent* of the direction of propagation of the sound wave. This "electrostrictive-fluctuation" result is in striking contrast to the "piezoelectric-relaxation" result of Geguzina and Krivoglaz (1968). It indicates that an investigation of the directional dependence of α could be used to separate the two effects. Waves propagating parallel to the polar axis in a uniaxial ferroelectric should have no anomaly due to piezoelectric coupling, but will have an electrostrictive anomaly due to thermal fluctuations in P. For waves propagating perpendicular to the polar axis, the piezoelectric effect will dominate below T_c (due to the presence of spontaneous polarization) or even above T_c for crystals which are piezoelectric in the paraelectric phase.

Let us next consider the effect on the ultrasonic velocity and attenuation in uniaxial ferroelectrics of applying an external electric field E along the polar axis. Minaeva and Strukov (1966) were the first to deal theoretically with this problem, but their treatment was restricted to nonpiezoelectric crystals. Recently, Geguzina and Timan (1968) have treated both Rochelle salt, which is piezoelectric in the paraelectric phase, and TGS, which is not. Their theory is based directly on the Landau–Khalatnikov approach, and the result for the field dependence of u and α can be cast in the relaxational form of Eqs. (11) and (12), where the relaxation time τ has an explicit dependence on E. In general, it was found that τ will decrease as the field increases; in particular, it was predicted that $\tau \propto E^{-2/3}$ at $T = T_c$.

For a crystal which is piezoelectric, the attenuation and velocity anomalies are appreciably reduced as the field is increased. For a crystal which is not piezoelectric, the situation is more complicated above the Curie point. The relaxation strength $[(u_\infty{}^2 - u_0{}^2)/2u^3$ in Eq. (12)] is proportional to $P_0{}^2$, where P_0 is the equilibrium value of the polarization. In this case, the Geguzina–Timan model does not contain any mechanism for acoustic loss in the paraelectric phase when $E = 0$. However, when an electric field is applied, ultrasonic attenuation can occur via interaction of the sound wave with the *induced* polarization. (Below T_c, there is, of course, a large spontaneous polarization even when $E = 0$.) As a result of the behavior in the relaxation strength, Geguzina and Timan predict that TGS-type ferroelectrics will show an *increase* in α when a field is applied above T_c.

There do not appear to be any theoretical papers on the behavior of sound waves in antiferroelectric crystals, and Table III shows that there have been only four, somewhat limited experimental studies. Thus, we will restrict our subsequent discussion to the ferroelectric case. Section V,B will present results for a typical piezoelectric crystal, and Section V,C will be concerned with a typical nonpiezoelectric crystal.

B. Potassium Dihydrogen Phosphate

Potassium dihydrogen phosphate (KDP) is a uniaxial ferroelectric with a Curie temperature of $\sim 122°K$. The completely analogous transition in KD_2PO_4 (KD*P) occurs at $\sim 222°K$, which shows that there is an unusually large isotope effect. In its paraelectric phase, KDP is tetragonal ($\bar{4}2m$) and the x_y mechanical strain is piezoelectrically coupled to the polarization along the ferroelectric z axis. Therefore, a transverse ultrasonic wave propagating in the [100] direction with its polarization in the [010] direction is the shear wave of interest. The elastic constant related to this shear is called c_{66}, but there are two limiting values of this constant, depending on the electrical boundary conditions. One can specify the elastic properties at constant polarization (c^P) or at constant electric field (c^E). Low-frequency resonance measurements by Mason (1946) on bare and on plated crystals show that c_{66}^P exhibits a normal linear temperature dependence, whereas c_{66}^E drops toward zero at the Curie point.

The ultrasonic velocity and attenuation are determined from the complex stiffness at constant field. For KDP, Eq. (35) becomes

$$c_{66}^* = c_{66}^P - a_{36}^2 \chi_{33}^x(\omega) \tag{38}$$

where a_{36} is the piezoelectric stress constant $(\partial E_3/\partial x_6)_P$ and $\chi_{33}^x(\omega)$ is the complex linear susceptibility of the clamped crystal. At ultrasonic frequencies, $\omega\tau \ll 1$ for KDP and the susceptibility can be represented in the form $\chi(\omega) = \chi(0) - i\chi(0)\omega\tau$, where $\chi(0)$ is the static value and the relaxation time τ is given by either Eq. (36) or (37). Taking the real part of Eq. (38), one obtains

$$c_{66}^{P,S} - c_{66}^{E,S} = a_{36}^2 \chi^{x,S}(0) \tag{39}$$

or the thermodynamically equivalent form

$$s_{66}^{E,S} - s_{66}^{P,S} = b_{36}^2 \chi^{X,S}(0) \tag{40}$$

where b_{36} is the piezoelectric strain constant, $\chi^{X,S}$ is the free susceptibility, and s_{66} is the elastic compliance ($=1/c_{66}$). The superscript S has been added to remind one that the elastic constant determined from the ultrasonic velocity is an adiabatic value. This distinction becomes important near T_c because of the piezocaloric effect. From the imaginary part of Eq. (38), one finds

$$\mathrm{Im}(c^*) \approx 2\rho u^3 \alpha/\omega = a_{36}^2 \chi^{x,S}(0)\omega\tau_{S,x} \tag{41}$$

which can be rewritten, with the help of Eq. (39), as

$$\alpha = [(c_{66}^P - c_{66}^E)/2\rho u^3]\omega^2 \tau_{S,x} \tag{42}$$

This expression is in agreement with the low-frequency limit of Eq. (12) if $c^P = \rho u_\infty^2$ and $c^E = \rho u_0^2$. As we shall see below, the velocity u at ultrasonic frequencies is indeed the zero-frequency limiting value for KDP. The relation $c^P = \rho u_\infty^2$ is also clearly valid, since, at infinite frequency, the polarization would not be able to follow the applied stress, and the effective (or "frozen") elastic constant would correspond to a constant polarization.

The dielectric susceptibilities are known to follow a Curie–Weiss law above T_c:

$$\chi^X = C/(T - T_c), \qquad \chi^x = C/(T - T_0) \tag{43}$$

where the Curie constant $C = 259°\mathrm{K}$ for KDP. In Eq. (43), T_c is the actual temperature at which spontaneous polarization appears, and T_0 is used to denote the lower temperature at which a *clamped* crystal would become ferroelectric. Combining Eqs. (39), (40), and (43), one obtains

$$(s_{66}^E - s_{66}^P)^{-1} = (T - T_c)/D, \qquad (c_{66}^P - c_{66}^E)^{-1} = (T - T_0)/D_0 \tag{44}$$

where $D = Cb_{36}^2$ and $D_0 = Ca_{36}^2$. Either form of Eq. (44) can be called an elastic Curie–Weiss law, where D or D_0 are the elastic Curie constants, and these are convenient ways to test the temperature dependence of the velocity in the paraelectric phase. The appropriate plots are shown in Fig. 21, where the Mason (1946) values of c^P and the Garland and Novotny (1969) values of c^E have been used. It is clear that the elastic Curie–Weiss law is well obeyed; moreover, the slopes in Fig. 21 are in excellent agreement with the values calculated from the known values of C, a, and b (Garland and Novotny, 1969).

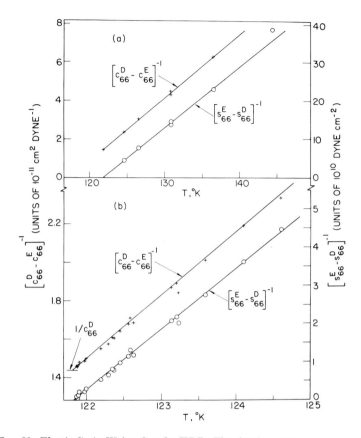

FIG. 21. Elastic Curie–Weiss plots for KDP. The elastic constant at constant diele-tric displacement c^D is essentially identical to c^P in materials with a high dielectric con-stant such as KDP. Part (a) shows data over a wide temperature interval, whereas much more detailed data close to T_c are shown in part (b). The corresponding lines in parts (a) and (b) are identical, i.e., drawn with the same slopes and intercepts (Garland and Novotny, 1969).

Evidence that velocity data at 10 MHz correspond to the zero-frequency limit is provided by Fig. 22, which shows the very good agreement between ultrasonic c^E_{66} values and those obtained from Brillouin scattering at fre-quencies ranging from 0.5 to 5 GHz (Brody and Cummins, 1968). In addi-tion, the τ values obtained below from an analysis of the attenuation indicate that velocity dispersion should be negligible at 10^7 Hz and very small even at 5×10^9 Hz. This is by no means always the case; in many materials, τ is large enough so that there is appreciable dispersion between ultrasonic and hypersonic frequencies (see Sections III,E, V,C, and VII,B). Another interesting feature of Fig. 22 is the very rapid rise in c^E just below T_c.

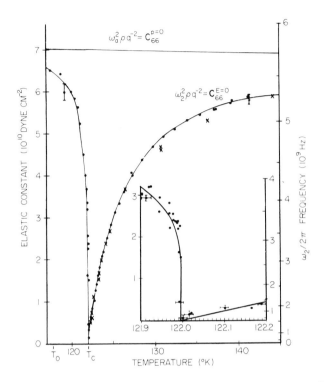

FIG. 22. Elastic constants c_{66}^E and c_{66}^P for KDP near its Curie point. Solid points were obtained from Brillouin shifts, and the crosses from ultrasonic measurements (Brody and Cummins, 1968).

Ultrasonic data cannot be obtained below T_c at zero applied field, because of domain scattering, but, recently, measurements have been made on single-domain KDP crystals which were poled with external fields $E_z = 1, 2, 3,$ and $4\ \text{kV cm}^{-1}$ (Litov and Garland, 1969). Figure 23 shows the effect of such applied fields on the temperature dependence of c_{66}. Again, there is good agreement with data obtained from Brillouin shifts measured in the presence of an applied field (Brody, 1969). Note from Fig. 23 that the effect of the field is to progressively smooth out and shift the minimum in c_{66}. This velocity behavior is in qualitative agreement with the theoretical predictions of Geguzina and Timan (1968). In the case of KD*P, Litov and Uehling (1968) measured c_{66}^E as a function of temperature at $E = 1.5\ \text{kV cm}^{-1}$. Although the general behavior is very similar to that in KDP, there is an apparently discontinuous jump in the velocity at T_c (in spite of the applied field). This strongly supports the now accepted view that there is a small, but definite first-order transition in KD*P.

Let us now consider the attenuation data in KDP and KD*P. The excellent data of Litov and Uehling (1968) on the variation of α in KD*P

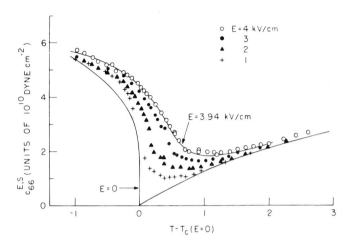

FIG. 23. Values of $c_{66}^{E,S}$ for KDP as a function of temperature for several constant values of the applied field. Ultrasonic data points were obtained at 15 MHz, and the two lines represent Brillouin data at zero field and at 3.94 kV cm^{-1} (Litov and Garland, 1969).

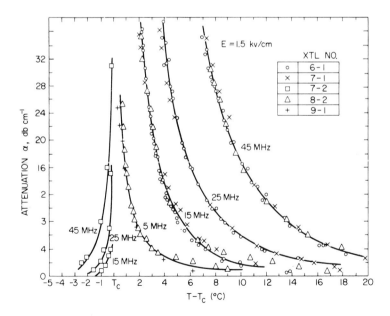

FIG. 24. Attenuation in KD*P versus temperature at several ultrasonic frequencies. All data were obtained at a constant field of 1.5 kV cm^{-1} (Litov and Uehling, 1968).

as a function of temperature and frequency at a field of 1.5 kV cm^{-1} are shown in Fig. 24. Analysis of these data shows that α varies as ω^2, and, consequently, verifies that $\omega\tau \ll 1$. It should also be mentioned that the magnitude of α for $|\Delta T| > 1°$K was insensitive to the value of the applied field as long as E was large enough to pole the crystal into a single domain. For an analysis of the temperature dependence of α and a determination of the polarization relaxation time, it is convenient to rewrite Eq. (42) as

$$\alpha = [(c_{66}^{P} - c_{66}^{E})/2uc^{P}]\omega^2\tau_{S,X} \tag{45}$$

where $\tau_{S,X}$ is the relaxation time at constant (zero) stress. This change is justified by the fact that the two relaxation times are simply related by $\tau_X/\tau_x = c^P/c^E = \chi^X/\chi^x$. From Eqs. (37) and (43), it can be seen that τ_X will diverge at the Curie point according to $CL^{-1}/(T - T_c)$, whereas τ_x varies as $CL^{-1}/(T - T_0)$ and will have a finite value at T_c [see also Geguzina and Krivoglaz (1968)]. By utilizing Eq. (45), Litov and Uehling (1968) obtained the τ_X and L values shown in Fig. 25. As expected from Landau theory, L is a slowly varying function of temperature, but τ_X^{-1} varies approximately as $|T - T_c|$. The interpretation of these results in terms of the microscopic Silsbee–Uehling–Schmidt (SUS) theory of KDP will not be given here.

The KDP attenuation results of Garland and Novotny (1969), obtained at zero applied field, were restricted to the paraelectric phase. An analysis

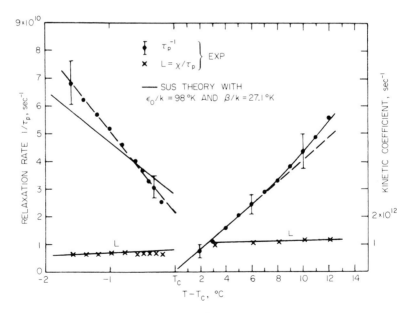

FIG. 25. Temperature variation of τ_X^{-1} and the kinetic coefficient L for KD*P in the vicinity of the Curie point. [Note: τ_p is used in this figure to denote τ_X.] (Litov and Uehling, 1968.)

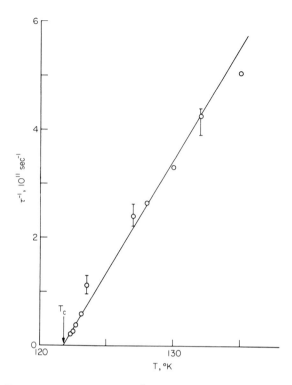

FIG. 26. Temperature variation of τ_x^{-1} for KDP in its paraelectric phase (Garland and Novotny, 1969).

identical to that described above for KD*P gave the τ_X values shown in Fig. 26. A comparison of Figs. 25 and 26 shows that there is a tenfold decrease in the paraelectric relaxation rate on deuteration (the kinetic coefficient L is $\sim 11 \times 10^{12}$ sec^{-1} for KDP and $\sim 1.1 \times 10^{12}$ sec^{-1} for KD*P). This seems quite reasonable, in view of the proton-tunneling motion which is associated with this transition. Very recently, Litov and Garland (1969) have measured the attenuation in KDP both above and below T_c at applied fields of 3 and 4 kV cm^{-1}. Although there is some small effect of the field on the τ_X values in the paraelectric phase, the most important feature of this work concerns the values of α below T_c. The ferroelectric τ_X values for KDP were found to be very similar to those for KD*P, which implies a very substantial decrease in the value of L when KDP becomes ordered. A comparison of these results with recent Brillouin results in the presence of an applied field (Brody, 1969) raises the question as to whether the relaxation formalism is a completely adequate description for KDP. Brillouin measurements on KD*P are currently in progress (Cummins, 1969) and should provide some additional insight into this system.

C. TRIGLYCINE SULFATE

Triglycine sulfate (TGS) is a uniaxial ferroelectric with a Curie temperature of about 49°C. The crystal is monoclinic both above and below T_c. In the paraelectric phase, TGS is centrosymmetric and nonpiezoelectric; in the ferroelectric phase, there is a spontaneous polarization directed along the y axis and the crystal becomes piezoelectric. Longitudinal strains along all three principal axes are coupled with P_2, but there is considerable anisotropy in the dispersion and attenuation of sound waves. The piezoelectric, electrostrictive, and low-frequency elastic properties of TGS have been thoroughly investigated by Ikeda *et al.* (1962).

O'Brien and Litovitz (1964) have combined the Landau approach to cooperative phase transitions with the Devonshire free-energy expansion for ferroelectrics in order to explain the velocity dispersion and ultrasonic attenuation in TGS. Let us first consider the behavior of the velocity. The velocity dispersion can be represented by the relaxation form given in Eq. (11), where O'Brien and Litovitz show that

$$(u_\infty{}^2 - u_0{}^2) = 2g_{k2}^2/\rho\xi \tag{46}$$

and

$$\tau = [2L\xi P_2{}^2]^{-1} \tag{47}$$

In the above equations, g_{k2} is the electrostrictive constant coupling a longitudinal strain x_k with the spontaneous polarization P_2, $\xi/4$ is the coefficient of $P_2{}^4$ in the Devonshire expansion, and L is the Landau kinetic coefficient. It is assumed that u_∞, g, ξ, and L are all temperature-independent quantities, and $P_2{}^2$ varies linearly with $(T_c - T)$ in this model. Thus, τ will behave as in Eq. (16): $\tau \propto (T_c - T)^{-1}$ below T_c and $\tau = \infty$ (since $P_2 = 0$) above T_c. As shown by Eq. (46), the quantity $(u_\infty{}^2 - u_0{}^2)$ is small and *temperature independent* for TGS: This is in marked contrast to the behavior of a piezoelectric type of ferroelectric such as KDP.

From Eqs. (11), (46), and (47), one can predict the expected velocity variation with temperature. For $T < T_c$, $u \approx u_0$ as long as $\omega\tau \ll 1$. As $T \to T_c$, τ increases, and u will increase rapidly toward u_∞ as $\omega\tau$ becomes comparable to and then greater than unity. Above T_c, $u = u_\infty$ at all temperatures. At the ultrasonic frequencies used by O'Brien and Litovitz (1964), the jump from u_0 to u_∞ is abrupt, since $\omega\tau$ does not become appreciable until ΔT is very small (less than 0.1°C). At the hypersonic frequencies (7–18 GHz) involved in Brillouin scattering, the dispersion is spread out over several degrees and can be analyzed. Gammon and Cummins (1966) have observed the Brillouin frequency shifts (which are directly proportional to phonon velocities) in TGS at two scattering angles; see Fig. 27. The orientations were chosen so that the direction of the acoustic phonon was the same in both cases, but the wavelengths, and thus frequencies, differed. Note that the predicted velocity behavior is observed for both transverse and longitudinal waves. Gammon and Cummins have taken the temperature at which $\omega\tau = 1$

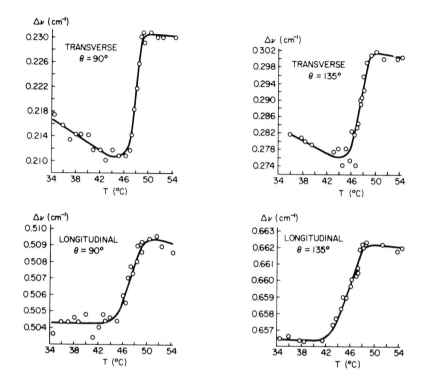

FIG. 27. Brillouin shift $\Delta\nu$ versus temperature for longitudinal and transverse components in TGS. The shift $\Delta\nu$ is proportional to the acoustic velocity. The frequency of the transverse waves is ~ 7.5 GHz and that of the longitudinal waves is ~ 18 GHz (Gammon and Cummins, 1966).

to be the temperature at which u is halfway between u_∞ and u_0, which is justified, since the dispersion is relatively small. On this basis, it was deduced that $\tau = (2.9 \pm 0.3) \times 10^{-11}/(T_c - T)$ sec. An additional observation in this Brillouin work was the marked anisotropy of the velocity dispersion. No dispersion was found for the sound velocity in the [010] direction ($\mathbf{q}$ parallel to the ferroelectric axis). For $\mathbf{q}$ perpendicular to [010], the magnitude of the dispersion depends strongly on the direction of propagation, and actually vanishes for some directions.

Let us now turn to a consideration of the attenuation. As discussed in Section V,A, there are two contributions to α in TGS: a "relaxation" term and a "fluctuation" term. The former arises from the coupling between the spontaneous polarization and the strain associated with the sound wave. This contribution is described by Eq. (12) together with the expressions for $(u_\infty{}^2 - u_0{}^2)$ and τ given by Eqs. (46) and (47). Thus, α_{relax} increases dramatically as $T \to T_c$ from below, and drops to zero above T_c. It also depends strongly on the direction of propagation (Geguzina and Krivoglaz, 1968). The second contribution is due to coupling between the sound wave and spatially inhomogeneous thermal fluctuations in the polarization. The α_{fluct} contribution is symmetric about T_c, is appreciable only close to T_c, and does not depend on the direction of the sound wave. These differences between α_{relax} and $\alpha_{\text{f uct}}$ are the explanation of the anisotropy in the attenuation shown in Fig. 28 (Minaeva *et al.*, 1969). In the case of α_y, the observed attenuation is

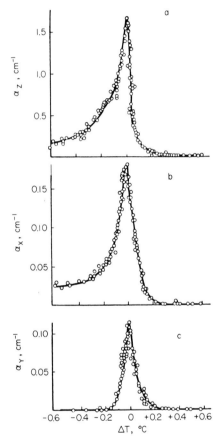

FIG. 28. Temperature dependence of the attenuation of 10-MHz longitudinal waves propagating parallel to (a) the z, (b) the x, and (c) the y axes in a TGS single crystal (Minaeva *et al.*, 1969).

completely due to the fluctuation contribution. In the case of α_z, the relaxation contribution dominates below T_c, while the tail above T_c is due to fluctuations. For both α_x and α_z, the data *below* T_c can be represented at low frequencies by

$$\alpha_{\text{relax}} = B\omega^2\tau \propto \omega^2(T_c - T)^{-1} \tag{48}$$

where $B = g^2/\rho\xi u^3$ is a different temperature-independent quantity for each direction, but τ is essentially the same relaxation time. Minaeva *et al.* (1969) report that $\tau^- = (3.5 \pm 1.0) \times 10^{-11}/(T_c - T)$ sec, in good agreement with

Gammon and Cummins (1966). The value of α_{fluct} is also predicted to vary as $\omega^2(T - T_c)^{-1}$ both above and below T_c, but this has not been explicitly tested.

O'Brien and Litovitz (1964) measured α_z (although the [001] direction is called [$\bar{1}$01] in their paper) as a function of frequency as well as temperature. They verified the ω^2 dependence in Eq. (48) and also found that $\tau = A/(T_c - T)$ for $T < T_c$, but their value of A (22.5×10^{-11}) is not in good agreement with those cited above. The reason for this discrepancy is not clear, but may involve the effect of domains or stray electric fields.

The effect of applied electric fields on the ultrasonic attenuation in TGS has been investigated by Gammon and Verdieck (1969). For measurements below T_c, the crystal was initially poled by applying a 10 V cm^{-1} field while cooling through T_c. In the single-domain ferroelectric phase, the field dependence of B in Eq. (48) is negligible for small fields and the behavior of α is directly determined by τ alone. When a field was applied in the direction of the crystal polarization, the increase in P_2 caused a decrease in τ, and thus a decrease in α. When the field was applied opposite to the crystal polarization, P_2 decreased and α increased. Above T_c, where the "relaxation" loss is due to coupling to the induced polarization, Eq. (48) can still be used, but B is both temperature and field dependent. Thus, α is observed to increase with the field until a saturation point is reached, at which the increase in B is offset by a decrease in τ. At higher fields, α decreases as the field is increased, since B approaches a constant value and τ continues to decrease. All of these field effects are in good qualitative agreement with a Landau-type theory, such as that of Geguzina and Timan (1968), which was discussed in Section V,A.

D. OTHER SYSTEMS

Rochelle salt is a system that has been frequently studied by Russian investigators (see Table III). This material is piezoelectric in its paraelectric phase, and its behavior is, in many ways, similar to that described for KDP. The principal complication is the existence of two Curie points. Since the ultrasonic properties of Rochelle salt have been described elsewhere in some detail (Sannikov, 1962; Geguzina and Timan, 1968; Geguzina and Krivoglaz, 1968), they will not be discussed here.

Another class of crystals which have not yet been mentioned are displacive ferroelectrics. The prototype, BaTiO$_3$, has been discussed by Sannikov (1962), by Dvorak (1968), and by Tani and Tsuda (1969). Unfortunately, much of the experimental data on BaTiO$_3$ has been obtained on polycrystalline samples and is of limited usefulness. Two other displacive ferroelectrics—SrTiO$_3$ and KTaO$_3$—will be discussed below. Neither of these crystals undergo a typical displacive transition, but there are interesting special effects in each case.[6]

---

[6] Note added in proof: See the recent review article of Barrett (1970).

Strontium titanate is a cubic crystal with the perovskite structure, and its dielectric constant follows a Curie–Weiss law with a Curie temperature around $0°K$. However, $SrTiO_3$ undergoes a subtle phase transition at $\sim 108°K$ which is *not* associated with any anomaly in the dielectric constant. This phase transition involves a very small tetragonal distortion, but remarkably large changes in the elastic constants. Bell and Rupprecht (1963) measured c_{11}, c_{12}, and c_{44} from $300°K$ down to $108°K$. They observed appreciable attenuation in addition to pronounced changes in velocity below $\sim 120°K$. Only for longitudinal waves propagating in a [111] direction was it possible to make accurate velocity measurements below the transition, and their results for this wave are given in Fig. 29. Note the behavior of the dielectric constant, which is also shown in this figure.

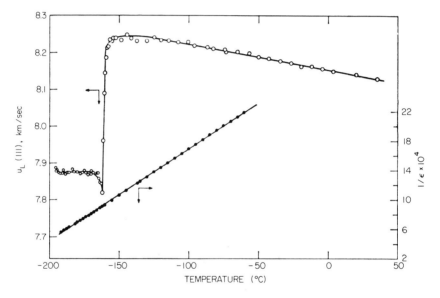

Fig. 29. Temperature dependence of the inverse dielectric constant $(1/\varepsilon)$ and longitudinal [111] ultrasonic velocity in $SrTiO_3$ (Bell and Rupprecht, 1963).

The $SrTiO_3$ transition at $108°K$ is of special interest because it does *not* fit in with the general ideas concerning the interaction of acoustic and polarization modes at $q = 0$. Cowley (1964) proposed a lattice-dynamic model in which the phase transition and the anomalous elastic properties were due to an accidental degeneracy of the LA and TO phonon branches over a wide rane of q values. However, a definitive neutron study by Shirane and Yamada (1969) has recently established that this transition is caused by a soft phonon mode at the [111] zone boundary. Thus, anomalous elastic properties in ferroelectrics are not always caused by, or even associated with, an anomalous susceptibility.

Another perovskite with interesting acoustic properties is $KTaO_3$. Although the dielectric constant above $\sim 40°K$ can be fitted fairly well by a Curie–Weiss law with a T_c value of a few degrees Kelvin, the paraelectric phase is stabilized at low temperatures by the anharmonic effect of zero-point fluctuations, and the crystal never becomes ferroelectric. Barrett (1968, 1969) has studied both longitudinal and transverse ultrasonic waves in $KTaO_3$ at frequencies in the 180–900-MHz range. The behavior of transverse waves is normal, but the longitudinal velocity and attenuation are both anomalous at low temperatures. The longitudinal velocity exhibits an unusual temperature dependence between 2 and 20°K, and the attenuation shows a large peak at $\sim 30°K$ (see Fig. 30). These observations were inter-

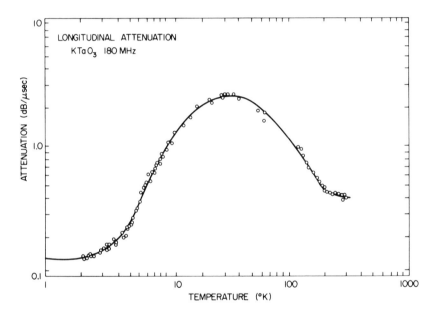

Fɪɢ. 30. Temperature dependence of α_{long} in $KTaO_3$ at 180 MHz (Barrett, 1969).

preted in terms of an Akhieser-type interaction between the sound wave and the soft optical mode. The formalism of Woodruff and Ehrenreich (1961) was used to develop an explicit theory of this interaction, in which the ultrasonic wave is treated as an early static strain which modulates the equilibrium populations of the thermal phonons. Since the modes require a finite relaxation time τ to adjust to the new equilibrium populations, there is a phase lag with respect to the strain and an acoustic loss. The frequency dependence of the attenuation data at 4.2°K, shown in Fig. 31, is consistent with a relaxation model with $\tau \approx 1.3 \times 10^{-10}$ sec. Barrett (1969) has given an extensive discussion of his data in terms of Nettleton and parabolic dispersion models for the soft optical mode.

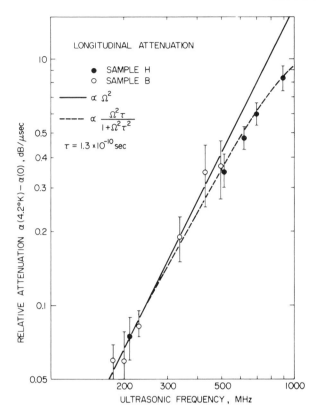

FIG. 31. The excess longitudinal attenuation $\alpha(4.2°\text{K}) - \alpha(0°\text{K})$ in $KTaO_3$ as a function of frequency (Barrett, 1969).

VI. Ferromagnetic and Antiferromagnetic Transitions

A. THEORY

Very little experimental[7] or theoretical work has been done on the behavior of the sound velocity near magnetic Curie or Néel points. Although Tanaka *et al.* (1962) developed an expression for the velocity dispersion in terms of Landau theory (see Section II,D), there do not seem to be any experimental results on dispersion in single-domain crystals. Vlasov (1966) has discussed a special kind of dispersion which occurs in ferromagnets with domain structure, and Kashcheev (1967b) has related the singularity in the velocity of "nontransversal" sound at T_c with the singularity in the spin

---

[7] Note added in proof: Very recently, Luthi *et al.* (1969) have presented a wide variety of results on ultrasonic propagation near phase transitions in both ferromagnets and antiferromagnets.

heat capacity. In neither case is the result applied to any experimental system. For the case of weak coupling between the strain and the exchange energy J, the compressible Ising model of Renard and Garland (1966a) can be used to predict the temperature dependence of the low-frequency elastic constants. This model has been applied to the antiferromagnet UO_2 by Brandt and Walker (1968). The behavior of c_{44} above T_N in UO_2 has also been discussed by Allen (1968) in terms of the molecular-field approximation and spin–lattice interactions involving nearest-neighbor exchange and local quadrupoles.

For models in which the spin–phonon interaction is due to magneto-striction, the effective elastic coupling will depend on the type of interaction (Bennett and Pytte, 1967). If the coupling is due primarily to volume magnetostriction, longitudinal velocities should be anomalous and shear velocities should not. On the other hand, for single-ion (linear) magneto-striction, there should be coupling to the transverse as well as the longitu-dinal modes. Unfortunately, no explicit theoretical predictions for ultrasonic velocities near a critical point have yet been developed on the basis of a magnetostrictive model. However, Melcher and Bolef (1969a,b) have very recently developed a magnetostrictive model to explain the magnetic-field and temperature dependence of the elastic constants in the antiferromagnetic state. In this model, low-frequency ultrasonic waves are coupled to anti-ferromagnetic resonance modes. The field dependence is due to changes in the equilibrium orientation of the sublattice magnetization which cause changes in the effective coupling constant. This model has been applied in detail to the antiferromagnetic ordered phase of $RbMnF_3$.

The critical ultrasonic anomaly of greatest interest in magnetic systems is the attenuation, which has been the subject of very active theoretical investigation. As described in Section II,D, Tanaka et al. (1962) presented a modified Landau treatment of the attenuation in both ferromagnets and antiferromagnets. Papoular (1965) has also used the basic Landau approach to develop a relaxation theory for the internal friction in the case of piezomagnetic coupling (linear in the long-range order parameter) and mag-netostrictive coupling (quadratic in the order parameter). He applied the resulting magnetostrictive equations to an analysis of the antiferromagnetic transition in MnO.

In the past few years, the Landau approach to the ultrasonic attenuation near magnetic critical points has been completely supplanted by theories based on spin-fluctuation correlation functions. A general background dis-cussion of such theoretical ideas is given in Section II,E. The basic physical idea involves the fact that a sound wave will be perturbed by the thermal fluctuations of the spins whenever there is significant spin–phonon coupling. Since the thermal spin fluctuations increase very rapidly near the critical temperature, there will also be an anomalous increase in α. The first theory of ultrasonic attenuation based on spin correlations seems to have been presented by Papoular (1964), but an independent formulation by Tani and Mori (1966) has had greater influence on subsequent developments. The mathematical formulation involves the use of an interaction Hamiltonian to calculate the rate of phonon annihilation minus the rate of phonon creation. As a result, the ultrasonic attenuation is found to be proportional to the double Fourier transform of a four-spin correlation function. In order to proceed further with the calculation, it is assumed that the four-spin correlation can be factored into products of two-spin correlations. In many cases, it is also assumed that the Fourier transform of the two-spin correlation function can be evaluated in the hydrodynamic limit (i.e., short-wavelength fluctuations are neglected).

The results of such fluctuation theories in both ferromagnets and antiferromagnets can be represented in the form

$$\alpha \propto \omega^2 (T - T_c)^{-\theta} \tag{49}$$

when $(T - T_c)/T_c$ is small. The only theoretical treatment of the ordered phase seems to be that of Tani and Tanaka (1968), who predict that $\theta = 4$ for isotropic systems and $\theta = 1$ for anisotropic ones. For the paramagnetic phase, the critical exponents θ obtained by various investigators are listed in Table IV. Kawasaki (1968c) has presented theoretical arguments to show that the dynamic behavior of the correlation functions is quite well described by molecular-field theory when there is a single easy axis of magnetization or strong anisotropic exchange. However, molecular-field theory is not valid for the case of isotropic exchange. For this reason, the theoretical results in Table IV have been listed in two classes—isotropic (Heisenberg) and uniaxial (molecular field).

TABLE IV

VARIOUS THEORETICAL VALUES FOR THE CRITICAL EXPONENT θ IN THE $|T - T_c|^{-\theta}$ DEPENDENCE OF THE LONGITUDINAL ULTRASONIC ATTENUATION IN THE PARAMAGNETIC PHASE

Ferromagnet		Antiferromagnet		
Isotropic	Uniaxial	Isotropic	Uniaxial	Reference
—	0.5	—	—	Papoular (1964)
—	1	—	1	Tani and Mori (1966, 1968)
—	49/9	—	—	Kashcheev (1967a)
—	2.5	—	1.5	Okamoto (1967)
~2.3	2.5	—	—	Bennett and Pytte (1967)
—	—	—	1.5	Pytte and Bennett (1967)
2.2	2.5	1.6	1.5	Kawasaki (1968a)
~5/3	—	~1	—	Kawasaki (1968b)
—	~4/3	1	~4/3	Laramore and Kadanoff (1969)

As can be seen from Table IV, there is considerable disparity among the various theoretical predictions. The last two entries are not only the most recent, but they should also be considered as the most reliable. Kawasaki (1968b) has used scaling-law ideas to find the asymptotic critical behavior of α in isotropic Heisenberg magnets. In this treatment, he has avoided the hydrodynamic assumption mentioned above and has included short-wavelength spin correlations. The theory of Laramore and Kadanoff (1969) is based on a mode–mode coupling scheme in which a sound mode breaks up into two dynamic spin fluctuations. The result for the longitudinal ultrasonic attenuation is

$$\alpha_L \propto \omega^2 \left[\frac{\partial(\ln J)}{\partial(\ln a)} \right]^2 \frac{C_p T}{\rho} \tau^* \tag{50}$$

where J is the magnetic coupling constant, a is the lattice parameter, and τ^* is a characteristic spin relaxation time for wave vector $q \approx \xi^{-1}$. For a uniaxial ferromagnet or antiferromagnet, τ^* was found to be proportional to the magnetic susceptibility χ: thus, $\alpha_L \propto C_p \chi$.

Two additional aspects of the behavior of α in ferromagnets are discussed in the Russian literature. Tarasov and Taborov (1966) appear to have discussed the frequency dependence of α, but their paper is in Ukrainian and has not been translated. Belov *et al.* (1959) have discussed the effect of an external magnetic field in terms of the Landau approach. This paper is very similar to those mentioned in Section V,A in connection with the effect of an electric field on ferroelectric crystals. It appears to be the only paper dealing with external magnetic field effects.

B. RARE-EARTH METALS

Gadolinium is a hexagonal ferromagnet with a Curie temperature of $\sim 290°$K. Since long-range indirect exchange is involved, Gd is not a typical Heisenberg ferromagnet. Indeed, Kadanoff (1969) classified gadolinium as a uniaxial ferromagnet in spite of the fact that the anisotropy in the spin–spin interaction is small. Luthi and Pollina (1968b) have measured the ultrasonic attenuation in gadolinium single crystals as a function of temperature, frequency, and direction of propagation for both longitudinal and transverse waves. The data for 50-MHz waves propagating along the hexagonal axis are shown in Fig. 32. The large difference between α_{long} and α_{trans} along the c axis (as well as the data in other directions) indicates that the spin–phonon coupling in Gd is due to volume magnetostriction rather than "single-ion" magnetostriction. This is confirmed by the anomalous variation in the longitudinal stiffness c_{33} and the normal behavior of the shear stiffness c_{44} (Long *et al.*, 1969). Luthi and Pollina restricted the analysis of their attenuation data to the paramagnetic phase. For all ΔT values from $+1.5$ to $+15°$C, it was shown that $\alpha \propto \omega^2$ from 36 to 180 MHz, as expected from Eq. (49). Figure 33 shows a log-log plot of α_L versus $(T - T_c)$, which indicates that the critical exponent $\theta = 1.2 \pm 0.1$. This value is considerably smaller than

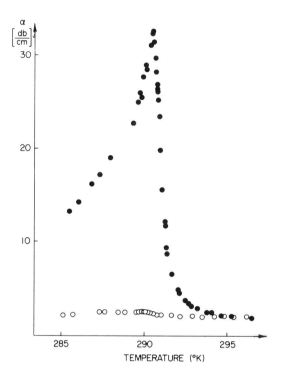

Fɪɢ. 32. Ultrasonic attenuation at 50 MHz for waves propagating parallel to the *ɔ* axis in Gd. Solid points are for the longitudinal wave; open circles are for the transverse wave. The very slight anomaly for α_{trans} is largely due to a 2° misalignment of the crystal (Luthi and Pollina, 1968b).

the Kawasaki (1968b) value of 1.66 for an isotropic ferromagnet, but is in fair agreement with the Laramore and Kadanoff (1969) value of ∼1.33 for a uniaxial ferromagnet. This would seem to support the latter classification for this sample of Gd. Recent measurements by Luthi *et al.* (1969) on another Gd single crystal of better quality indicate that $\theta = 1.6$, which suggests a more isotropic sample.

Pollina and Luthi (1969) have also investigated three antiferromagnetic rare-earth metals—terbium, dysprosium, and holium. In all three cases, $\alpha \propto \omega^2$ from 50 to 170 MHz in the paramagnetic phase. A log-log plot of α_{L} versus $(T - T_{\text{N}})$ for Ho is shown in Fig. 34. The linearity is excellent over almost three decades, and $\theta = 1.0 \pm 0.1$, which is in very good agreement with the theoretical value for an isotropic antiferromagnet such as Ho (see Table IV). The other two metals are anisotropic antiferromagnets, and the critical exponents for their longitudinal attenuation are $\theta = 1.24$ for Tb and $\theta = 1.37$ for Dy. These values are in reasonably good agreement with the Laramore and Kadanoff (1969) value of ∼1.33 for a uniaxial antiferromagnet. In all three metals, Pollina and Luthi show that the longitudinal attenuation is primarily due to volume magnetostrictive coupling.

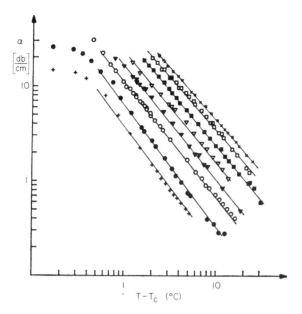

FIG. 33. Temperature dependence of the *c*-axis longitudinal attenuation in the paramagnetic phase of Gd. Frequency values are 36, 50, 70, 90, 108, 130, 150, and 180 MHz (Luthi and Pollina, 1968b).

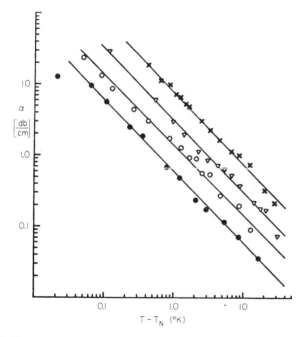

FIG. 34. Temperature dependence of the *c*-axis longitudinal attenuation in the paramagnetic phase of Ho. Frequency values are 50, 70, 110, and 170 MHz. (Pollina and Luthi, 1969).

TABLE V

ULTRASONIC INVESTIGATIONS OF FERROMAGNETIC AND ANTIFERROMAGNETIC SYSTEMS

Substance	T_c (°K)	Temperature range (°K)	Magnetic field	Frequency[a]	Quantity measured	Reference
Ferromagnets						
Gd	290	4.2–300	—	10	$u_L, u_T, \alpha_L, \alpha_T$	Rosen (1968c)
		285–305	—	36–180	α_L, α_T $\theta = 0$ (c axis), 53, 90 (a axis)	Luthi and Pollina (1968a, b)
		77–345	0–8 kOe	5	c_{33}, c_{44}	Long et al. (1969)
Ni	631	77–650	—	10	$\alpha_L(110)$	West (1958)
		4.2–760	10 kOe	10?	$c_{44}, C', c_L(110)$	Alers et al. (1960)
Fe–30% Ni alloy	~450	290–650	10 kOe	10?	$c_{44}, C', c_L(110)$	Alers et al. (1960)
Invar alloy	384	273–473	0–500 Oe	1 kHz, 0.1, 5	E, Q^{-1}	Belov et al. (1960)
RbFeF$_3$	45, 86, 101	4.2–300	—	20	u_L, u_T	Testardi et al. (1967)
Antiferromagnets						
Cr	310 (also 120)	77–500	0, 10 kG	10	c_{11}, c_{44}, C'	Bolef and de Klerk (1963)
		77–425	—	1 Hz	G, log dec δ	DeMorton (1963)
		90–450	0, 15 kOe	35 kHz	E, log dec δ	Street (1963)
		310–312	—	15–75	$\alpha_L(100)$	O'Brien and Franklin (1966)
α-Mn	96	4.2–300	—	10	$u_L, u_T, \alpha_L, \alpha_T$	Rosen (1968a)
Ho	132	4.2–300	—	10	$u_L, u_T, \alpha_L, \alpha_T$	Rosen (1968c)
		129–135	—	50–170	α_L(c axis)	Pollina and Luthi (1969)
Dy	177 (also 86)	4.2–300	—	10	$u_L, u_T, \alpha_L, \alpha_T$	Rosen (1968c)
		173–182	—	30–150	α_L, α_T (c axis)	Pollina and Luthi (1969)

Substance		Temperature range	Field	Frequency[a]	Quantities measured	References
Tb	228 (also 2222?)	4.2–300	—	10	$u_L, u_T, \alpha_L, \alpha_T$	Rosen (1968c)
		214–236	—	50–150	α_L (c axis)	Pollina and Luthi (1969)
Eu, Er, Pr, Nd, Sm		4.2–300	—	10	$u_L, u_T, \alpha_L, \alpha_T$	Rosen (1968b, c, 1969)
NiO	~507	430–555	—	Few kHz?	E, log dec δ	Street and Lewis (1951)
CoO	~289	225–315	—	Few kHz?	E	Street and Lewis (1951)
		77–300	—	Low	E, Q^{-1}	Fine (1953)
		243–303	—	30	$c_{11}, c_{44}, C', C_L(110), \alpha_L(100), \alpha_T(100), \alpha_T(110)$	Aleksandrov et al. (1968)
MnO	113	93–248	—	0.14	E, Q^{-1}	Belov et al. (1960)
		77–275	—	~85 kHz	G, Q^{-1}	Yevtushchenko and Levitin (1961)
MnTe	306.7	290–325	0, 11 kOe	50–190	u_L, α_L	Walther (1967)
UO$_2$	30.8	4.2–300	0–35 kG	20–90	c_{11}, c_{44}, C'	Brandt and Walker (1967, 1968)
		15–50	0–35 kG	50	$\alpha_L, \alpha_T(100), \alpha_T(110)$	Brandt and Walker (1967, 1968)
MnF$_2$	67.3	58–84	0, 3.6 kG	8–65	$u, \alpha_L(110)$	Neighbours et al. (1963)
		62–72	—	10–70	$\alpha_L(001), (110)$	Neighbours and Moss (1968)
Cr$_2$O$_3$	310	273–373	—	~85 kHz	G, Q^{-1}	Yevtushchenko and Levitin (1961)
RbMnF$_3$	83	4.2, 50, 80–84	0–8 kOe	10	$u_L, \alpha_L(100)$	Melcher et al. (1967)
		4.2–300	—	10–70	$c_{11}, c_L(110)$	Golding (1968)
		~82–~90	—	30–150	$\alpha_L(001)$	Golding (1968)
		4.2–300	0, 7.5 kOe	30	$c_{11}, c_{44}, C', c_L(110)$	Melcher and Bolef (1969a)

[a] In megahertz, except where otherwise noted.

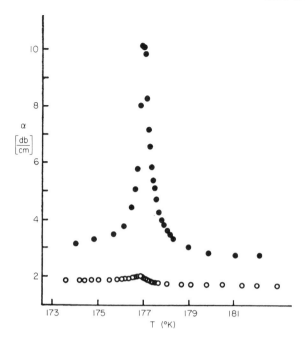

FIG. 35. Ultrasonic attenuation at 50 MHz for waves propagating along the c axis in Dy. Solid points are for the longitudinal wave, and open circles are for the transverse wave (Pollina and Luthi, 1969).

In the case of dysprosium, there is a weak, but definite anomaly in the shear-wave attenuation along the c axis (see Fig. 35). These data ca￼ ￼lso be fit (over about a single decade) by Eq. (49) with a critical exp￼ ￼ent $\theta = 0.8 \pm 0.15$. Pollina and Luthi state that α_{trans} arises from linear (si￼ ￼gle-ion-type) magnetostrictive coupling, and suggest that this θ value may be explained by including higher-order terms in the Hamiltonian.

Rosen (1968b,c, 1969) has investigated a variety of rare-earth metals from 4.2 to 300°K (see Table V). In these investigations, longitudinal and transverse velocities and attenuations were measured in polycrystalline samples at a single frequency of 10 MHz. Although such investigations provide an interesting survey of the elastic behavior, they are not valuable as detailed single-crystal work near the phase transition.

C. RUBIDIUM MANGANESE FLUORIDE

$RbMnF_3$ has an ideal perovskite structure (O_h point group). The thermal expansion anomaly near T_N is remarkably small, and cubic symmetry appears to be maintained at all temperatures. This implies that the spin system is very weakly coupled to the lattice. Furthermore, no lattice instability is observed near T_N and the antiferromagnetic phase transition is of the lambda type.

In addition to these structural properties, the magnetic anisotropy is very low ($H_A \approx 4$ Oe). Thus, RbMnF$_3$ should be well represented by an isotropic antiferromagnetic Heisenberg model.

Golding (1968) has made ultrasonic measurements on single-crystal RbMnF$_3$ at frequencies from 30 to 150 MHz. The behavior of [100] longitudinal waves near T_N is shown in Fig. 36. It was observed that α is quadratic in frequency for temperatures close to T_N, and that the temperature of the attenuation maximum (and the velocity minimum) was independent of frequency. The attenuation data in the paramagnetic phase were well represented by Eq. (49) over the region $3 \times 10^{-4} \leq (T - T_N)/T_N \leq 2 \times 10^{-2}$. The resulting exponent was $\theta = 0.32 \pm 0.02$. In contrast to the rare-earth metals discussed in Section VI,B, this exponent is in definite disagreement with all the theoretical predictions in Table IV. The reason for this disagreement is at present unclear. One possible source of theoretical trouble lies in approximating the four-spin correlation function by two-spin correlation functions. This factorization may overemphasize the critical spin fluctuations near T_N, which would enhance the strength of the divergence in α. A similar low value of θ is also observed in MnF$_2$, which is briefly mentioned in Section VI,D.

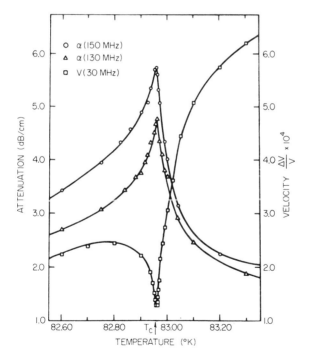

Fɪɢ. 36. Velocity and attenuation variations for [100] longitudinal waves in RbMnF$_3$. The Néel temperature of 82.96°K is indicated by the arrow (Golding, 1968).

In addition to the attenuation results described above, there has been an extensive study of the elastic constants of $RbMnF_3$ over the temperature range from 4.2 to 300°K (Melcher and Bolef, 1969a). This investigation included a study of the dependence of the velocity on the applied magnetic field and was principally aimed at elucidating the magnetoelastic coupling in the antiferromagnetic phase. In the critical region just below T_N, the longitudinal velocity is strongly field dependent (see Fig. 37), but it is

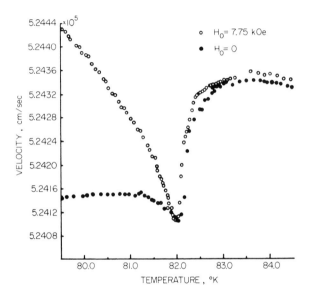

FIG. 37. Velocity of [100] longitudinal waves in $RbMnF_3$. The ultrasonic frequency is 10 MHz, and the applied field of 7.75 kOe is aligned parallel to [100] (Melcher *et al.*, 1967).

necessary to differentiate carefully between true critical effects and magneto-elastic effects due to antiferromagnetic resonance. As shown by Fig. 37, c_{11} has a sharp minimum at T_N, and the same is true of $c_L = (c_{11} + c_{12} + 2c_{44})/2$. However, neither c_{44} nor $C' = (c_{11} - c_{12})/2$ shows any effect of the long-range magnetic ordering which occurs at T_N. These results are consistent with a critical spin–phonon interaction due to volume magnetostriction, and are not compatible with a model based on single-ion magnetostriction.

D. OTHER SYSTEMS

Among the investigations listed in Table V, the one most closely related to those described above is the study of tetragonal MnF_2 by Neighbours and Moss (1968). They observed a lambda-type peak in the attenuation of longitudinal waves at the antiferromagnetic Néel point, but no anomaly in the shear-wave attenuation. The α_L values for propagation in the [001] direction were consistently greater than those for the [110] direction, and the peak attenuation in either direction was "roughly proportional to the square of the frequency" over the 10–70-MHz range. Attenuation data in the [001] direction conformed to Eq. (49) over about two decades ($0.01 < \Delta T < 2°K$) both above and *below* T_N. The resulting exponents were $\theta^- \approx 0.18$ for $T < T_N$ and $\theta^+ = 0.40 \pm 0.02$ for $T > T_N$. This θ^- value does not agree with the theoretical prediction of Tani and Tanaka (1968) for the antiferromagnetic state, and the θ^+ value does not agree with any of the theoretical values in Table IV. However, the θ^+ value is quite similar to that in $RbMnF_3$, although MnF_2 is a uniaxial antiferromagnet and $RbMnF_3$ is isotropic. Although volume magnetostriction appears to be the dominant interaction in MnF_2 as in $RbMnF_3$, the velocity behavior in these two crystals is quite different. Within an experimental precision of 2 parts in 10^4, there is no anomaly in the velocity near T_N in MnF_2. In spite of the apparent disagreement between θ^+ and the predictions of various fluctuation theories, Tani and Mori (1968) state that their theory is in good agreement with the preliminary MnF_2 data reported by Neighbours *et al.* (1963). The absence of any velocity anomaly and the presence of anisotropy in the attenuation were shown to be consequences of their model, and the temperature dependence of α was represented by

$$\alpha T \propto [1 + 0.48(T/T_c)^2 \varepsilon^{-1/2} \exp(-2.8\varepsilon^{1/2})]^2 \qquad (51)$$

where $\varepsilon = (T - T_c)/T_c$. Although it can be seen that Eq. (51) will have the asymptotic form of Eq. (49) with the exponent $\theta = 1$ when ε is sufficiently small, the α variation for $\varepsilon \gtrsim 10^{-3}$ will be more gradual. This points out the difficulty of determining the extent of the "critical region."

There are two other investigations which have shown interesting anomalies in the temperature dependence of the single-crystal elastic constants. Bolef and de Klerk (1963) have studied chromium, with emphasis on the detailed behavior near the Néel point ($T_N = 310°K$) and a low-temperature transition ($\sim 120°K$) between two states of different antiferromagnetic ordering. Near T_N, there is a sharp dip in c_{11}, a very small change in C', and essentially no anomaly in c_{44}. Near the 120°K transition, all three elastic constants behave anomalously, but the most pronounced change occurs in C', which increases by $\sim 5\%$ on cooling through the transition. The other investigation was concerned with high-field elastic constants of nickel and a Fe–Ni alloy (Alers *et al.*, 1960). By applying a field of 10 kOe, the "ΔE effect" associated with domain-wall motion was eliminated and the intrinsic effect of magnetic ordering was clearly observed on going through the Curie temperature.

As a postscript to this section, it might be mentioned that Dietz and Jaumann (1962) have investigated a large series of polycrystalline ferromagnetic metals and alloys (Ni, Fe, Fe–Ni alloys, Mn–Zn–Ferrite, etc.). They measured u and α at 1, 2.5, and 5 MHz from 110 to 460°K and from 0 to 16 kG. These results are difficult to summarize and are *not* listed in Table V.

VII. Order–Disorder Lambda Transitions

A. HELIUM

The lambda-transition line in liquid He[4] extends from 2.17°K at the saturation vapor pressure of 0.05 atm to 1.76°K at 29.8 atm (at which point, the λ line intersects the liquid–solid phase line). Most investigations have been made on liquid helium in equilibrium with its vapor, and it is traditional to call this low-pressure end of the transition line the λ point. As indicated in Table VI, there have been quite a few measurements of the variation in ultrasonic velocity near the λ point. In Fig. 38, a pulse measurement at

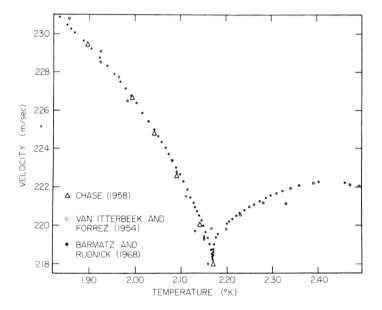

FIG. 38. Variation of the acoustic velocity over an extended temperature range near the λ point in liquid helium (Barmatz and Rudnick, 1968).

1 MHz (Chase, 1958) and an interferometer measurement in the 0.2–0.8-MHz range (van Itterbeek and Forrez, 1954) are compared with a recent resonance measurement at 22 kHz (Barmatz and Rudnick, 1968). Atkins and Stasior (1953), by measuring u as a function of temperature at various constant pressures, were able to follow the character of the anomaly all the way along the λ line. Their data indicate that the dip in u is somewhat more pronounced at the high-pressure end than that shown in Fig. 38.

The velocity data of Barmatz and Rudnick (1968) have been obtained with exceptionally high precision and temperature resolution. Because of this and because they were obtained at a low frequency, they are ideal for an analysis of the temperature behavior of u. As discussed in Section II,A, one can derive, from the Pippard equations, an exact thermodynamic expression for $(u_\lambda^{-2} - u^{-2})$, where the velocities are the static (zero-frequency) limiting values. Close to T_λ, where $(u - u_\lambda)/u_\lambda \ll 1$, Eq. (4) can be rewritten in the form $(u - u_\lambda) = A/C_p$, with the behavior of A determined primarily by $(dS/dT)_t^2$. Figure 39, a plot of velocity versus C_p^{-1}, shows that $(dS/dT)_t^2$ is

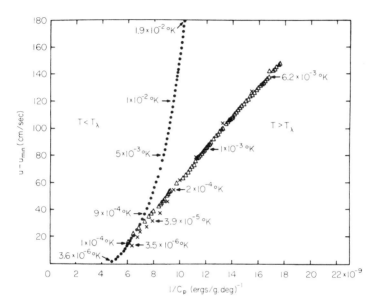

FIG. 39. Plot of $(u - u_{\min})$ versus C_p^{-1}; $u_{\min}$ is the minimum velocity value measured in He at 22 kHz (Barmatz and Rudnick, 1968).

essentially constant over a reasonable range above T_λ, but not below. A least-squares fit to the data above T_λ gave

$$u - u_{\min} = (12.0 \times 10^9/C_p) - 58.6 \quad \text{cm sec}^{-1} \tag{52}$$

which indicates that the minimum experimental velocity at 22 kHz is 58.6 cm sec^{-1} greater than the expected static u_λ value. It is possible that this discrepancy is due to velocity dispersion, since there is a difference between $u(44 \text{ kHz})$ and $u(17 \text{ kHz})$ within ± 20 μdeg of T_λ (see Fig. 40). Gravity effects may also be important.

Ultrasonic attenuation near the λ point has been measured over an unusually wide range of frequency. Let us first consider the data in the "low"-frequency range from 22 kHz to 12 MHz. Barmatz and Rudnick (1968) have analyzed their α values at 22 kHz (see Fig. 41) to show that $\alpha \propto (T_\lambda - T)^{-1}$ below T_λ and $\alpha \propto (T - T_\lambda)^{-1/2}$ above T_λ. This temperature dependence is confirmed in the 1–12-MHz range by Chase (1958). In both

TABLE VI

ULTRASONIC INVESTIGATIONS OF LAMBDA-TYPE TRANSITIONS

Substance	T_λ (°K)	Temperature range[a]	Pressure range	Frequency[b]	Quantity measured	Reference
He	2.17 at satn v.p.	1.57–4.5	Satn	15	u, α	Pellam and Squire (1947)
		0.85–4.2	Satn	2–12	u, α	Chase (1953)
		1.2–4.2	1–69 atm	12	u	Atkins and Stasior (1953)
		1.1–4.2	Satn	0.2–0.8	u	van Itterbeck and Forrez (1954)
		1.3–2.2	Satn	1	u, α	Chase (1958)
		$\Delta T < 15 \times 10^{-3}$	Satn	9.75 kHz	u	Rudnick and Shapiro (1965)
		1.8–2.5	Satn	17–44 kHz	u, α	Barmatz and Rudnick (1968)
		1.2–4.2	Satn	653	α	Heinicke et al. (1969)
		1.1–4.2	Satn	1 GHz	α	Imai and Rudnick (1969)
CH_4	20.49 (also ~8)	14.4–21.2, 63.5–77.4	Satn	?	u_L, u_T	Bezuglyi et al. (1966)
NH_4Cl	~241.9 at 1 atm	150–300	1 atm	5–55	c_{11}, c_{44}, C'	Garland and Jones (1963)
		200–270	1 atm	5–55	$\alpha_L(100)$	Garland and Jones (1965)
		150–320	1 atm	20, 60	c_{11}, c_{44}, C'	Garland and Renard (1966b)
		250–310	0–12 kbar	20	c_{11}, c_{44}, C'	Garland and Renard (1966b)
		215–300	1 atm	10–60	$\alpha_L(100)$	Garland and Yarnell (1966b)
		238–245	1 atm	5, 15	$\alpha_L(100)$	Shustin et al. (1967)
		223–323	1 atm	~9 to 19 GHz	$c_{11}, c_{44}, c_L(110)$	Lazay (1969)
		240–270	0–3.5 kbar	10–30	$\alpha_L(100)$	Garland and Snyder (1969)
NH_4Br	234.5 at 1 atm	235–333	1 atm	~13	c_{11}, c_{44}, C'	Haussühl (1960)

Material	T_c	Temp. range	Pressure	Frequency	Constants	Reference
		235–320	1 atm	20	c_{11}, C'	Garland and Yarnell (1966a)
		100–320	1 atm	20	c_{44}	Garland and Yarnell (1966a)
		255–315	0–12 kbar	20	c_{11}, c_{44}, C'	Garland and Yarnell (1966a)
		180–240	0–6 kbar	20	c_{11}, c_{44}, C'	Garland and Young (1968b)
Quartz	~574°C	0–580°C	1 atm	~0.5	c_{11}, c_{14}, c_{44}, c_{66}	Atanasoff and Hart (1941)
		0–650°C	1 atm	2–11	c_{44}	Atanasoff and Kammer (1941)
		580–800°C	1 atm	~0.5	c_{11}, c_{33}, c_{13}, c_{44}, c_{66}	Kammer et al. (1948)
		−195 to 700°C	1 atm	~75 kHz	Six c_{ij}, Q^{-1} (45° cut)	Mayer (1960)
		20–650°C	1 atm	?	Six c_{ij}	Zubov and Firsova (1962)
		500–600°C	1 atm	20–60	c_{11}, c_{33}, $\alpha_L(1\bar{1}0)$	Snyder (1968)
		30–600°C	1 atm	~20 GHz	c_{11}	Shapiro and Cummins (1968)
β-Brass (50.0% Zn)	~735	4–800	1 atm	10	c_{11}, c_{44}, C'	McManus (1963)
α-Uranium	~42	30–77	1 atm	35–60	Nine c_{ij}	Fisher and McSkimin (1961)
$NaNO_3$	549.9	293–573	1 atm	~150 Hz	s_{11}, s_{33}, s_{12}, s_{44}	Kornfeld and Chubinov (1958)
		323–573	1 atm	5–60	c_{33}, c_{44}, $\alpha_L(001)$	Craft and Slutsky (1968), Craft et al. (1969)
$KMnF_3$	198	200–290	1 atm	5, 30	c_{11}, c_{44}, c_{12}	Aleksandrov et al. (1966, 1967)
Poly-L-lysine (aqueous soln)	$f(pH)$	293–309	1 atm	3–100	u, α	Parker et al. (1968)

[a] In degrees Kelvin, except where otherwise noted.
[b] In megahertz, except where otherwise noted.

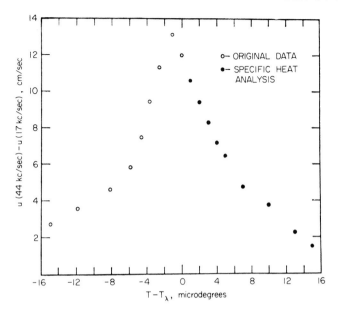

FIG. 40. Velocity dispersion in helium near T_λ (Barmatz and Rudnick, 1968).

cases, the data were discussed in terms of the relaxation model of Landau and Khalatnikov (1954), which was described in Section II,D. However, this model does not provide a good representation for the data very close to T_λ (where $\omega\tau > 1$). Ferrell *et al.* (1968) have developed a dynamic scaling expression for the low-frequency attenuation which does not involve any adjustable parameters:

$$\alpha/\omega^2 = 1.8\,|\Delta T|^{-1}\,\ln^{-1/2}(2/|\Delta T|) \times 10^{-18}\quad \text{cm}^{-1}\text{ sec}^2 \tag{53}$$

This result is a consequence of the considerable mixing between first and second sound modes near T_λ; for a general discussion of dynamic scaling ideas, see Section II,E. A comparison of Eq. (53) with the experimental α/ω^2 values is shown in Fig. 42. The line labeled "scaling" does not extend beyond $\Delta T \approx 10^{-2}\,°\text{K}$, since Eq. (53) is not valid for temperatures farther away from the λ point. The lines labeled "Landau–Khalatnikov" represent the best fit of that model to the data of Chase (1958). Note that the logarithmic factor in Eq. (53) somewhat reduces the $|\Delta T|^{-1}$ dependence, which corresponds to the Landau result at small $\omega\tau$ values. The dynamic scaling result is confirmed at low frequencies by the mode–mode coupling calculation of Swift and Kadanoff (1968). They show that there are two characteristic frequencies near the lambda point: $\omega_2{}^* \propto |\Delta T|$ for second sound, and the larger quantity $\omega_1{}^* \propto |\Delta T|^{2/3}$ for first sound. The scaling result is obtained for $\omega < \omega_2{}^*$, while it is predicted that α will vary like $|\Delta T|^{-2/3}$ when $\omega_2{}^* \ll \omega \lesssim \omega_1{}^*$. This breakdown of Eq. (53) at high frequencies and small $|\Delta T|$ values can be seen from the 1-MHz data in Fig. 42 and from the hypersonic data described below. Despite their success below T_λ, neither dynamic scaling nor mode–mode coupling models have yet been able to explain the $(T - T_\lambda)^{-1/2}$ dependence of α above T_λ. Another important feature of the data above the λ point is the considerable deviation from a quadratic (hydrodynamic) frequency dependence (Barmatz and Rudnick, 1968).

Let us now consider the behavior of the attenuation at high frequencies. Heinicke *et al.* (1969) have used a novel Brillouin technique to obtain data at 0.65 GHz, while Imai and Rudnick (1969) have used a pulsed acoustic interferometer to obtain even more detailed data at 1 GHz. Neither set of data can be represented by the Landau–Khalatnikov relaxation theory, which predicts that the maximum value of α should occur when $\omega\tau = 1$. On the basis of $\tau = 1.5 \times 10^{-11}(T_\lambda - T)^{-1}$, used by Barmatz and Rudnick (1968) to obtain the best L–K fit to their He(II) data, the attenuation peak at 1 GHz should occur $\sim 0.09°$K below T_λ. Figure 43 clearly shows that the attenuation peak is actually within a few millidegrees of T_λ. The smooth curves in Fig. 43 represent the best fits to the power law $\alpha \propto |T - T_\lambda|^{-\theta}$, with $\theta^+ = 0.41 \pm 0.05$ for $T > T_\lambda$ and $\theta^- = 0.52 \pm 0.05$ for $T < T_\lambda$. Thus, at high frequencies, there is greater symmetry about T_λ, which is in agreement with the general expectations of the mode–mode coupling theory. However, the experimental results in helium still provide a considerable challenge to our theoretical understanding.

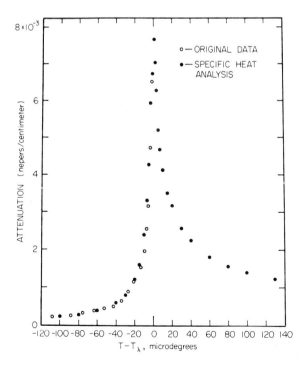

FIG. 41. Ultrasonic attenuation at 22 kHz in helium. The data have been corrected for a background attenuation of $\sim 1.5 \times 10^{-3}$ cm^{-1} (Barmatz and Rudnick, 1968).

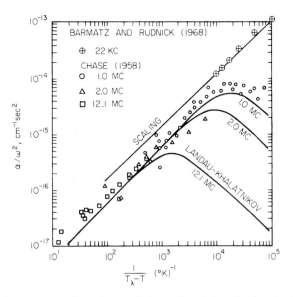

FIG. 42. Temperature dependence of α/ω^2 values below the λ point in He (Ferrell *et al.*, 1968).

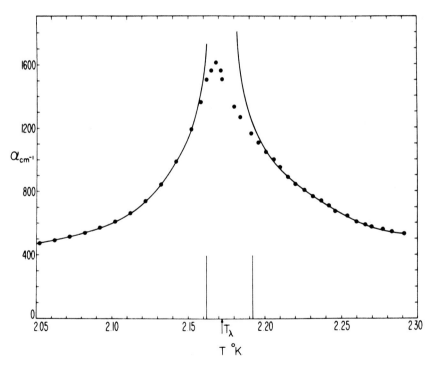

FIG. 43. Excess attenuation at 1 GHz in He near its λ point. The classical attenuation due to shear viscosity and thermal conductivity has been subtracted (Imai and Rudnick, 1969).

128

B. Ammonium Chloride

Ammonium chloride undergoes a lambda transition of the order–disorder type which involves the relative orientations of the tetrahedral ammonium ions in a CsCl-type structure. The most stable orientation of the NH_4^+ ion in the cubic unit cell is for the hydrogen atoms to point toward the nearest-neighbor Cl^- ions. Thus, there are two possible positions for the ammonium ion. In the completely ordered state, all NH_4^+ tetrahedra have the same relative orientation with respect to the crystallographic axes; in the completely disordered state, the orientations are random with respect to these two positions. It is clear that the orientational ordering is completely analogous to the spin ordering of a simple-cubic ferromagnet in zero external field. The difference in interaction energy between parallel and antiparallel NH_4^+ ions is very largely due to octopole–octopole terms between nearest NH_4^+ neighbors, and thus the Ising model is quite a good approximation. Furthermore, the ordering process should have little effect on the dynamics of such an ionic lattice.

Garland and Renard (1966b) have measured both longitudinal and transverse ultrasonic velocities in NH_4Cl over a wide range of temperature (150–320°K) and pressure (0–12 kbar). Special emphasis was given to "anomalous" behavior near the lambda line. As shown in Fig. 44, the

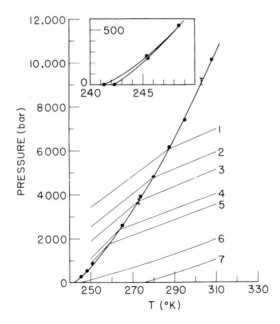

FIG. 44. Phase diagram for NH_4Cl. The high-pressure, low-temperature field corresponds to the ordered phase. The data points were obtained from the abrupt "break" in the ultrasonic shear velocities at the λ-transition point (see Fig. 46). The light lines numbered 1–7 represent isochores at various volumes; see legend of Fig. 46 for values of V_i (Garland and Renard, 1966b).

transition temperature is a fairly strong function of pressure, and some hysteresis occurs at low pressures. The adiabatic elastic constants c_{11} and c_{44} are shown in Figs. 45 and 46 as functions of temperature at 1 atm. Note the discontinuous jump in the value of c_{44}. On cooling, this change occurred at $241.4 \pm 0.1°K$; on warming, the break occurred at $242.3 \pm 0.1°K$. The same effect was observed for $C' = (c_{11} - c_{12})/2$, but strong attenuation of longitudinal waves precluded any such observation for c_{11}. Both the jump and the hysteresis can be explained on the basis of a compressible Ising model. Garland and Renard (1966a) showed that such a model is unstable in the

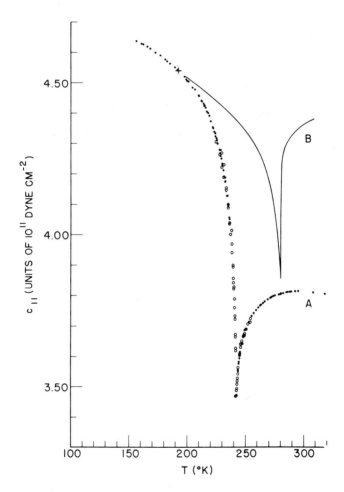

Fig. 45. Variation of c_{11} in NH_4Cl with temperature. Curve A: data at 1 atm. Curve B: calculated curve at constant volume $V_2 = 34.15$ cm^3 mole^{-1}; V_2 corresponds to V_λ at $280°K$ (Garland and Renard, 1966b).

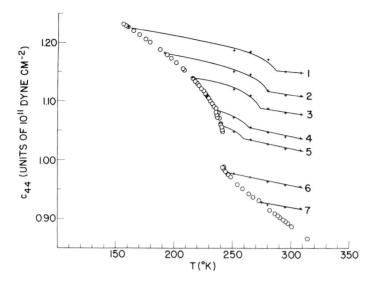

FIG. 46. Variation of c_{44} in NH_4Cl with temperature on cooling. Open circles are experimental data at 1 atm. Curves 1–7 are calculated for various constant volumes: $V_1 = 34.002$; $V_2 = 34.150$; $V_3 = 34.266$; $V_4 = 34.428$; $V_5 = 34.507$; $V_6 = 34.768$; $V_7 = 34.928$ cm^3 mole^{-1} (Garland and Renard, 1966b).

immediate vicinity of the critical point and will undergo a first-order transition. Thus, the transition in NH_4Cl is primarily lambda-like in character, but there is a small first-order change at low pressures.

The pressure dependences of the effective elastic constants c_{11} and c_{44} are shown in Figs. 47 and 48, and one can see the progressive change in the anomalous behavior as the transition occurs at higher and higher pressures. These high-pressure data can also be used to evaluate the effective elastic constants at constant volume. In Fig. 46, the temperature dependence of c_{44} is shown along the seven isochores indicated in Fig. 44. To avoid confusion in Fig. 45, only the c_{11} variation along the V_2 isochore is shown. The behavior of these constant-volume elastic constants can be understood in terms of the compressible Ising model discussed in Section II,B.

One can predict directly from Eqs. (6) and (7) the qualitative behavior of the shear constants c_{44} and C'. Both "disordered-lattice" contributions should show a slow, smooth (almost linear) increase as the temperature is decreased; this is based on the behavior of any normal ionic crystal. The term $-lU_1(0,H)/NJ$ in Eq. (7) increases from zero in the completely disordered state to a constant positive value at temperatures quite a bit below T_λ. This increase is especially rapid as the temperature is decreased through the lambda point (which depends on the volume, since J is a function of V). The constant-volume c_{44} curves in Fig. 46 show excellent agreement with this prediction. The elastic constant C' should have very much the same behavior

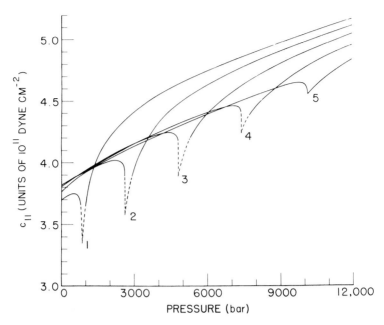

FIG. 47. Dependence of c_{11} in NH_4Cl on pressure at various temperatures: $T_1 =$ 250.72°K; $T_2 = 265.00$°K; $T_3 = 280.05$°K; $T_4 = 295.02$°K; $T_5 = 308.04$°K. Dashed portions of the curves indicate regions where data are less accurate are or missing due to high attenuation (Garland and Renard, 1966b).

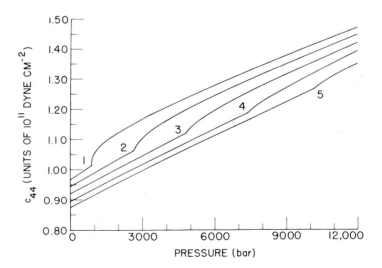

FIG. 48. Dependence of c_{44} in NH_4Cl on pressure at various temperatures; see legend of Fig. 47 for T values (Garland and Renard, 1966b).

as c_{44}, although they are not identical, because Eq. (6) contains the term $-mG(0, H)$. In order to discuss c_{11}, let us consider the appropriate linear combination of Eqs. (5) and (6). Since the configurational heat capacity has a sharp maximum at T_λ, the term $-(vT/J^2)(C_1/N)(dJ/dv)^2$ will dominate the temperature dependence of c_{11}^T. Hence, c_{11}^T should display a very pronounced minimum at the lambda point. From ultrasonic data, we obtain c_{11}^S rather than c_{11}^T. Although these isothermal and adiabatic stiffnesses differ considerably very close to T_λ, the difference between c_{11}^T and c_{11}^S is less than 10% when $|T - T_\lambda| \sim 1°\text{K}$, and this difference decreases as $|T - T_\lambda|$ increases. Thus, c_{11}^S should follow the predicted behavior of c_{11}^T quite closely. From Fig. 45, we see that the shape of c_{11}^S at constant volume is strikingly similar to the shape expected from the heat-capacity curve.

The compressible Ising model has also been used by Renard and Garland (1966b) to analyze the behavior of the "constants" in the Pippard equations and the substantial variation in the slope of the transition line. Garland and Young (1968a) later used this model to show that the anomalous changes in the volume of NH_4Cl are essentially identical to those in the shear constant c_{44}. Thus, shear-velocity measurements may provide an experimentally attractive method (especially at high pressures) of obtaining information about volume changes due to cooperative ordering phenomena.

Hypersonic velocities for both longitudinal and transverse modes in NH_4Cl have been determined from Brillouin shifts by Lazay (1969). For the shear wave corresponding to c_{44}, there was no velocity dispersion between this measurement at ~ 9 GHz and the Garland and Renard (1966b) data at 20 MHz. This agrees with the absence of anomalous ultrasonic attenuation for shear waves. For longitudinal waves in the [100] and [110] directions, corresponding to c_{11} and to $c_L = (c_{11} + c_{12} + 2c_{44})/2$, respectively, the Brillouin velocity values at ~ 18 GHz differ markedly from the 20-MHz ultrasonic values. Indeed, the variations in u_L at hypersonic frequencies are very similar to those for the shear velocity—both c_{11} and c_L show a normal linear variation above T_λ and a rapid change just below T_λ. A dispersion $(u_{\text{hyper}} - u_{\text{ultra}})/u_{\text{hyper}}$ equal to ~ 0.1 was observed at T_λ; this value drops to ~ 0.01 at $220°\text{K}$ and at $310°\text{K}$. The presence of such dispersion is consistent with the large longitudinal attenuation near T_λ in NH_4Cl.

The 1-atm attenuation data of Garland and Yarnell (1966b) for longitudinal waves propagating along the [100] direction are shown in Fig. 49. The frequency dependence of α is clearly quadratic for the 10–60-MHz range, and the temperature dependence is approximately given by $\alpha \propto |T - T_\lambda|^{-1.15}$ both above and below T_λ. The solid lines shown in Fig. 49 represent an attempt to fit the data by the Landau expression $\alpha \propto |\Delta T|^{-1}$ over a range of about $\pm 10°\text{K}$. It is not clear which the of the exponents, $\theta = 1$ or $\theta = 1.15$, is a better choice for NH_4Cl, since the extent of the "critical" region is unknown. In general, one would prefer to fit the data closest to T_λ. Unfortunately, the behavior of α_L very close to T_λ (i.e., when $|\Delta T| < 0.7°\text{K}$) is complicated by metastability and the occurrence of a first-order transition. This region has been studied with special care (Garland

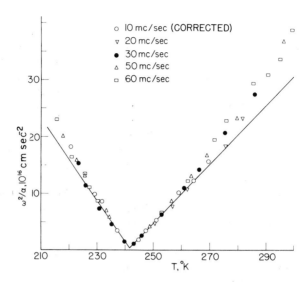

FIG. 49. Plot of ω^2/α versus temperature for NH_4Cl. Frequency values in MHz: (O) 10, (▽) 20, (●) 30 , (△) 50, (□) 60 (Garland and Yarnell, 1966b).

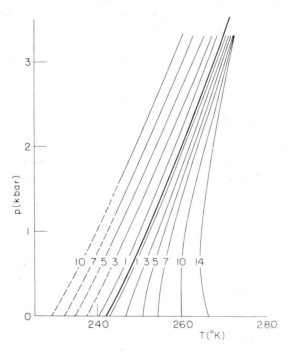

FIG. 50. Lines of constant (ω^2/α) for NH_4Cl. The units of the ω^2/α values are 10^{16} cm sec^{-2}, and the heavy line represents the λ line as shown in Fig. 44 (Garland and Snyder, 1969).

and Yarnell, 1966b), and the attenuation data can be interpreted in terms of a compressible Ising model.

The attenuation coefficient α of longitudinal ultrasonic waves propagating in the [100] direction has also been measured as a function of pressure in the vicinity of the lambda line (Garland and Snyder, 1969). For seven different temperatures between 241 and 270°K, measurements were carried out at 10, 20, and 30 MHz as the pressure was varied from 1 to 3500 bar. In all cases, α was quadratic in the frequency. Figure 50 shows that lines of constant attenuation lie parallel to the lambda line in the ordered phase. Note, however, that such constant-attenuation lines in the disordered phase converge toward the lambda line as the pressure is increased. In both phases, isobaric values of α^{-1} vary approximately like $|T - T_\lambda(p)|$ near the lambda line, where $T_\lambda(p)$ is the transition temperature at pressure p.

C. AMMONIUM BROMIDE

Ammonium bromide and ammonium chloride crystals are structurally very closely related. At room temperature, both crystals have disordered CsCl-type structures with the NH_4^+ ions distributed at random with respect to two equivalent orientations. At very low temperatures, both have "parallel" ordered CsCl-type structures. Both also undergo lambda transitions of the order–disorder type. However, there are major differences between the types of ordering observed in the chloride and in the bromide. In the case of NH_4Cl, there is a single lambda line marking the transition between the disordered cubic phase and the parallel ordered cubic phase (see Section VII,B). In NH_4Br, there are two additional ordered phases—an ordered tetragonal phase which is stable at low pressures, and a high-pressure ordered phase denoted as O_{II}. Indeed, the latter phase was discovered in a recent ultrasonic investigation by Garland and Young (1968b), who established the phase diagram shown in Fig. 51.

Garland and Young (1968b) measured ultrasonic velocities in single-crystal NH_4Br at pressures from 0 to 6 kbar and at temperatures between 180 and 240°K. Extensive measurements were made in the new high-pressure ordered phase O_{II} and in the disordered cubic phase D, as well as along the lambda line between these phases. Domain formation takes place in the ordered tetragonal phase O_T, and the resulting attenuation makes velocity measurements difficult. Various attempts were made to align the tetragonal axes and obtain a single-domain crystal, but these were unsuccessful. Measurements were made, however, in this phase for the "average" c_{44} shear constant which results from a random orientation of the tetragonal axes of the domains along the directions of the three equivalent axes in the disordered cubic phase. A few measurements were also made in the low-temperature ordered phase O_I. This was very difficult, since single crystals crack upon undergoing the first-order phase change from O_{II} to O_I. Although there are many similarities in the acoustic behavior along the D–O_{II} lambda line and along the lambda line in NH_4Cl, it can be proved acoustically that the O_{II} phase is *not* analogous to the parallel-ordered cubic phase of NH_4Cl.

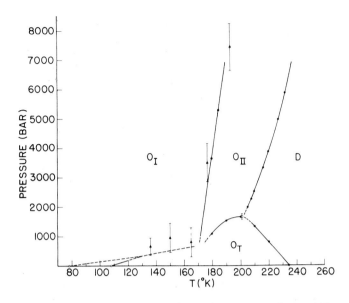

FIG. 51. Phase diagram for NH_4Br. Solid circles represent ultrasonic data, and the other symbols represent earlier volume and specific-heat measurements. The phases shown are the disordered cubic (D), ordered tetragonal (O_T), the new ordered phase (O_{II}), and the low-temperature cubic ordered phase (O_I) (Garland and Young, 1968b).

The effective elastic constants c_{11} and c_{44} are shown as functions of pressure at various constant temperatures in Figs. 52 and 53. Figure 53 illustrates very clearly the various anomalous elastic changes which are associated with changes in ordering. The behavior of c_{44} at temperature $T_5 = 240°K$ is essentially the normal behavior expected of any crystal. The pressure dependence at $240°K$ is, in fact, in excellent agreement with those measured by Garland and Yarnell (1966a) at 255, 275, and $295°K$. At 210 and $220°K$, one can see the effects of the two separate lambda transitions, while, at lower temperatures, there is a first-order transition between the ordered tetragonal and the O_{II} phases. Note that c_{44} is anomalously *small* (compared to the value in the disordered phase) in the tetragonal phase and anomalously large in the O_{II} phase. This difference is basically due to the behavior of the volume. On cooling NH_4Br, there is an anomalous lattice *expansion* as the crystal undergoes the transition from the disordered cubic to ordered tetragonal phase. On the other hand, NH_4Br contracts when the disordered phase transforms into the O_{II} phase. This anomalous contraction is five times larger than the similar change which occurs in NH_4Cl, and it dominates the variation of c_{44} near the $D–O_{II}$ lambda line. Indeed, an analysis of all the available elastic data shows that the *constant-volume* values of c_{44} are actually smaller in the O_{II} phase than in the disordered phase (Garland and Young, 1968b). This is in marked contrast to the situation in NH_4Cl (see Fig. 46).

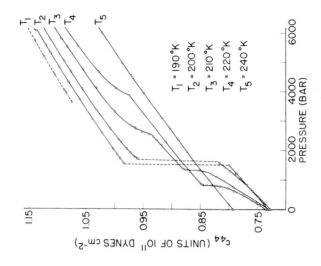

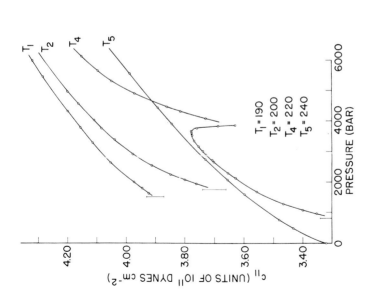

Fig. 52. Variation of c_{11} with pressure in NH_4Br. No data could be obtained in the ordered tetragonal phase. Here, $T_1 = 190°K$, $T_2 = 200°K$, $T_4 = 220°K$, $T_5 = 240°K$ (Garland and Young, 1968b).

Fig. 53. Variation of c_{44} with pressure in NH_4Br. The dashed curve near the top represents data taken in phase O_I at 180°K, where the lattice parameter is not well known. Here, $T_1 = 190°K$, $T_2 = 200°K$, $T_3 = 210°K$, $T_4 = 220°K$, $T_5 = 240°K$ (Garland and Young, 1968b).

VIII. First-Order Phase Transitions

As indicated by Table VII, there have been relatively few ultrasonic investigations of first-order phase transitions *per se*. Instabilities lead to small first-order discontinuities in KD_2PO_4 (see Section V,B) and NH_4Cl (see Section VII,B), but these transitions are essentially of the cooperative order–disorder type. Only in connection with CO_2 (see Fig. 10) and NH_4Br (see Fig. 53) have we presented data involving purely first-order changes. The first case obviously involves a liquid–vapor transition, and the latter involves the transformation of an ordered tetragonal structure into a differently ordered high-pressure structure. Of the investigations listed in Table VII, about one half concern melting, and the other half concern structural transformations in solids.

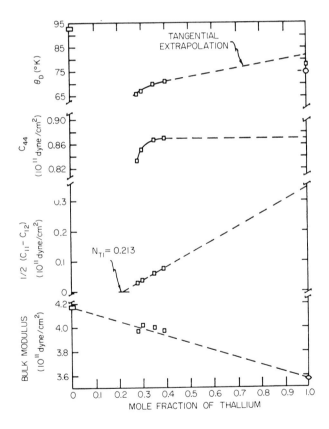

Fig. 54. Elastic constants and Debye temperatures of Tl–In alloys at 300°K. The points at $X = 1$ represent values for hcp thallium, the points at $X = 0$ represent values for face-centered-tetragonal indium, and the points in the range $X = 0.28$–0.4 are values for the fcc phase of the alloy (Novotny and Smith, 1965).

ULTRASONIC INVESTIGATIONS OF FIRST-ORDER PHASE TRANSITIONS

Substance	T_t	Temperature range	Pressure range	Frequency[a]	Quantity measured	Reference
Cd	594°K	300–575°K	1 atm	10	Five c_{ij}	Chang and Himmel (1966)
Zn	692°K	4.2–670°K	1 atm	10	Five c_{ij}	Alers and Neighbours (1958)
H_2O	0°C	0–60°C	1 atm	7.4–66	α	Pinkerton (1947)
		30°C, 50°C	0–6 kbar	15	u	Holton (1951)
		−12 to 129°C	0–10 kbar	12	u	Smith and Lawson (1954)
		0°C	0–2 kbar	25	u, α	Litovitz and Carnevale (1955)
		30°C	0–2 kbar	45	u, α	Litovitz and Carnevale (1955)
Benzene	5.5°C	0–7°C	1 atm	1–23	α	Gorbunov et al. (1966)
Paraffin	~53°C	48–55°C	1 atm	2–22	α	Gorbunov et al. (1966)
Liquid crystals[b]	70, 72.5, 75.5°C	70–80°C	1 atm	2–15	u_L, α_L	Kapustin and Zvereva (1966)
RbCl	$p_t \approx 5.5$ kbar at 298°K	298°K	0–20 kbar	3–5	u_L, u_T	Voronov and Goncharova (1966)
RbI	$p_t \approx 3.9$ kbar at 298°K	298°K	0–12 kbar	3–5	u_L, u_T	Voronov and Goncharova (1966)
In–Tl alloy (27–40 at% Tl)	f (compos)	200–350°K	1 atm	10	c_{11}, c_{44}, C'	Novotny and Smith (1965)
		300°K	0–3 kbar	10	c_{11}, c_{44}, C'	Hill and Smith (1968)
Cu_3Au	387.5°C	20–450°C	1 atm	Low?	c_{11}, c_{44}, C'	Siegel (1940)
Na	~30°K	4.2–200°K	Low	~20 kHz	f_R, Q^{-1}	Verdini (1961)
Li	~77°K	4.2–200°K	Low	~26 kHz	f_R, Q^{-1}	Verdini (1961)
Martensitic transition in Ru, Rh, Co, Dy, Er					$c_{ij}(T)$	Fisher and Dever (1967)

[a] In megahertz, except where otherwise noted.
[b] p,p'-Nonoxybenzaltoluidine.

Most of the studies of liquid–solid transitions were motivated, to some extent, by Born's theory of melting. A conspicuous difference between the properties of liquids and solids is the fact that liquids flow under a shear stress. Thus, Born proposed that the melting point of a crystalline solid represents the upper temperature limit of its shear stability; i.e., the velocity of at least one of the shear modes in a solid should vanish at the melting point. However, none of the systems—Cd, Zn, H_2O, benzene, or paraffin—shows such behavior. In each case, there is a discontinuous break in all the elastic constants at the melting point.

The measurements on liquid water are of special interest because of "structural" changes involving extensive hydrogen bonding. Since there is an excellent general discussion of the anomalous properties of water (Lawson and Hughes, 1963) which includes a detailed presentation of ultrasonic data, no discussion of H_2O will be given here.

The first-order transitions studied in solids include: martensitic (bcc → hcp) transformations in alkali metals and in transition metals, the change from the NaCl-structure to the CsCl-structure in rubidium halides, the discontinuous order–disorder transition in Cu_3Au, and a fcc → fct transformation in thallium–indium alloys. In the case of the rubidium halides, Hardy and Karo (1965) have made the theoretical prediction that the phase transition is associated with [100] phonon modes, whose frequency goes to zero at the edge of the Brillouin zone. If this is so, then the situation will be much like that in $SrTiO_3$ (see Section V,D). In the case of Tl–In alloys, as the indium concentration increases, the face-centered-cubic alloy becomes unstable with respect to a $C' = (c_{11} - c_{12})/2$ shear deformation and transforms into a face-centered-tetragonal phase. For a given temperature, this instability will occur at a corresponding critical value of the composition. An extrapolation of C' versus mole fraction at 300°K is shown in Fig. 54, and one see that $C' \to 0$ at $X_{Tl} = 0.213$. This value is within ~ 1 at.% of the actual transformation composition at this temperature. The fact that C' vanishes in the vicinity of the fcc–fct phase boundary is confirmed by the high-pressure ultrasonic investigation of Hill and Smith (1968).

ACKNOWLEDGMENT

The author wishes to thank D. Eden and E. Litov for helpful discussions and a critical reading of the manuscript.

REFERENCES

Aleksandrov, K. S., Reshchikova, L. M., and Beznosikov, B. V. (1966). *Phys. Status Solidi* 18, K17.

Aleksandrov, K. S., Reshchikova, L. M., and Beznosikov, B. V. (1967). *Soviet Phys.-Solid State (English Transl.)* 8, 2904.

Aleksandrov, K. S., Shabanova, L. A., and Reshchikova, L. M. (1968). *Soviet Phys.-Solid State (English Transl.)* 10, 1316.

Alers, G. A., and Neighbours, J. R. (1958). *Phys. Chem. Solids* 7, 58.

Alers, G. A., Neighbours, J. R., and Sato, H. (1960). *Phys. Chem. Solids* 13, 40.

Alfrey, G. F., and Schneider, W. G. (1953). *Discussions Faraday Soc.* **15**, 218.

Allen, S. J. (1968). *Phys. Rev.* **167**, 492.

Anantaraman, A. V., Walters, A. B., Edmonds, P. D., and Pings, C. J. (1966). *J. Chem. Phys.* **44**, 2651.

Anderson, N. S., and Delsasso, L. P. (1951). *J. Acoust. Soc. Am.* **23**, 423.

Andrews, T. (1869). *Proc. Roy. Soc.* **18**, 42.

Atanasoff, J. V., and Hart, P. J. (1941). *Phys. Rev.* **59**, 85.

Atanasoff, J. V., and Kammer, E. (1941). *Phys. Rev.* **59**, 97.

Atkins, K. R., and Stasior, R. A. (1953). *Can. J. Phys.* **31**, 1156.

Baranskii, K. N., Shustin, O. A., Velichkina, T. S., and Yakovlev, I. A. (1963). *Soviet Phys. JETP (English Transl.)* **16**, 518.

Barmatz, M. (1970). *Phys. Rev. Letters* **24**, 651.

Barmatz, M., and Rudnick, I. (1968). *Phys. Rev.* **170**, 224.

Barrett, H. H. (1968). *Phys. Letters A* **26**, 217.

Barrett, H. H. (1969). *Phys. Rev.* **178**, 743.

Barrett, H. H. (1970). *In* " Physical Acoustics " (W. P. Mason and R. N. Thurston, eds.), Vol. 6, Chapter 2. Academic Press, New York and London.

Bell, R. O., and Rupprecht, G. (1963). *Phys. Rev.* **129**, 90.

Belov, K. P., Katayev, G. I., and Levitin, R. Z. (1959). *Soviet Phys. JETP (English Transl.)* **10**, 670.

Belov, K. P., Katayev, G. I., and Levitin, R. Z. (1960). *J. Appl. Phys. Suppl.* **31**, 153S.

Benedek, G. B. (1969). Private communication.

Bennett, H. S., and Pytte, E. (1967). *Phys. Rev.* **155**, 553.

Berlincourt, D., and Jaffe, H. (1958). *Phys. Rev.* **111**, 143.

Bezuglyi, P. A., Burma, N. G., and Minyafaev, R. Kh. (1966). *Soviet Phys.-Solid State (English Transl.)* **8**, 596.

Bhatia, A. B. (1959). *J. Acoust. Soc. Am.* **31**, 16.

Bolef, D. I., and de Klerk, J. (1963). *Phys. Rev.* **129**, 1063.

Botch, W., and Fixman, M. (1965). *J. Chem. Phys.* **42**, 199.

Brandt, O. G., and Walker, C. T. (1967). *Phys. Letters* **18**, 11.

Brandt, O. G., and Walker, C. T. (1968). *Phys. Rev.* **170**, 528.

Breazeale, M. A. (1962). *J. Chem. Phys.* **36**, 2530.

Breazeale, M. A. (1963). *J. Chem. Phys.* **38**, 1786.

Brody, E. M. (1969). Ph.D. Thesis, Dept. of Phys., Johns Hopkins Univ., Baltimore, Maryland.

Brody, E. M., and Cummins, H. Z. (1968). *Phys. Rev. Letters* **21**, 1263.

Brown, A. E. (1967). *Acustica* **18**, 169.

Brown, A. E., and Richardson, E. G. (1959). *Phil. Mag.* **4**, 705.

Buckingham, M. J., and Fairbank, W. M. (1961). *Progr. Low Temp. Phys.* **3**, 80.

Cevolani, M., and Petralia, S. (1952). *Atti Accad. Nazl. Lincei* **2**, 674.

Cevolani, M., and Petralia, S. (1958). *Nuovo Cimento* [10], **7**, 866.

Chang, Y. A., and Himmel, L. (1966). *J. Appl. Phys.* **37**, 3787.

Chase, C. E. (1953). *Proc. Roy. Soc.* **A220**, 116.

Chase, C. E. (1958). *Phys. Fluids* **1**, 193.

Chase, C. E. (1959). *Phys. Rev. Letters* **2**, 197.

Chase, C. E., Williamson, R. C., and Tisza, L. (1964). *Phys. Rev. Letters* **13**, 467.

Chernov, L. (1960). "Wave Propagation in a Random Medium," p. 55. McGraw-Hill, New York.

Chynoweth, A. G., and Schneider, W. G. (1951). *J. Chem. Phys.* **19**, 1566.

Chynoweth, A. G., and Schneider, W. G. (1952). *J. Chem. Phys.* **20**, 1777.

Cowley, R. A. (1964). *Phys. Rev.* **134**, A981.

Craft, W. L., and Slutsky, L. J. (1968). *J. Chem. Phys.* **49**, 638.
Craft, W. L., Eckhardt, R., and Slutsky, L. J. (1969). *J. Phys. Soc. Japan Suppl.* **26**, 184.
Cummins, H. Z. (1969). Private communication.
D'Arrigo, G., and Sette, D. (1968). *J. Chem. Phys.* **48**, 691.
D'Arrigo, G., Mistura, L., and Tartaglia, P. (1970). *Phys. Rev.* **A1**, 286.
DeMorton, M. (1963). *Phys. Rev. Letters* **10**, 208.
deSobrino, L. (1968). *Can. J. Phys.* **46**, 2821.
Deutch, J. M., and Zwanzig, R. (1967). *J. Chem. Phys.* **46**, 1612.
Dietz, G., and Jaumann, J. (1962). *Z. Angew Phys.* **14**, 222.
Dvorak, V. (1967). *Can. J. Phys.* **45**, 3903.
Dvorak, V. (1968). *Phys. Rev.* **167**, 525.
Feke, G. T. (1969). M. S. Thesis, John Carroll Univ., Cleveland, Ohio.
Ferrell, R. A., Menyhard, N., Schmidt, H., Schwabl, F., and Szepfalusy, P. (1968). *Ann. Phys. (N.Y.)* **47**, 565.
Fine, M. E. (1953). *Rev. Mod. Phys.* **25**, 158.
Fisher, E. S., and Dever, D. (1967). *Trans. AIME* **239**, 48.
Fisher, E. S., and McSkimin, H. J. (1961). *Phys. Rev.* **124**, 67.
Fisher, M. E. (1967). *Rept. Progr. Phys.* **30**, 615.
Fixman, M. (1962). *J. Chem. Phys.* **36**, 1961.
Fixman, M. (1964). *Advan. Chem. Phys.* **4**, 175–228.
Ford, Jr., N. C., Langley, K. H,, and Puglielli, V. G., (1968). *Phys. Rev. Letters* **21**, 9.
Fushimi, S., and Ikeda, T. (1966). *Rev. Elec. Commun. Lab. (Tokyo)* **14**, 161.
Gammon, R. W., and Cummins, H. Z. (1966). *Phys. Rev. Letters* **17**, 193.
Gammon, R. W., and Verdieck, M. J. (1969). To be published.
Gammon, R. W., Swinney, H. L., and Cummins, H. Z. (1967). *Phys. Rev. Letters* **19**, 1467.
Garland, C. W. (1964a). *J. Chem. Phys.* **41**, 1005.
Garland, C. W. (1964b). *Phys. Rev.* **135**, A1696.
Garland, C. W., and Jones, J. S. (1963). *J. Chem. Phys.* **39**, 2874.
Garland, C. W., and Jones, J. S. (1965). *J. Chem. Phys.* **42**, 4194.
Garland, C. W., and Novotny, D. B. (1969). *Phys. Rev.* **117**, 9711.
Garland, C. W., and Renard, R. (1966a). *J. Chem. Phys.* **44**, 1120.
Garland, C. W., and Renard, R. (1966b). *J. Chem. Phys.* **44**, 1130.
Garland, C. W., and Snyder, D. D. (1969). *Phys. Chem. Solids.* To be published.
Garland, C. W., and Yarnell, C. F. (1966a). *J. Chem. Phys.* **44**, 1112.
Garland, C W., and Yarnell, C. F. (1966b). *J. Chem. Phys.* **44**, 3678.
Garland, C. W., and Young, R. A. (1968a). *J. Chem. Phys.* **48**, 146.
Garland, C. W., and Young, R. A. (1968b). *J. Chem. Phys.* **49**, 5282.
Geguzina, S. Ya., and Krivoglaz, M. A. (1968). *Soviet Phys.-Solid State (English Transl.)* **9**, 2441.
Geguzina, S. Ya., and Timan, B. L. (1968). *Soviet Phys.-Solid State (English Transl.)* **9**, 1702.
Ginsberg, D. M., and Hebel, L. C. (1969). *In* "Superconductivity" (R. D. Parks, ed.), Vol. I, Chapter 4. Dekker, New York.
Giterman, M. Sh., and Kontorovich, V. M. (1965). *Soviet Phys. JETP (English Transl.)* **20**, 1433.
Glinskii, A. A. (1965). *Soviet Phys. Acoust. (English Transl.)* **11**, 87.
Golding, B. (1968). *Phys. Rev. Letters* **20**, 5.
Golubeva, O. N., and Shustin, O. A. (1968). *JETP Letters (English Transl.),* **7**, 358.

Gorbunov, M. A., Koshkin, N. I., and Sheloput, D. V. (1966). *Soviet Phys. Acoust. (English Transl.)* **12**, 20.

Grechkin, V. I., and Nozdrev, V. F. (1964). *Soviet Phys. Acoust. (English Transl.)* **9**, 304.

Halperin, B. I., and Hohenberg, P. C. (1969). *Phys. Rev.* **177**, 952.

Hamano, K., Negishi, K., Marutake, M., and Nomura, S. (1963). *Japan J. Appl. Phys.* **2**, 83.

Hardy, J. R., and Karo, A. M. (1965). *Lattice Dyn. Proc. Intern. Conf., Copenhagen*, 1963, p. 195. Pergamon, New York.

Haussühl, S. (1960). *Acta Cryst.* **13**, 685.

Heims, S. P. (1966). *J. Chem. Phys.* **45**, 370.

Heinicke, W., Winterling, G., and Dransfeld, K. (1969). *Phys. Rev. Letters* **22**, 170.

Herget, C. M. (1940). *J. Chem. Phys.* **8**, 537.

Herzfeld, K. F., and Litovitz, T. A. (1959). "Absorption and Dispersion of Ultrasonic Waves," Secs. 31 and 104. Academic Press, New York.

Hill, E. R., and Smith, C. S. (1968). A.E.C. Tech. Rept. Contract, No. AT(11-1)-623, Case-Western Reserve Univ., Cleveland, Ohio.

Holton, G. (1951). *J. Appl. Phys.* **22**, 1407.

Hueter, T. F., and Neuhaus, D. P. (1955). *J. Acoust. Soc. Am.* **27**, 292.

Huibregtse, E. J., Bressey, W. H., and Drougard, M. E. (1959). *J. Appl. Phys.* **30**, 899.

Ikeda, T. (1957). *J. Phys. Soc. Japan* **13**, 809.

Ikeda, T., Tanaka, Y., and Toyodo, H. (1962). *Japan J. Appl. Phys.* **1**, 13.

Imai, J. S., and Rudnick, I. (1969). *Phys. Rev. Letters* **22**, 694.

Inoue, M. (1969). *J. Phys. Soc. Japan* **26**, 420.

Janovec, V. (1966). *J. Chem. Phys.* **45**, 1874.

Kadanoff, L. P. (1968). *Comments Solid State Phys.* **1**, 5.

Kadanoff, L. P. (1969). *J. Phys. Soc. Japan Suppl.* **26**, 122.

Kadanoff, L. P., and Martin, P. C. (1963). *Ann. Phys. (N.Y.)* **24**, 419.

Kadanoff, L. P., and Swift, J. (1968). *Phys. Rev.* **166**, 89.

Kadanoff, L. P., Götze, W., Hamblen, D., Hecht, R., Lewis, E. A. S., Palciauskas, V. V., Rayl, M., Swift, J., Aspnes, D., and Kane, J. (1967). *Rev. Mod. Phys.* **39**, 395.

Kalianov, B. I., and Nozdrev, V. F. (1958). *Soviet Phys. Acoust. (English Transl.)* **4**, 198.

Kammer, E. W., Pardue, T. E., and Frissel, H. F. (1948). *J. Appl. Phys.* **19**, 265.

Kapustin, A. P., and Zvereva, G .E. (1966). *Soviet Phys. Cryst. (English Transl.)* **10**, 603.

Kashcheev, V. N. (1967a). *Phys. Letters A* **24**, 627.

Kashcheev, V. N. (1967b). *Phys. Letters A* **25**, 71.

Kawasaki, K. (1968a). *Solid State Commun.* **6**, 57.

Kawasaki, K. (1968b). *Phys. Letters A* **26**, 543.

Kawasaki, K. (1968c). *Progr. Theoret. Phys. (Kyoto)* **39**, 285.

Kawasaki, K. (1968d). *Progr. Theoret. Phys. (Kyoto)* **39**, 1133.

Kawasaki, K. (1968e). *Progr. Theoret. Phys. (Kyoto)* **40**, 11.

Kawasaki, K. (1968f). *Progr. Theoret. Phys. (Kyoto)* **40**, 706.

Kawasaki, K. (1968g). *Progr. Theoret. Phys. (Kyoto)* **40**, 930.

Kawasaki, K. (1970). To be published.

Kawasaki, K., and Tanaka, M. (1967). *Proc. Phys. Soc. (London)* **90**, 791.

Kendig, A. P., Bigelow, R. H., Edmonds, P. D., and Pings, C. J. (1964). *J. Chem. Phys.* **40**, 1451.

Kikuchi, R. (1960). *Ann. Phys. (N.Y.)* **10**, 127.

Kornfeld, M. I., and Chubinov, A. A. (1958). *Soviet Phys. JETP (English Transl.)* **6**, 26.

Krasnyi, Yu. P., and Fisher, I. Z. (1967). *Ukranian Phys. J.* (*English Transl.*) **12**, 462.

Kravtosov, V. M. (1963). *Soviet Phys. Acoust.* (*English Transl.*) **9**, 193.

Kruus, P. (1964). *Can. J. Chem.* **42**, 1712.

Kruus, P., and Bak, T. A. (1966). *Acta Chem. Scand.* **20**, 231.

Kubo, R. (1966). *Rept. Progr. Phys.* **29**, 255.

Landau, L. D., and Khalatnikov, I. M. (1954). *Dokl. Akad. Nauk SSSR* **96**, 469.

Laramore, G., and Kadanoff, L. (1969). *Phys. Rev.* **187**, 619.

Lawson, A. W., and Hughes, A. J. (1963). *In* "High Pressure Physics and Chemistry " (R. S. Bradley, ed.), Vol. 1, Academic Press, New York.

Lazay, P. (1969). Ph.D. Thesis, Phys. Dept. M.I.T., Cambridge, Massachusetts.

Lefkowitz, I., and Hazony, Y. (1968). *Phys. Rev.* **169**, 441.

Levanyuk, A. P. (1966). *Soviet Phys. JETP* (*English Transl.*) **22**, 901.

Levanyuk, A. P., Minaeva, K. A., and Strukov, B. A. (1969). *Soviet Phys. Solid State* (*English Transl.*) **10**, 1919.

Litov, E., and Garland, C. W. (1969). To be published.

Litov, E., and Uehling, E. A. (1968). *Phys. Rev. Letters* **21**, 809.

Litovitz, T. A., and Carnevale, E. H. (1955). *J. Appl. Phys.* **26**, 816.

Long, Jr., M., Wazzan, A. R., and Stern, R. (1969). *Phys. Rev.* **178**, 775.

Luthi, B., and Pollina, R. J. (1968a). *J. Appl. Phys.* **39**, 718.

Luthi, B., and Pollina, R. J. (1968b). *Phys. Rev.* **167**, 482.

Luthi, B., Moran, T. J., and Pollina, R. J. (1969). *Phys. Chem. Solids* (to be published).

Lynton, E. A. (1969). "Superconductivity," 3rd ed., pp. 149–152. Methuen, London.

McManus, G. M. (1963). *Phys. Rev.* **129**, 2004.

Madigosky, W. M., and Litovitz, T. A. (1961). *J. Chem. Phys.* **34**, 489.

Makhanko, I. G., and Nozdrev, V. F. (1964). *Soviet Phys. Acoust.* (*English Transl.*) **10**, 207.

Mason, W. P. (1946). *Phys. Rev.* **68**, 173.

Mason, W. P., and Matthias, B. T. (1952). *Phys. Rev.* **88**, 477.

Mayer, G. (1960). *Centre Etudes Nucl. de Saclay, Rapport* No. 1330.

Melcher, R. L., and Bolef, D. I. (1969a). *Phys. Rev.* **178**, 864.

Melcher, R. L., and Bolef, D. I. (1969b). *Phys. Rev.* **186**, 491.

Melcher, R. L., Bolef, D. I., and Stevenson, R. W. H. (1967). *Solid State Commun.* **5**, 735.

Merkulov, L. G., and Sokolova, E. S. (1962). *Soviet Phys. Acoust.* (*English Transl.*) **7**, 401.

Minaeva, K. A., and Levanyuk, A. P. (1965). *Bull. Acad. Sci. USSR, Phys. Ser.* **29**, 978.

Minaeva, K. A., and Strukov, B. A. (1966). *Soviet Phys.-Solid State* (*English Transl.*). **8**, 24.

Minaeva, K. A., Strukov, B. A., and Koptsik, V. A. (1966). *Soviet Phys.-Solid State* (*English Transl.*) **8**, 1299.

Minaeva, K. A., Levanyuk, A. P., Strukov, B. A., and Koptsik, V. A. (1967). *Soviet Phys.-Solid State* (*English Transl.*) **9**, 950.

Minaeva, K. A., Strukov, B. A., and Varnstorff, K. (1969). *Soviet Phys.-Solid State* (*English Transl.*) **10**, 1665.

Mnatsakanyan, A. V., Shuvalov, L. A., Zheludev, I. S., and Gavrilova, I. V. (1966) *Soviet Phys. Cryst.* (*English Transl.*) **11**, 412.

Mountain, R. D. (1968). *J. Chem. Phys.* **48**, 2189.

Mountain, R. D., and Zwanzig, R. (1968). *J. Chem. Phys.* **48**, 1451.

Mueller, P. E., Garland, C. W., and Eden, D. (1969). To be published.

Neighbours, J. R., and Moss, R. W. (1968). *Phys. Rev.* **173**, 542.

Neighbours, J. R., Olivre, R. W., and Stillwell, C. H. (1963). *Phys. Rev. Letters* **11**, 125.

Noury, J. (1951). *Compt. Rend.* **233**, 516.

Novotny, D. B., and Smith, J. F. (1965). *Acta Met.* **13**, 881.

Nozdrev, V. F. (1955). *Soviet Phys. Acoust.* (*English Transl.*) **1**, 249.

Nozdrev, V. F., and Sobolev, V. D. (1956). *Soviet Phys. Acoust.* (*English Transl.*) **2**, 408.

Nozdrev, V. F., and Stepanov, N. G. (1968). *Soviet Phys. Acoust.* (*English Transl.*) **13**, 538.

Nozdrev, V. F., and Tarantova, G. D. (1962). *Soviet Phys. Acoust.* (*English Transl.*) **7**, 402.

Nozdrev, V. F., and Yashina, L. S. (1966). *Soviet Phys. Acoust.* (*English Transl.*) **11**, 339.

Nozdrev, V. F., Osadchii, A. P., and Rubstov, A. S. (1962). *Soviet Phys. Acoust.* (*English Transl.*) **7**, 305.

O'Brien, E. J., and Franklin, J. (1966). *J. Appl. Phys.* **37**, 2809.

O'Brien, E. J., and Litovitz, T. A. (1964). *J. Appl. Phys.* **35**, 180.

Okamoto, H. (1967). *Progr. Theoret. Phys.* (*Kyoto*) **37**, 1348.

Papoular, M. M. (1964). *Compt. Rend.* **258**, 4446.

Papoular, M. M. (1965). *Phys. Letters* **16**, 259.

Parbrook, H. D. (1953). *Acustica*, **3**, 49.

Parbrook, H. D., and Richardson, E. G. (1952). *Proc. Phys. Soc.* (*London*) **B65**, 437.

Parker, R. C., Slutsky, L. J., and Applegate, K. R. (1968). *J. Phys. Chem.* **72**, 3177.

Pellam, J. R., and Squire, C. F. (1947). *Phys. Rev.* **72**, 1245.

Pinkerton, J. M. M. (1947). *Nature* **160**, 128.

Pippard, A. B. (1951). *Phil. Mag.* [7] **42**, 1209.

Pippard, A. B. (1956). *Phil. Mag.* **1**, 473.

Pollina, R. J., and Luthi, B. (1969). *Phys. Rev.* **177**, 841.

Price, W. (1949). *Phys. Rev.* **75**, 946.

Pytte, E., and Bennett, H. S. (1967). *Phys. Rev.* **164**, 712.

Renard, R., and Garland, C. W. (1966a). *J. Chem. Phys.* **44**, 1125.

Renard, R., and Garland, C. W. (1966b). *J. Chem. Phys.* **45**, 763.

Rosen, M. (1968a). *Phys. Rev.* **165**, 357.

Rosen, M. (1968b). *Phys. Rev.* **166**, 561.

Rosen, M. (1968c). *Phys. Rev.* **174**, 504.

Rosen, M. (1969). *Phys. Rev.* **180**, 540.

Rudnick, I., and Shapiro, K. A. (1965). *Phys. Rev. Letters* **15**, 386.

Sannikov, D. G. (1962). *Soviet Phys.-Solid State* (*English Transl.*) **4**, 1187.

Schacher, G. E. (1967). *J. Chem. Phys.* **46**, 3565.

Schneider, W. G. (1951). *Can. J. Chem.* **29**, 243.

Schneider, W. G. (1952). *J. Chem. Phys.* **20**, 759.

Sette, D. (1955). *Nuovo Cimento* [X], **1**, 800.

Shapiro, S. M., and Cummins, H. Z. (1968). *Phys. Rev. Letters* **21**, 1587.

Shimakawa, S. (1961). *J. Phys. Soc. Japan* **16**, 113.

Shirane, G., and Yamada, Y. (1969). *Phys. Rev.* **177**, 858.

Shirokov, A. M., and Shuvalov, L. A. (1964). *Soviet Phys. Cryst.* (*English Transl.*) **8**, 586.

Shustin, O. A., Velichkina, T. S., Baranskii, K. N., and Yakovlev, I. A. (1961). *Soviet Phys. JETP* (*English Transl.*) **13**, 683.

Shustin, O. A., Yakovlev, I. A., and Velichkina, T. S. (1967). *JETP Letters* (*English Transl.*) **5**, 3.

Shuvalov, L. A., and Likhacheva, Yu. S. (1960). *Bull. Acad. Sci. USSR, Phys. Ser.* **24**, 1219.

Shuvalov, L. A., and Minaeva, K. A. (1963). *Soviet Phys. "Doklady" (English Transl.)* **7**, 906.

Shuvalov, L. A., and Mnatsakanyan, A. V. (1965). *Bull. Acad. Sci. USSR, Phys. Ser.* **25**, 1809.

Shuvalov, L. A., and Mnatsakanyan, A. V. (1966). *Soviet Phys. Cryst. (English Transl.)* **11**, 210.

Shuvalov. L. A., and Pluzhnikov, K. A. (1962). *Soviet Phys. Cryst. (English Transl.)* **6**, 555.

Siegel, S. (1940). *Phys. Rev.* **57**, 537.

Singh, R. P., and Verma, G. S. (1968). *J. Phys. C. (Proc. Phys. Soc.)* [2], **1**, 1476.

Singh, R. P., Darbari, G. S., and Verma, G. S. (1966). *Phys. Rev. Letters* **16**, 1150.

Smith, A. H., and Lawson, A. W. (1954). *J. Chem. Phys.* **22**, 351.

Snyder, D. D. (1968). Ph.D. Thesis, M.I.T., Cambridge, Massachusetts.

Steinemann, S. (1952). Diplomarbeit, Swiss Federal Inst. of Technol.

Street, R. (1963). *Phys. Rev. Letters* **10**, 210.

Street, R., and Lewis, B. (1951). *Nature* **168**, 1036.

Swift, J. (1968). *Phys. Rev.* **173**, 257.

Swift, J., and Kadanoff, L. P. (1968). *Ann. Phys. (N.Y.)* **50**, 312.

Tanaka, T., Meijer, P. H. E., and Barry, J. H. (1962). *J. Chem. Phys.* **37**, 1397.

Tani, K., and Mori, H. (1966). *Phys. Letters* **19**, 627.

Tani, K., and Mori, H. (1968). *Progr. Theoret. Phys. (Kyoto)* **39**, 876.

Tani, K., and Tanaka, H. (1968). *Phys. Letters* A**27**, 25.

Tani, K., and Tsuda, N. (1969). *J. Phys. Soc. Japan* **26**, 113.

Tanneberger, H. (1959). *Z. Physik* **153**, 445.

Tarasov, B. F., and Taborov, B. F. (1966). *Ukr. Fiz. Zh.* **11**, 570.

Testardi, L. R., Levinstein, H. J., and Guggenheim, H. J. (1967). *Phys. Rev. Letters* **19**, 503.

Tielsch, H., and Tanneberger, H. (1954). *Z. Physik* **137**, 256.

Trelin, Yu. S., and Sheludyakov, E. P. (1966). *JETP Letters (English Transl.)* **3**, 63.

van Dael, W., van Itterbeek, A., and Thoen, J. (1967). *Advan. Cryog. Eng.* **12**, 754.

van Itterbeek, A., and Forrez, G. (1954). *Physica* **20**, 133.

Verdini, L. (1961). *Proc. Intern. Congr. Acoust. 3rd, Stuttgart,* 1959, p. 480. Elsevier, Amsterdam.

Vlasov, K. B. (1966). *Bull. Acad. Sci. USSR Phys. Ser.* **30**, 985.

Voronov, F. F., and Goncharova, V. A. (1966). *Soviet Phys. JETP (English Transl.)* **23**, 777.

Walther, K. (1967). *Solid State Commun.* **5**, 399.

West, F. G. (1958). *J. Appl. Phys.* **29**, 480.

Widom, B. (1965). *J. Chem. Phys.* **43**, 3898.

Williamson, R. C. (1970). To be published.

Williamson, R. C., and Chase, C. E. (1968). *Phys. Rev.* **176**, 285.

Woodruff, T. O., and Ehrenreich, H. (1961). *Phys. Rev.* **123**, 1553.

Yakovlev, I. A., and Velichkina, T .S. (1957). *Usp. Fiz. Nauk* **63**, 411 (*Usp. Adv. Phys. Science (English Transl.)* **63**, 552).

Yakovlev, I. A., Velichkina, T. S., and Baranskii, K. A. (1957). *Soviet Phys. JETP (English Transl.)* **5**, 762.

Yakovlev, I. A., Velichkina, T. S., and Baranskii, K. A. (1958). *Soviet Phys. JETP (English Transl.)* **6**, 830.

Yevtushchenko, L. A., and Levitin, R. Z. (1961). *Phys. Metals Metallog. (USSR) (English Transl.)* **12**, 139.

Zubov, V. G., and Firsova, M. M. (1962). *Soviet Phys. Cryst. (English Transl.)* **7**, 374.

ADDITIONAL REFERENCES

In the six months between June 1969 and January 1970, there has been a considerable volume of new work which has come to the author's attention. In a few cases, it has been possible to incorporate references to such very recent papers into the text, but many important contributions could not be included. Such papers (with their full titles) are listed below. No attempt has been made to carry out a complete literature search of very recent papers (i.e., those published after June 1969), but it seemed worthwhile to list those of which the author was aware.

Ahlers, G. (1969a). "Thermodynamics of the isentropic sound velocity near the superfluid transition in He⁴," **182**, 352 (see also erratum **187**, 397).

Ahlers, G. (1969b). "On the attenuation and dispersion of first sound near the superfluid transition in He⁴," *J. Low Temp. Phys.*, to be published.

Berre, B., Fossheim, K., and Müller, K. A. (1969). "Critical attenuation of sound by soft modes in $SrTiO_3$," *Phys. Rev. Letters* **23**, 589.

Edmonds, P. E., and Orr, D. A. (1967). "Ultrasonic absorption and dispersion at phase transitions in liquid crystalline compounds." *In* "Liquid Crystals" (G. H. Brown, G. J. Dienes, and M. M. Labes, eds.). Gordon and Breach, New York.

Evans, R. G., and Cracknell, M. F. (1969). "Ultrasonic attenuation in MnO in the vicinity of the Néel point," *Phys. Chem. Solids*, to be published.

Fisher, E., and Manghnani, M. (1969). "Pressure coefficients of the single crystal elastic coefficients in ferromagnetic and paramagnetic gadolinium." Colloque Intern. CNRS: Les Proprietes Physiques des Solids Sous Pression, Grenoble, September 8–10, 1969.

Giterman, M. Sh., and Gorodetskii, E. E. (1969). "Behavior of kinetic coefficients near the critical point of pure liquids," *Soviet Phys. JETP* **29**, 347.

Golding, B., and Barmatz, M. (1969). "Ultrasonic propagation near the magnetic critical point of nickel," *Phys. Rev. Letters* **23**, 223.

Golding, B., and Buehler, E. (1969). "Ultrasonic propagation in EuSe," *Solid State Commun.* **7**, 747.

Harnik, E., and Shimshoni, M. (1969). "Sound propagation in the polar phase of KH_2PO_4," *Phys. Letters* **29A**, 620.

Hatta, I., Ishiguro, T., and Mikoshiba, N. (1969a). "Ultrasonic attenuation near the critical points in $NaNO_2$," *Phys. Letters* **29A**, 421.

Hatta, I., Ishiguro, T., and Mikoshiba, N. (1969b). "Ultrasonic attenuation near the transition points in $NaNO_2$," *Proc. Intern. Meeting on Ferroelectricity, Kyoto*, September 4–9, 1969. *Suppl. J. Phys. Soc. Japan*, to be published.

Ichiyanagi, M. (1969). "Attenuation of first sound waves near the λ point of liquid helium," *Progr. Theoret. Phys. (Kyoto)* **42**, 147.

Ikushima, A. (1969a). "Sound velocity near the Néel point of MnF_2," *Phys. Letters* **29A**, 364.

Ikushima, A. (1969b). "Ultrasonic attenuation near the antiferromagnetic critical point of CoO," *Phys. Letters* **29A**, 417.

Kawasaki, K. (1969a). "A note on the ultrasonic attenuation near the magnetic critical points," *Phys. Letters* **29A**, 406.

Kawasaki, K. (1969b). "Transport coefficients of van der Waals fluids and fluid mixtures," *Progr. Theoret. Phys. (Kyoto)* **41**, 1190.

Kawasaki, K., and Ikushima, A. (1970). "Sound velocity of MnF_2 near the Néel temperature." To be published.

Leisure, R. G., and Moss, R. W. (1969). "Ultrasonic velocity in MnF$_2$ near the Néel temperature," *Phys. Rev.* **188**, 840.

Levanyuk, A. P., Strukov, B. A., and Minaeva, K. A. (1969). "Anisotropy of ultrasonic attenuation in uniaxial ferroelectrics," *Proc. Intern. Meeting on Ferroelectricity, Kyoto,* September 4–9, 1969. *Suppl. J. Phys. Soc. Japan,* to be published.

Liebermann, R. C., and Banerjee, S. K. (1970). "Anomalies in the compressional and sheer properties of hematite in the region of the Morin transition." To be published.

Litov, E., and Uehling, E. A. (1970). "Polarization relaxation and susceptibility in the ferroelectric transition region of KD$_2$PO$_4$". *Phys. Rev.,* to be published.

Litster, J. D., and Stinson, T. W. III. (1970). "Critical slowing of fluctuations in a nematic liquid crystal," *J. Appl. Phys.,* to be published.

Luthi, B., Papon, P., and Pollina, R. J. (1969). "Ultrasonic attenuation at magnetic phase transitions," *J. Appl. Phys.* **40**, 1029.

Marutoke, M. (1969). "Quasi-phenomenological theory of elastic anomaly in ferroelectric crystals," *Proc. Intern. Meeting on Ferroelectricity, Kyoto,* September 4–9, 1969. *Suppl. J. Phys. Soc. Japan,* to be published.

Meincke, P. P. M., and Litva, J. (1969). "Velocity of sound in invar at low temperatures," *Phys. Letters* **29 A**, 390.

Mohr, R., Langley, K. H., and Ford, N. C. Jr. (1970). "Brillouin scattering from SF$_6$ in the vicinity of the critical point." To be published.

Moran, T. J., and Luthi, B. (1969). "Critical changes in sound velocity near a magnetic phase transition," *Phys. Letters,* to be published.

Mountain, R. D. (1969). "Dynamical model for Brillouin scattering near the critical point of a fluid," *J. Res. Natl. Bur. Std. (U.S.)* **73A**, 593.

Postnikov, V. S., Pavlov, V. S., and Turkov, S. K. (1969). "Internal friction in ferroelectrics due to interaction of domain boundaries and point defects," *Phys. Chem. Solids,* to be published.

Romanov, V. P., and Solov'ev, V. A. (1968). "Sound absorption near the critical point," *Soviet Phys.–Acoustics (English Transl.)* **14**, 213.

Rosen, M. (1969a). "Effect of the low-temperature phase transformations on the elastic behavior of cerium," *Phys. Rev.* **181**, 932.

Rosen, M. (1969b). "Elastic properties of rare-earth single crystals," *Colloque Intern. CNRS: Les Proprietes Physiques des Solides Sous Pression, Grenoble,* September 8–10, 1969.

Sawamoto, K., Ashida, T., Omachi, Y., and Uno, T. (1969). "Behavior of LiTaO$_3$ single crystal near its Curie point. Part II. Dielectric and ultrasonic properties," *Proc. Intern. Meeting on Ferroelectricity, Kyoto,* September 4–9, 1969. *Suppl. J. Phys. Soc. Japan,* to be published.

Shapira, Y., Foner, S., and Misetich, A. (1969). "Magnetic phase diagram of MnF$_2$ from ultrasonic and differential magnetization measurements," *Phys. Rev. Letters* **23**, 98.

Sorge, G., Hegenbarth, E., and Schmidt, G. (1969). "Mechanical relaxation and nonlinearity in strontium titanate single crystals," *Proc. Intern. Meeting on Ferroelectricity, Kyoto,* September 4–9, 1969. *Suppl. J. Phys. Soc. Japan,* to be published.

Tani, K. (1969). "Ultrasonic attenuation in magnetics at low temperatures," *Progr. Theoret. Phys. (Kyoto)* **41**, 891.

Turik, A. V. (1969). "Dielectric, elastic and piezoelectric properties of BaTiO$_3$ single crystals," *Proc. Intern. Meeting on Ferroelectricity, Kyoto,* September 4–9, 1969. *Suppl. J. Phys. Soc. Japan,* to be published.

Wright, P. G. (1969). "Note on departures from Garland's relation between the elastic constants of a cubic crystal near a λ-transition," *J. Phy. C (Proc. Phys. Soc.)* [2] **2**, 1352.

—3—

Ultrasonic Attenuation in Normal Metals and Superconductors: Fermi-Surface Effects

J. A. RAYNE

Carnegie-Mellon University
Pittsburgh, Pennsylvania

and

C. K. JONES

Westinghouse Research Laboratories
Pittsburgh, Pennsylvania

I. Introduction

The attenuation of an acoustic wave propagating in a pure metal at low temperatures is significantly affected by the direct interaction between the resulting lattice vibrations and the conduction electrons. This effect, which was first observed in normal and superconducting lead by Bömmel (1954) and Mackinnon (1954), has been extensively investigated both theoretically and experimentally. From these investigations, much useful information has been obtained both about the electronic properties of metals and the magnitude and anisotropy of the electron–phonon interaction.

It is the purpose of this chapter to review the current status of both the theoretical and experimental aspects of low-temperature ultrasonic attenuation measurements in metals. The discussion is restricted to the case of zero magnetic field, since the magnetoacoustic effect has been considered in previous volumes of this series by Peverly (1966) and Roberts (1968). For the same reason, the discussion of attenuation due to dislocation motion and the effects of electron damping is confined to only those aspects affecting the reduction of experimental data. The general plan of the chapter is, first, to consider normal metals, and then to extend the discussion to superconductors.

II. Theory of Attenuation in Normal Metals

A. FREE-ELECTRON MODEL

Many features of the attenuation of ultrasonic waves propagating in a normal metal are given by treating the latter in the free-electron approximation, i.e., as a regular array of point ions imbedded in a compensating sea of conduction electrons. Purely quantum-mechanical solutions of this problem have been given by Morse (1959) and Kittel (1955). However, since the acoustic wavelengths are much larger than those of the conduction electrons, it is possible to give a semiclassical treatment, as shown by Pippard (1955) and Holstein (1956). As the physical principles involved in the attenuation mechanism are somewhat clearer in this approach, it will be considered first.

Accordingly, let us consider a plane acoustic wave with propagation vector $\mathbf{q}$ traveling in a metal with N electrons per unit volume. The resulting ionic motion is characterized by a sinusoidally varying velocity $\mathbf{u}$ given by

$$\mathbf{u} = \mathbf{u}_0 \exp[i(\omega t - \mathbf{q} \cdot \mathbf{r})] \tag{1}$$

Associated with the wave are harmonically varying electromagnetic fields which cause a transfer of acoustic energy to the electron assembly. Collisions of the electrons with impurities cause this energy to be returned to the lattice in a random form. There is, therefore, an irreversible flow of energy from the sound wave, which is consequently attenuated.

At the frequencies usually employed in these experiments (1 GHz or less), the rate at which the electron assembly loses energy can be calculated from the Joule heating term alone (Holstein, 1956). This term can be written, per unit volume, in the form

$$\dot{Q} = \tfrac{1}{2} \operatorname{Re}(\mathbf{J}_{\text{el}}^{*} \cdot \mathbf{E}) \tag{2}$$

where $\mathbf{J}_{\text{el}}$ is the electronic current density and $\mathbf{E}$ the local electric field. The total energy of the sound wave per unit volume is $\tfrac{1}{2}\rho u^2$, where ρ is the density of the metal. Hence, if v_{s} is the acoustic-wave velocity, the attenuation constant per unit length is given by

$$\alpha = 2\dot{Q}/\rho v_{\text{s}} u^2 \tag{3}$$

The total electric current density **J** is the sum of the electronic and ionic current densities, so that, taking e to be the electronic charge and N to be the charge density, we have

$$\mathbf{J} = \mathbf{J}_{el} - Ne\mathbf{u} \tag{4}$$

This current density **J** is related to the electric field **E** and the associated magnetic field **H** by Maxwell's equations. In the case of longitudinal waves, these can be solved to give, together with Eq. (4),

$$\mathbf{J}_{el} = Ne\mathbf{u} - (i\omega/4\pi)\mathbf{E} \tag{5}$$

The second term is negligible compared to the first if the acoustic frequency is much lower than the plasma frequency, a condition easily satisfied in all experiments involving metals. Thus, the total current in the metal must vanish, and the electric fields, set up in this case by the minute charge imbalance, cause the electrons to move in such a way that they cancel the ionic current.

In the case of transverse waves, it is easily shown that there is no space charge and that the electric fields are induced by the magnetic fields associated with the current imbalance between the electrons and ions. The solution of Maxwell's equations together with Eq. (4) now gives

$$\mathbf{J}_{el} = Ne\mathbf{u} + (iq^2c^2/4\pi\omega)\mathbf{E} \tag{6}$$

If the acoustic wavelength is much greater than the skin depth for electromagnetic waves of frequency ω, the second term on the right-hand side is negligible compared with the first, and again the electronic and ionic currents cancel. It is to be noted that the condition for quasi-current neutrality is more stringent in this case, and, in fact, may not be satisfied at the highest acoustic frequencies now used. Nevertheless, in what follows, we shall always assume that the total current is zero.

To evaluate the attenuation, we must now calculate the electronic current density self-consistently from the electric field and from the microscopic distribution function. It has been shown by Holstein (1956) that the latter relaxes to the equilibrium distribution function corresponding to the local lattice velocity and the Fermi energy for the disturbed electron density. Expanding to first order in the lattice velocity **u**, we thus have that the distribution relaxes to

$$\bar{f}_0(\mathbf{v}, \mathbf{r}, t) = f_0(\mathbf{v}, E_F) - \frac{\partial f_0}{\partial E}\left(m\mathbf{v} \cdot \mathbf{u} + \frac{2}{3}E_F \frac{\mathbf{u} \cdot \mathbf{q}}{\omega}\right) \tag{7}$$

where $f_0(\mathbf{v}, E_F)$ is the Fermi function for the equilibrium value of E_F, and the last term on the right-hand side represents the effects of the change in electron density due to the ultrasonic wave. It is to be noted that this term is zero for shear waves, for which **u** is perpendicular to **q**.

Using the kinetic method of Chambers (1952), or by solving the Boltzmann equation directly, it can then be shown that the change in distribution function contributing to the electron current is given by

$$\delta f(\mathbf{v}, \mathbf{r}, t) = -e\tau \frac{\partial f_0}{\partial E} \mathbf{v} \cdot \frac{\mathbf{E} + (m u/e\tau)[1 + (iav/3v_s)]}{(1 + i\omega\tau - i\mathbf{q} \cdot \mathbf{v}\tau)} \tag{8}$$

Here, τ is the assumed relaxation time and $a = ql/(1 + i\omega\tau)$, where $l = v\tau$ is the electron mean free path. Thus, if propagation is along the x direction, the components of electronic current are given by

$$
\begin{aligned}
(J_{el})_x &= \sigma_{xx}\{E_x + (m u_x/e\tau)[1 + (iav/3v_s)]\} \\
(J_{el})_y &= \sigma_{yy}\{E_y + (m u_y/e\tau)\} \\
(J_{el})_z &= \sigma_{zz}\{E_z + (m u_z/e\tau)\}
\end{aligned} \tag{9}
$$

where σ_{xx}, σ_{yy}, and σ_{zz} are the diagonal components of the conductivity tensor

$$\sigma_{ij} = -\frac{e^2}{4\pi^3} \int d\mathbf{k} \, \frac{v_i v_j}{1 + i\omega\tau - i\mathbf{q} \cdot \mathbf{v}\tau} \frac{\partial f_0}{\partial E} \tag{10}$$

From Eqs. (2)–(4), (6), and (9), we can solve for the attenuation to obtain, for longitudinal and transverse waves, respectively,

$$
\begin{aligned}
\alpha_l &= \frac{Nm}{\rho v_{sl} \tau} \operatorname{Re}\left(\frac{\sigma_0}{\sigma_{xx}} - 1 - \frac{a^2}{3}\right) \\
\alpha_t &= \frac{Nm}{\rho v_{st} \tau} \operatorname{Re}\left(\frac{\sigma_0}{\sigma_{yy}} - 1\right)
\end{aligned} \tag{11}
$$

where we now write $a = ql$, since $\omega\tau \ll 1$, and where $\sigma_0 = Ne^2\tau/m$ is the static conductivity.

In the limit $ql \gg 1$, we see from Eq. (10) that the only contribution to the conductivity comes from those electrons for which $\mathbf{q} \cdot \mathbf{v} = 0$. Thus, only those electrons lying on the effective zone with velocities perpendicular to the propagation direction contribute to the attenuation. This result can readily be understood in physical terms, since it is only the electrons lying on the effective zone which can stay in synchronism with the ultrasonic wave and can absorb energy from it.

For arbitrary ql, the integrals involved in evaluating the conductivity tensor can be solved, only the diagonal terms being nonzero. Since $\omega\tau \ll 1$, their real parts are given by

$$
\begin{aligned}
\sigma_{xx} &= \sigma_0 (3/a^3)(a - \tan^{-1} a) \\
\sigma_{yy} &= \sigma_{zz} = \sigma_0 (3/2a^3)[(1 + a^2)(\tan^{-1} a) - a]
\end{aligned} \tag{12}
$$

Substituting these results in Eq. (11), we finally obtain

$$
\begin{aligned}
\alpha_l &= \frac{Nm}{\rho v_{sl} \tau} \left[\frac{a^2 \tan^{-1} a}{3(a - \tan^{-1} a)} - 1\right] \\
\alpha_t &= \frac{Nm}{\rho v_{st} \tau} \left[\frac{2}{3} \frac{a^3}{(1 + a^2)(\tan^{-1} a) - a} - 1\right]
\end{aligned} \tag{13}
$$

For $a \ll 1$, as in the case of low frequencies or low purity, Eq. (13) can be expanded in powers of a to give

$$\alpha_1 = \tfrac{4}{15}(Nmv/\rho v_{s1})q^2l$$
$$\alpha_t = \tfrac{1}{5}(Nmv/\rho v_{s1})q^2l \tag{14}$$

Thus, in this limit, the attenuation varies linearly with the electron mean free path l and exhibits a quadratic dependence on the acoustic frequency. These results can, of course, be understood from purely classical arguments in which the electron gas is treated as a viscous elastic medium. For this case, the attenuation coefficient for longitudinal waves is given by

$$\alpha_1 = \tfrac{4}{3}(\eta/\rho v_{s1})q^2 \tag{15}$$

where η is the viscosity of the electron gas. Assuming a constant relaxation time, we can write for this viscosity

$$\eta = \tfrac{1}{5}Nmvl \tag{16}$$

which, on substitution in Eq. (15), gives the same result as before. It is also possible to derive Eqs. (14) by a relaxation argument, as shown by Akhiezer *et al.* (1957).

For $ql \gg 1$, Eqs. (14) can be expanded in powers of $1/a$ to give

$$\alpha_1 = \frac{\pi}{6} \frac{Nmv}{\rho v_{s1}} q = \frac{\pi^2}{3} \frac{Nmv}{\rho v_{s1}^2} f$$
$$\alpha_t = \frac{4}{3\pi} \frac{Nmv}{\rho v_{st}} q = \frac{8}{3} \frac{Nmv}{\rho v_{st}^2} f \tag{17}$$

From these expressions, we see that, in this limit, the attenuation is independent of the electron mean free path l and exhibits a linear variation with frequency. The value of α/f depends only on the Fermi momentum $p = mv$ and the appropriate elastic modulus ρv_s^2. Table I and Fig. 1 give the variation of α and α/a, normalized to their limiting high-frequency behavior, as functions of a.

As noted previously, the attenuation can also be calculated quantum-mechanically for the free-electron model. If we adopt a simple deformation potential to describe the electron–phonon interaction, the perturbing Hamiltonian for longitudinal waves is given by

$$H' = -i\tfrac{2}{3}E_F \sum_{\mathbf{k},\mathbf{q}} (\hbar/2\rho\omega_q)^{1/2}q(a_q c^*_{\mathbf{k}+\mathbf{q}}c_{\mathbf{k}} - a_q{}^* c^*_{\mathbf{k}-\mathbf{q}}c_{\mathbf{k}}) \tag{18}$$

where a_q and $a_q{}^*$ are the destruction and creation operators for phonons of wave vector $\mathbf{q}$, and $c_{\mathbf{k}}$ and $c_{\mathbf{k}}{}^*$ are the corresponding operators for an electron of wave vector $\mathbf{k}$. It is to be noted that, with this form of interaction, there is no attenuation for shear waves.

The probability per unit time that a phonon in state $\mathbf{q}$ will be absorbed in scattering an electron from $\mathbf{k}$ to $\mathbf{k} + \mathbf{q}$ is

$$w_- = (2\pi/\hbar)|\langle \mathbf{k}+\mathbf{q}, n_q - 1|H'|\mathbf{k}, n_q\rangle|^2 \, \delta(E_{\mathbf{k}} + \hbar\omega_q - E_{\mathbf{k}+\mathbf{q}}) \tag{19}$$

TABLE I

ACOUSTIC ATTENUATION FOR LONGITUDINAL
WAVES AS A FUNCTION OF $a = ql$ ACCORD-
ING TO THE FREE-ELECTRON MODEL

a	$\alpha/\lim_{a \to \infty}(\alpha/a)$	$(\alpha/a)/\lim_{a \to \infty}(\alpha/a)$
0.0	0.00000	0.0000
0.2	0.02166	0.1008
0.4	0.07838	0.1959
0.6	0.16893	0.2815
0.8	0.28491	0.3561
1.0	0.42004	0.4200
2.0	1.2478	0.6239
4.0	3.4102	0.7850
6.0	5.1020	0.8503
8.0	7.0827	0.8853
10.0	9.0711	0.9071
15.0	14.055	0.9370
20.0	19.048	0.9524
25.0	24.043	0.9617

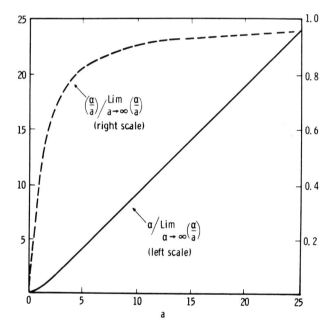

FIG. 1. Dependence of the normalized electronic attenuation α and α/a on the parameter $a = ql$, where q is the phonon wave number and l is the electronic mean free path.

On averaging over an electron ensemble in thermal equilibrium, we have

$$w_- = (4\pi E_F{}^2 q/9\rho v_s \hbar) n_q f_0(\mathbf{k})[1 - f_0(\mathbf{k} + \mathbf{q})]\, \delta(E_{\mathbf{k}} + \hbar\omega_{\mathbf{q}} - E_{\mathbf{k}+\mathbf{q}}) \quad (20)$$

Similarly, for phonon absorption, we find

$$w_+ = (4\pi E_F{}^2 q/9\rho v_s \hbar)(n_q + 1) f_0(\mathbf{k})[1 - f_0(\mathbf{k} - \mathbf{q})]\, \delta(E_{\mathbf{k}} - \hbar\omega_{\mathbf{q}} - E_{\mathbf{k}-\mathbf{q}}) \quad (21)$$

Since $q \ll k$, terms involving q^2 in the delta functions may be neglected, so that, allowing for both spin orientations, we obtain from Eqs. (20) and (21)

$$\alpha = (16\pi E_F{}^2 q/9\rho v_s{}^2\hbar) \sum_{\mathbf{k}} f_0(\mathbf{k})[\mathbf{q} \cdot \nabla_k\, f_0(\mathbf{k})]\, \delta[(\hbar^2/m)\mathbf{k} \cdot \mathbf{q} - \hbar\omega_{\mathbf{q}}] \quad (22)$$

The summation in Eq. (22) may be readily evaluated to give

$$\alpha_1 = (\pi/6)(Nmv/\rho v_s)q \quad (23)$$

which is identical to the first of Eqs. (17). It is to be noted that, since $v \gg v_s$, the delta function in Eq. (22) means that only those electrons for which $\mathbf{q} \cdot \mathbf{k} \approx 0$ contribute to the attenuation. Thus, as before, these electrons must lie on an effective zone perpendicular to the propagation direction.

B. REAL METALS

Akhiezer *et al.* (1957), Blount (1959), Pippard (1960), and Chambers (1961) have calculated the ultrasonic attenuation in a real metal, taking into account the effect of band structure. In the outline to be given here, we shall again adopt a quasiclassical approach, since it is simpler to understand physically. A quantum-mechanical treatment and a comparison with other solutions to this problem have recently been given by Smilowitz (1965).

The analysis is most conveniently carried out in a frame of reference moving with the local lattice velocity. In this system, the condition of quasi-current neutrality is given by the equation

$$\mathbf{J}_{\text{el}} = 0 \quad (24)$$

since the ionic current is, by hypothesis, zero. The electric field due to the sound wave must be calculated self-consistently so as to satisfy Eq. (24). However, since the frame of reference is no longer inertial, there are fictitious forces on an electron. The first of these simply results from the change in the effective wave-number vector $\mathbf{k}$ of an electron due to relative motion of various parts of the lattice. If it is assumed that this change occurs adiabatically with the strain due to the ultrasonic wave, then we have

$$\dot{k}_i = -\dot{\varpi}_{ij} k_j \quad (25)$$

Here, ϖ_{ij} is the unsymmetrized strain tensor with components $\partial \xi_j / \partial x_i$, where ξ_j is the displacement corresponding to the velocity given by Eq. (1). From the latter equation, we obtain

$$\dot{k}_i = i q_i u_j k_j \quad (26)$$

which is equivalent to a fictitious force acting on the electron

$$\Pi_i^{(1)} = \hbar \dot{k}_i = i\hbar q_i u_j k_j \tag{27}$$

In what follows, we shall only be concerned with the component of force normal to the Fermi surface, i.e., with $\Pi_i n_i$, where the n_i are the components of the normal. Thus, we have that, due to this effect, there is an effective force

$$\Pi^{(1)} = i\hbar q_i n_i u_j k_j \tag{28}$$

If the Fermi surface had the same shape at every point, $\Pi^{(1)}$ would describe how an electron originally on the Fermi surface would leave it as a consequence of the different relative velocities of the various regions. The deformed shape of the surface, and hence the current relative to the lattice, could then be easily calculated. However, in general, the Fermi surface changes shape when the lattice is strained. For a *static* strain given by ϖ_{ij}, the change can be specified by the equation

$$\Delta k_n = \varpi_{ij} K_{ij} \tag{29}$$

where Δk_n is the normal displacement of the Fermi surface and K_{ij} is a generalized deformation tensor, as defined by Pippard (1960). An electron moving through such a strain field is constrained to remain on the Fermi surface by forces which are instantaneously set up, and which are due, in part, to any space charge set up by the strain. In the moving strain field due to the sound wave, these forces are unable to accomplish this constraint in a complete fashion. If we consider an electron originally on the Fermi surface, then, at time dt later, the sound wave has moved a distance $v_s \, dt$, and the electron is now displaced from the Fermi surface by an amount corresponding to the differences of Δk_n computed for points in a statically strained lattice at a distance $v_s \, dt$ apart. Thus, from Eq. (29), we have

$$\Delta k_n = |\mathrm{grad}(\varpi_{ij} K_{ij})| v_s \tag{30}$$

which, from (1), reduces to

$$\Delta k_n = i q_i u_j K_{ij} \tag{31}$$

There is thus an effective force given by

$$\Pi^{(2)} = i\hbar q_i u_j K_{ij} \tag{32}$$

which, when combined with the other component, gives a total effective force on the electron

$$\Pi = \Pi^{(1)} + \Pi^{(2)} = i\hbar q_i u_j (K_{ij} + k_i n_j) \tag{33}$$

It is convenient to define a new tensor D_{ij} given by

$$D_{ij} = K_{ij} + k_i n_j \tag{34}$$

so that Eq. (33) may be written as

$$\Pi = i\hbar q_i u_j D_{ij} \tag{35}$$

The tensor D_{ij} is of fundamental importance in the theory of ultrasonic attenuation. From consideration of charge conservation, we have, for a static strain,

$$\varpi_{ij}\,\delta_{ij}\int d\mathbf{k} + \int \varpi_{ij} K_{ij}\,dS = 0 \tag{36}$$

so that

$$\int \varpi_{ij}(k_i n_j + K_{ij})\,dS = 0 \tag{37}$$

We thus obtain the condition on D_{ij}

$$\int D_{ij}\,dS = 0 \tag{38}$$

where the integration is over the free Fermi surface.

For propagation along the x direction, say, only the components ϖ_{xx}, ϖ_{xy}, and ϖ_{xz} of the tensor ϖ_{ij} do not vanish, and we may therefore specify the strain by a vector $\boldsymbol{\varpi}$. Correspondingly, the deformation tensor D_{ij} has only three nonvanishing components, which specify a deformation vector $\mathbf{D} = (D_{xx}, D_{xy}, D_{xz})$. Equation (33) may therefore be written in the more convenient form

$$\Pi = i\hbar q \mathbf{D} \cdot \mathbf{u} \tag{39}$$

where $\mathbf{D}$ satisfies the charge conservation condition

$$\int \mathbf{D}\,dS = 0 \tag{40}$$

By the introduction of the fictitious force Π, we have reduced the problem to a simple calculation in a static lattice. It is now necessary to determine the electric field $\mathbf{E}$ such that, under the combined influence of this field and the force Π, the total electronic current satisfies Eq. (24). Again by using the kinetic method or by solving directly the Boltzmann equation, we have for the change in distribution function due to Π

$$\delta f = -\tau v\,\frac{\partial f_0}{\partial E}\,\frac{\Pi}{1 - i\mathbf{q}\cdot\mathbf{v}\tau + i\omega\tau} \tag{41}$$

so that the related electron current is given by an integral over the Fermi surface

$$\mathbf{J} = \frac{e}{4\pi^3\hbar}\int \frac{\mathbf{v}\tau\Pi}{1 - i\mathbf{q}\cdot\mathbf{v}\tau + i\omega\tau}\,dS \tag{42}$$

Since $\omega\tau \ll 1$, we can write this equation in the form

$$\mathbf{J} = \frac{e}{4\pi^3\hbar} \int \frac{\Pi \mathbf{l}\, dS}{1 - ia\cos\varphi} \tag{43}$$

where $\mathbf{l} = \mathbf{v}\tau$ is the vector mean free path of an electron and φ is the angle between the propagation direction and the electron velocity. It is this current which is neutralized by the field $\mathbf{E}$. Thus, if ρ_{ij} is the appropriate resistivity tensor, Eqs. (39) and (43) give

$$E_j = \frac{e}{4\pi^3}\,\rho_{ij} \int \frac{a_i(\mathbf{D}\cdot\mathbf{u}\, a\cos\varphi)}{1 + a^2\cos^2\varphi}\, dS \tag{44}$$

using the fact that $\mathbf{D}(\mathbf{k}) = \mathbf{D}(-\mathbf{k})$. The component of $\mathbf{E}$ parallel to the electronic motion is $\mathbf{E}\cdot\mathbf{a}/a$, so that the total parallel force on an electron is

$$\Pi_{\text{total}} = i\hbar q\mathbf{D}\cdot\mathbf{u} + \frac{e^2}{4\pi^3 a}\,\rho_{ij}a_j \int \frac{a_i(\mathbf{D}\cdot\mathbf{u}\, a\cos\varphi)}{1 + a^2\cos^2\varphi}\, dS \tag{45}$$

To calculate the attenuation, we now calculate the excess energy due to the extra filled states resulting from the sound wave. For the element dS of the Fermi surface, there are $\Delta E\, dS/4\pi^3\hbar v$ additional states, where ΔE is given by

$$\Delta E = \Pi_{\text{total}}\, v\tau/(1 - i\mathbf{q}\cdot\mathbf{v}\tau + i\omega\tau) \tag{46}$$

The average energy exceeds the Fermi energy by $\Delta E/2$, so that the excess energy due to the element dS has a peak value $|\Delta E|^2\, dS/8\pi^3\hbar v$ and an average value one half this amount. It is this excess energy which is converted to heat with a time constant $\tau/2$, so that we have for the rate of heat production per unit volume

$$\dot{Q} = (1/8\pi^3\hbar) \int (|\Delta E|^2/l)\, dS \tag{47}$$

Thus, from Eq. (3), the attenuation coefficient can be written

$$\alpha = (1/4\pi^3\hbar\rho v_s u^2) \int (|\Delta E|^2/l)\, dS \tag{48}$$

Substituting for ΔE from Eq. (46) and again neglecting $\omega\tau$ in the denominator, we find

$$\alpha = \frac{1}{4\pi^3\hbar\rho v_s u^2} \int \frac{|\Pi_{\text{total}}|^2\, l\, dS}{1 + a^2\cos^2\varphi} \tag{49}$$

Since ρ_{ij} is real if $\omega\tau \ll 1$, Eqs. (45) and (49) may be combined to give

$$\alpha = \frac{\hbar q}{4\pi^3\rho v_s} \left[\int \frac{\mathscr{D}^2 a\, dS}{1 + a^2\cos^2\varphi} + \left(\frac{e^2}{4\pi^3\hbar q}\right)^2 \rho_{ij}\rho_{kl} I_i I_k A_{jl} \right] \tag{50}$$

Here, $\mathscr{D}$ is written for $\mathbf{D} \cdot \mathbf{u}/u$, while I_i and A_{jl} are integrals over the Fermi surface,

$$I_i = \int \frac{a_i(\mathscr{D}a \cos \varphi)}{1 + a^2 \cos^2 \varphi} \, dS, \qquad A_{jl} = \int \frac{a_j a_l \, dS}{a(1 + a^2 \cos^2 \varphi)} \tag{51}$$

It is easily shown that the conductivity tensor σ_{ij} is related to the integral A_{ij} by the equation

$$\sigma_{ij} = (e^2/4\pi^3\hbar q)A_{ij} \tag{52}$$

Thus, since $\sigma_{ij}\rho_{jk} = \delta_{ik}$, we may simplify Eq. (50) to obtain

$$\alpha = \frac{\hbar q}{4\pi^3\rho v_s} \left(\int \frac{\mathscr{D}^2 a \, dS}{1 + a^2 \cos^2 \varphi} + B_{ij} I_i I_j \right) \tag{53}$$

where B_{ij} is the inverse of A_{ij} and is equal to $e^2\rho_{ij}/4\pi^3\hbar q$.

1. Longitudinal Waves

For the case of pure longitudinal waves propagating along a direction of high symmetry, the foregoing expression may be simplified, since

$$\mathscr{D} = K_x + k_x \cos \varphi$$

$$I_x = \int \frac{\mathscr{D}a^2 \cos^2 \varphi \, dS}{1 + a^2 \cos^2 \varphi} = -\int \frac{\mathscr{D} \, dS}{1 + a^2 \cos^2 \varphi} \tag{54}$$

$$I_y = I_z = 0$$

where Eq. (40) has been used in deriving the expression for I_x. Thus, only the term involving $B_{xx} = A_{xx}^{-1}$ need be considered in evaluation of (53), which reduces to

$$\alpha_1 = \frac{\hbar q}{4\pi^3\rho v_s} \left\{ \int \frac{\mathscr{D}a^2 \, dS}{1 + a^2 \cos^2 \varphi} + \left[\left(\int \frac{\mathscr{D} \, dS}{1 + a^2 \cos^2 \varphi} \right)^2 \Big/ \int \frac{a \cos^2 \varphi}{1 + a^2 \cos^2 \varphi} \, dS \right] \right\} \tag{55}$$

where v_s is the longitudinal sound wave velocity.

If Eq. (55) is applied to the free-electron case, for which $K_x = -\frac{1}{3}k_0$, k_0 being the radius of the Fermi sphere, the integrals may be readily evaluated to give the same result as previously (cf. the first of Eqs. 13). For more general models, we note that the second term in (55) can be neglected in the limits $a \ll 1$ and $a \gg 1$. When $a \ll 1$, we thus have

$$\alpha_1 \approx (\hbar q^2/4\pi^3\rho v_s) \int \mathscr{D}^2 l \, dS \tag{56}$$

so that, just as in the free-electron model, the attenuation increases quadratically with frequency. It is to be noted that the above analysis is probably incorrect for a multiband metal, where interband relaxation effects could be important, as was indicated by Pippard (1960). Thus, in this case, the behavior of the attenuation at low values of ql cannot be used to infer the variation of l over the Fermi surface.

For $a \gg 1$, the presence of the term $a^2 \cos^2 \varphi$ in the denominator causes the first integral to be dominated by those regions of the Fermi surface for which $\cos \varphi$ is small. Thus, as in the free-electron case, the attenuation is governed by those electrons on the effective zone which move perpendicular to the propagation direction. The integral can then be written

$$\int \frac{\mathscr{D}^2 a \, dS}{1 + a^2 \cos^2 \varphi} \approx \int\int \frac{R K_x^2 a \, d\eta \, d\psi}{1 + a^2 \eta^2} \tag{57}$$

where $\eta = (\pi/2) - \varphi$, ψ is the azimuthal angle for the velocity as defined in Fig. 2, and R is the reciprocal of the Gaussian curvature of the Fermi surface, i.e., the product of the principal radii of curvature. On integrating over η, we obtain for $a \gg 1$

$$\alpha_1 \approx \frac{\hbar q}{4\pi^2 \rho v_s} \oint R K_x^2 \, d\psi = \frac{\hbar f}{2\pi \rho v_s^2} \oint R K_x^2 \, d\psi \tag{58}$$

where the line integral is taken around the effective zone. Thus, in the limit $a \gg 1$, the attenuation again varies linearly with frequency, just as in the

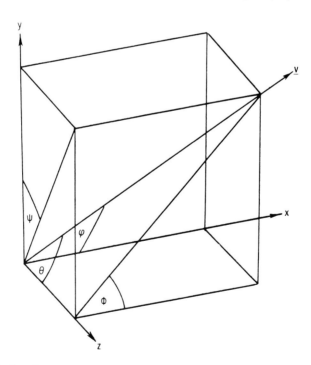

FIG. 2. Coordinate system used in the general discussion of the electronic attenuation in a metal with arbitrary Fermi surface.

free-electron model. The limiting value of α/f is independent of the mean free path l and involves only the elastic constant $\rho v_s{}^2$ for the propagation direction under consideration, together with an integral depending on the properties of the Fermi surface.[1] Thus, if the geometry of the latter is known, thereby determining R, measurements of the anisotropy of the electronic contribution to the attenuation should yield information about the deformation parameter K_x.

In this limit, the assumption of a relaxation time defined at all points of the Fermi surface is not at all crucial. For a model two-band Fermi surface consisting of spheres of radii k_1 and k_2, it has been shown by Pippard (1960) that cross-relaxation effects are negligible and that

$$\alpha_1 = (\hbar q/2\pi\rho v_s)(k_1{}^2 K_1{}^2 + k_2{}^2 K_2{}^2) \tag{59}$$

where K_1 and K_2 are the relevant deformation parameters. Thus, it is to be expected that Eq. (58) may be generalized to a multiband model

$$\alpha_1 \approx (\hbar q/4\pi^2 \rho v_s) \sum \oint R K_x{}^2 \, d\psi \tag{60}$$

where the sum extends over the effective zones on each sheet of the Fermi surface.

2. Shear Waves

For shear waves polarized along $0y$, Eq. (50) may again be simplified, since

$$\mathscr{D} = K_y + k_y \cos\varphi, \qquad I_x = I_z = 0$$

$$I_y = \int \frac{\mathscr{D}a^2 \sin\varphi \cos\varphi \cos\psi \, dS}{1 + a^2 \cos^2\varphi} \tag{61}$$

Thus, the only component of ρ_{ij} which need be evaluated is

$$\rho_{yy} = \frac{4\pi^3 \hbar q}{e^2} \bigg/ \int \frac{a \sin^2\varphi \cos^2\psi \, dS}{1 + a^2 \cos^2\varphi} \tag{62}$$

so that the attenuation is given by

$$\alpha_t = \frac{\hbar q}{4\pi^3 \rho v_s} \left\{ \int \frac{\mathscr{D}^2 a \, dS}{1 + a^2 \cos^2\varphi} + \left[\left(\int \frac{\mathscr{D}a^2 \sin\varphi \cos\varphi \cos\psi \, dS}{1 + a^2 \cos^2\varphi} \right)^2 \bigg/ \int \frac{a \sin^2\varphi \cos^2\psi \, dS}{1 + a^2 \cos^2\varphi} \right] \right\} \tag{63}$$

where v_s is the relevant shear-wave sound velocity.

[1] The transformation from a surface to a line integral in Eq. (58) is not valid when there is a flat spot (R infinite) on the effective zone. In this case α/f does not saturate and still varies as a.

For the free-electron model, $K_y = 0$, and Eq. (63) reduces to the same expression obtained previously. In the more general case, the second term can be neglected for $a \ll 1$ to give

$$a_t \approx (\hbar q^2 / 4\pi^3 \rho v_s) \int \mathscr{D}^2 l \, dS \tag{64}$$

Just as for longitudinal waves, this equation may be derived by a relaxation argument. The neglect of cross-relaxation effects again limit its validity to a single-band Fermi surface.

For $a \gg 1$, the second term in the above equation is not negligible, since, along a direction of high symmetry, $\mathscr{D}$ may vanish on the effective zone, and the first term becomes much smaller than that for a longitudinal wave. In this case, we may write Eq. (63) in the form

$$a_t \approx \frac{\hbar q}{4\pi^3 \rho v_s} \left\{ \int \frac{\mathscr{D}^2 \, dS}{a \cos^2 \varphi} + \frac{1}{\pi} \frac{(\int \mathscr{D} \tan \varphi \cos \psi \, dS)^2}{\oint R \cos^2 \psi \, d\psi} \right\} \tag{65}$$

Since the first term becomes negligible for $a \to \infty$, we then have

$$\alpha_t \approx \frac{\hbar q}{4\pi^4 \rho v_s} \frac{(\int \mathscr{D} \tan \varphi \cos \psi \, dS)^2}{\oint R \cos^2 \psi \, d\psi} \tag{66}$$

However, for some propagation directions (e.g., a trigonal axis), it is possible to have a pure shear mode, although the plane $k_x = 0$ is not one of reflection symmetry. In this case, the first term must be retained, and Eq. (63) gives

$$\alpha_t \approx \frac{\hbar q}{4\pi^2 \rho v_s} \left\{ \oint R K_y{}^2 \, d\psi + \frac{1}{\pi^2} \frac{(\int \mathscr{D} \tan \varphi \cos \psi \, dS)^2}{\oint R \cos^2 \psi \, d\psi} \right\} \tag{67}$$

3. *Arbitrary Direction of Propagation*

We shall now return to the general result given by Eq. (53) and examine its limiting form for $a \gg 1$. In this limit, some of the components may be expressed as line integrals around the effective zone, whose values are independent of a. The remainder are expressed as surface integrals tending to zero as $1/a$. Thus,

$$A_{xx} \approx \int dS/a, \qquad\qquad A_{xy} = A_{yx} \approx \int \tan \varphi \cos \psi \, dS/a$$

$$A_{yy} \approx \pi \oint R \cos^2 \psi \, d\psi, \qquad\qquad A_{xz} \approx \int \tan \varphi \sin \psi \, dS/a \tag{68}$$

$$A_{zz} \approx \pi \oint R \sin^2 \psi \, d\psi, \qquad\qquad A_{yz} = A_{zy} \approx \pi \oint R \cos \psi \sin \psi \, d\psi$$

The components of B_{ij}, retaining only the leading terms, are

$$B_{xx} = A_{xx}^{-1}, \qquad\qquad B_{xy} = B_{yx} = \frac{A_{xy}A_{yz} - A_{xy}A_{zz}}{A_{xx}(A_{yy}A_{zz} - A_{yz}^2)}$$

$$B_{yy} = \frac{A_{zz}}{A_{yy}A_{zz} - A_{zy}^2}, \qquad B_{xz} = B_{zx} = \frac{A_{xy}A_{yz} - A_{yy}A_{xz}}{A_{xx}(A_{yy}A_{zz} - A_{yz}^2)} \qquad (69)$$

$$B_{zz} = \frac{A_{yy}}{A_{yy}A_{zz} - A_{yz}^2}, \qquad B_{yz} = B_{zy} = \frac{-A_{yz}}{A_{yy}A_{zz} - A_{yz}^2}$$

so that all are independent of a, except B_{xx}, which is proportional to a.

Again using the charge conservation condition on $\mathscr{D}$, we have for the integrals I_i

$$I_x \approx -\pi \oint R\mathscr{D}\,d\psi/a, \qquad I_y \approx \int \mathscr{D}\tan\varphi\cos\psi\,dS, \qquad I_z \approx \int \mathscr{D}\tan\varphi\sin\psi\,dS \qquad (70)$$

It is thus clear that the terms in $B_{ij}I_iI_j$ which involve I_x all tend to zero as $1/a$, while the remainder tend to a constant limit. Thus, for $a \gg 1$, we have

$$\alpha \approx \frac{\hbar q}{4\pi^2\rho v_s}\left\{\oint R\mathscr{D}^2\,d\psi + \frac{A_{zz}I_y^2 - 2A_{yz}I_yI_z + A_{yy}I_z^2}{\pi(A_{yy}A_{zz} - A_{yz}^2)}\right\} \qquad (71)$$

so that α/f tends to a limiting value independent of the mean free path, just as found before for more special cases.

For an arbitrary propagation direction, pure longitudinal or pure transverse modes cannot be propagated. It is convenient to classify an elastic wave as a quasilongitudinal or quasitransverse mode, following Perz (1966), and to specify the angle Ω between the propagation direction Ox and the direction of particle motion given by the vector $\mathbf{u}$. If the y axis is chosen so that $\mathbf{u}$ is in the xy plane, then we have

$$\mathscr{D} = (K_x + k_x\cos\varphi)\cos\Omega + (K_y + k_y\cos\varphi)\sin\Omega \qquad (72)$$

whence, from Eq. (70),

$$I_y = \cos\Omega\int(K_x + k_x\cos\varphi)\tan\varphi\cos\psi\,dS$$

$$+ \sin\Omega\int(K_y + k_y\cos\varphi)\tan\varphi\cos\psi\,dS$$

$$I_z = \cos\Omega\int(K_x + k_x\cos\varphi)\tan\varphi\sin\psi\,dS \qquad (73)$$

$$+ \sin\Omega\int(K_y + k_y\cos\varphi)\tan\varphi\sin\psi\,dS$$

Thus, Eq. (71) may be written in the form

$$\alpha = (\hbar q/4\pi^2\rho v_s)(J_1 + J_2) \tag{74}$$

where J_1 is an effective zone integral given by

$$J_1 = \cos^2 \Omega \oint RK_x{}^2 \, d\psi + 2 \sin \Omega \cos \Omega \oint RK_x K_y \, d\psi + \sin^2 \Omega \oint RK_y{}^2 \, d\psi \tag{75}$$

The term J_2 is obtained by combining Eqs. (71) and (73) and may be written in terms of integrals over the Fermi surface

$$J_2 = L_1 \cos^2 \Omega + L_2 \cos \Omega \sin \Omega + L_3 \sin^2 \Omega \tag{76}$$

If the Fermi surface has reflection symmetry about xz and xy, then it is easily seen that $L_1 = L_2 = 0$, but that $L_3 \neq 0$. This result is also true if Ox is an axis of trigonal symmetry. In such cases, for pure longitudinal waves, $\Omega = 0$, $J_2 = 0$, and Eq. (74) becomes identical to Eq. (58). For pure transverse modes, $\Omega = \pi/2$, and we again recover Eq. (67). It is to be noted, however, that for almost pure quasilongitudinal modes (Ω small), the integral J_2 may still give a significant contribution to the attentuation if L_1 is large.

III. Theory of Attenuation in Superconductors

In the normal state, the attenuation for a superconductor behaves in exactly the same way as a normal metal. For the frequencies usually employed in ultrasonic experiments, the phonon energy is very much smaller than the BCS gap parameter Δ (i.e., $\hbar\omega/\Delta \ll 1$). Thus, at absolute zero, the attenuation goes to zero in the superconducting state, since there is insufficient energy to break any of the correlated electron pairs in the ground state.

To obtain an explicit expression for the attenuation in a superconductor, let us first consider the normal state. Following Eqs. (20) and (21), we can write, if the subscripts on the distribution functions are omitted,

$$\alpha_n = \int C_{\mathbf{kk'}} |M_{\mathbf{kk'}}|^2 \{f(E)[1 - f(E')] - f(E')[1 - f(E)]\} N(E)N(E') \, dE \, dS \tag{77}$$

where $N(E)$ and $N(E')$ give the densities of states at energies E and E', respectively, and where energy conservation requires that $E' = E + \hbar\omega$. The occupation factors take into account both absorption and induced emission of phonons of frequency ω. Both the matrix element $M_{\mathbf{kk'}}$ between initial and final states and the constant factor $C_{\mathbf{kk'}}$ involved in the transformation from a sum over $\mathbf{k}$, $\mathbf{k'}$ into the above integral are assumed to be independent of E. Thus, Eq. (77) can be separated in the form

$$\alpha_n = \int C_{\mathbf{kk'}} |M_{\mathbf{kk'}}|^2 \, dS \times \int [f(E) - f(E + \hbar\omega)] N(E)N(E + \hbar\omega) \, dE \tag{78}$$

so that, taking the energy zero at the Fermi level, we have

$$\alpha_n = \hbar\omega[N(0)]^2 \int C_{\mathbf{kk'}} |M_{\mathbf{kk'}}|^2 \, dS \tag{79}$$

In the superconducting state, it is assumed that the factors $C_{\mathbf{kk'}}$ and $M_{\mathbf{kk'}}$ remain unchanged from those in the normal state. The integral in Eq. (77) is, however, changed because of the presence of coherence effects involved in the scattering of an electron from a state $\mathbf{k}$ to a state $\mathbf{k'}$. Thus, in a normal metal, the transition from $\mathbf{k}, \sigma$ to $\mathbf{k'}, \sigma'$ is independent of the scattering from $-\mathbf{k'}, -\sigma'$ to $-\mathbf{k}, -\sigma$ as well as all other transitions. However, in the superconducting state, there are initial configurations in which, for example, the pair $\mathbf{k'}\uparrow, -\mathbf{k'}\downarrow$ is occupied as well as the single-particle state $\mathbf{k}\uparrow$. In this case, scattering of a particle from $-\mathbf{k'}\downarrow$ to $-\mathbf{k}\downarrow$ will give a final state in which there is a single excited electron in $\mathbf{k'}\uparrow$ and a ground-state pair $\mathbf{k}\uparrow, -\mathbf{k}\downarrow$. This scattering process is coherent with that in which the initial and final states only involve single-particle excitations $\mathbf{k}\uparrow$ and $\mathbf{k'}\uparrow$, respectively. It has been shown by Bardeen *et al.* (1957) that, for an ordinary potential interaction such as occurs in ultrasonic absorption, these processes interfere destructively and that the square of the matrix element occurring in Eq. (77) must be modified by inclusion of a coherence factor $[1 - (\Delta^2/EE')]$. If it is assumed that the gap parameter Δ is independent of position on the Fermi surface, then we can write, in analogy with Eq. (78),

$$\alpha_s = \int C_{\mathbf{kk'}} |M_{\mathbf{kk'}}|^2 \, dS \times \int [f(E) - f(E + \hbar\omega)] N(E) N(E')[1 - (\Delta^2/EE')] \, dE \tag{80}$$

where $N_s(E) = N(0)E/(E^2 - \Delta^2)^{1/2}$, and $E' = E + \hbar\omega$, as before.

Combining Eqs. (79) and (80), we then find for the ratio α_s/α_n

$$\frac{\alpha_s}{\alpha_n} = \frac{1}{\hbar\omega} \int \frac{[E(E + \hbar\omega) - \Delta^2][f(E) - f(E + \hbar\omega)]}{(E^2 - \Delta^2)^{1/2}[(E + \hbar\omega)^2 - \Delta^2]^{1/2}} \, dE \tag{81}$$

where the integration extends over all $|E| \geq \Delta$. For the case $\hbar\omega/\Delta \ll 1$, as is usually the case, the coherence factor in the numerator is canceled by the denominator, and the ratio α_s/α_n becomes

$$\alpha_s/\alpha_n = (1/\hbar\omega) \int [f(E) - f(E + \hbar\omega)] \, dE = \int (-\partial f/\partial E) \, dE \tag{82}$$

Thus, we have

$$\alpha_s/\alpha_n = 1 - f(-\Delta) + f(\Delta) = 2f(\Delta) \tag{83}$$

where $\Delta = \Delta(T)$ can be obtained from tables in the literature (for example, Mühlschegel, 1959). It is of interest to note that α_s/α_n drops with vertical tangent at T_c, because the gap is also changing with infinite slope at this

point. This behavior contrasts with the steep, but nonvertical, drop that
would be predicted by the simple two-fluid model for the superconducting
state developed by Gorter and Casimir (1934). Since it is only the normal
electrons that can absorb energy from low-frequency sound waves, α_s/α_n
should fall as $(T/T_c)^4$ according to the latter model.

For arbitrary values of phonon energy, Eq. (81) has been solved numerically by Bobetic (1964). The resulting curves of α_s/α_n versus reduced
temperature $t = T/T_c$, for various values of $\hbar\omega$ expressed as a function of
$\Delta(0) = 1.75\ k_B T_c$, are shown in Fig. 3. It should be noted that, for finite values
of $\hbar\omega/\Delta(0)$, there is a discontinuity in attenuation at the temperature T_d for
which the phonon energy equals the gap energy $2\Delta(T)$. The magnitude of this
discontinuity was first shown by Privorotskii (1962) to be

$$\delta(\alpha_s/\alpha_n) = (\pi/2)\tanh(\hbar\omega/4k_B\,T_d) \tag{84}$$

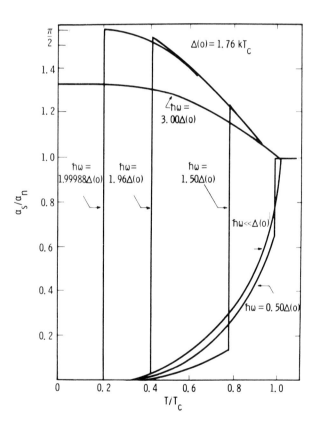

Fig. 3. Plot of normalized attenuation in the superconducting state α_s/α_n as a
function of reduced temperature $t = T/T_c$ for various values of phonon energy. The
phonon energy $\hbar\omega$ is expressed in terms of $\Delta(0)$, the gap parameter at absolute zero.

For low phonon frequencies, the discontinuity is very small and occurs so close to the transition temperature that it would not be observable experimentally. Another feature of the curves is the initial rise in α_s/α_n below the transition temperature. For $\hbar\omega > 2\Delta(0)$, that is, for phonon frequencies sufficient to span the energy gap at absolute zero, the ratio α_s/α_n is always greater than unity.

The above analysis is based on the original BCS theory of superconductivity, in which a constant effective electron–electron interaction and a spherical Fermi surface are assumed. Thus, the energy-gap parameter Δ is taken to be isotropic. An extension of the theory by Pokrovskii (1961) and Privorostskii (1962), taking into account gap anisotropy, gives the expression:

$$\alpha_s/p\alpha_n = 2f(\Delta_{min}) \tag{85}$$

where Δ_{min} is the minimum value of the gap on the effective zone, defined, as before, by the condition $\mathbf{q} \cdot \mathbf{v} = 0$. If the anisotropy on the effective zone is small, the weighing factor p changes little with temperature and is close to unity. For a strongly coupled superconductor with large gap anisotropy, the weighing factor can be written

$$p \sim (t/\delta\Delta)^{1/2} \tag{86}$$

where again t is the reduced temperature and $\delta\Delta = \Delta_{max} - \Delta_{min}$ is the change of Δ on the effective zone. The criterion for strong coupling is determined by the inequality $e^{\delta\Delta/t} \gg 1$, so that, in this case, we may write

$$\frac{\alpha_s}{\alpha_n} \sim \left(\frac{t}{\delta\Delta}\right)^{1/2} \exp\left\{-\frac{\Delta_{min}(0)}{t}f(t)\right\} \tag{87}$$

where $f(t) = \Delta(T)/\Delta(0)$ takes into account the temperature variation of $\Delta(T)$ and, with certain reservations, can be taken from the BCS theory.

Markowitz and Kadanoff (1963) and Clem (1966) have also considered the effects of gap anisotropy, assuming a matrix element for the attractive interaction of the form

$$V_{kk'} = \begin{cases} -(1 + a_k)V(1 + a_{k'}), & |\varepsilon_k|, |\varepsilon_{k'}| < \hbar\omega_D \\ 0, & \text{otherwise} \end{cases} \tag{88}$$

Here, ε_k is the Bloch energy of the state $\mathbf{k}$ referred to the Fermi level, ω_D is the Debye frequency, V is a positive interaction parameter, and a_k is a small number which depends only on direction and which is zero for the original BCS theory. For this form of the interaction, the energy-gap parameter exhibits an anisotropy of the form

$$\Delta_k(T) = \bar{\Delta}(T)(1 + a_k) \tag{89}$$

where $\bar{\Delta}(T)$ is the average of $\Delta_k(T)$ taken over the Fermi surface. The resulting form of the temperature dependence of the gap parameter is shown in Fig. 4 for a typical value of $\langle a^2 \rangle_{av} = 0.04$. Table II gives the values of

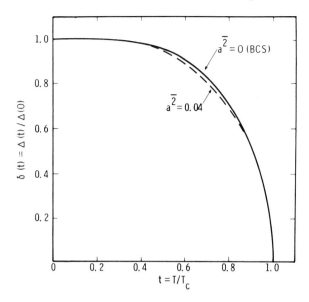

Fig. 4. Plot of reduced energy gap $\delta(t) = \Delta(t)/\Delta(0)$ as a function of reduced temperature for an anisotropic superconductor with $\langle a^2 \rangle_{av} = 0.04$. The full line gives the BCS relation with $\langle a^2 \rangle_{av} = 0$.

$\langle a^2 \rangle_{av}$ for various elements obtained from the variation of transition temperature with residual resistivity.

TABLE II

ANISOTROPY PARAMETER $\langle a^2 \rangle_{av}$ OF VARIOUS ELEMENTS
OBTAINED FROM THE VARIATION OF
TRANSITION TEMPERATURE WITH RESIDUAL RESISTIVITY

Element	$\langle a^2 \rangle_{av}$	Reference
V	0.016	Radebaugh and Keesom (1966)
Sn	0.019	Markowitz and Kadanoff (1963)
In	0.021	Markowitz and Kadanoff (1963)
Al	0.011	Markowitz and Kadanoff (1963)
Zn	0.047	Farrell *et al.* (1964)
Ta	0.011	Seraphim *et al.* (1961)

Quite apart from these considerations, Eq. (83) is only valid for longitudinal waves propagating along directions of high symmetry. For arbitrary polarization and propagation direction, the Meissner effect causes $M_{kk'}$ in the superconducting state to be different from that in the normal state, thus invalidating the analysis. To find the attenuation, however, it is not necessary to evaluate the integrals directly. Since $ql \gg 1$, they can be obtained from a modification of the treatment in Section IIB. In particular, the normal-state attenuation must be given by Eq. (74), i.e.,

$$\alpha_n = (\hbar q / 4\pi^2 \rho v_s)(J_1 + J_2) \tag{90}$$

where J_1 and J_2 are defined by Eqs. (75) and (76). In the superconducting state, the electromagnetic reactions normally responsible for transverse current neutrality are screened when the penetration depth is small compared to the sound wavelength, i.e., for all temperatures except those very near the transition temperature. Longitudinal current neutrality is, however, still assured by the more stringent requirement of local charge neutrality. Thus, associated with the sound wave there is now only a longitudinal electric field E_x given by Eq. (44) with the index i equal to x. Correspondingly, the superconductivity attentuation is now given by

$$\alpha_s = \alpha_s{}^b[2f(\Delta)] \tag{91}$$

where the basic attenuation corresponding to the superconducting matrix element $M_{\mathbf{kk}'}$ is given by Eq. (50) with the sum restricted to a single term, i.e.,

$$\alpha_s{}^b = \frac{\hbar q}{4\pi^3 \rho v_s} \left\{ \int \frac{\mathscr{D}^2 a \, dS}{1 + a^2 \cos^2 \varphi} + \left(\frac{e^2}{4\pi^3 \hbar q} \right)^2 \rho_{xx}^2 I_x{}^2 A_{xx} \right\} \tag{92}$$

For the limit $a \gg 1$, the second term becomes negligible and the first can be expressed as an effective zone integral,

$$\alpha_s{}^b = \frac{\hbar q}{4\pi^2 \rho v_s} \int R \mathscr{D}^2 \, d\psi = \frac{\hbar q}{4\pi^2 \rho v_s} J_1 \tag{93}$$

so that, from Eq. (91),

$$\alpha_s = (\hbar q / 4\pi^2 \rho v_s) J_1 [2f(\Delta)] \tag{94}$$

At the transition temperature, there is a discontinuity in $\alpha_n - \alpha_s$ associated with the J_2 integral of Eq. (90). In the limit $a \gg 1$, the residual superconducting attenuation below T_c given by Eq. (93) has the same frequency dependence as α_n. Thus, the ratio α_s / α_n is independent of frequency for arbitrary polarization and propagation direction. For the special case of shear waves, this result has been derived independently by Leibowitz (1964b). Since Ω is equal to $\pi/2$ for shear waves, the integral J_1 is given by

$$J_1 = \oint R K_y{}^2 \, d\psi \tag{95}$$

where K_y is the shear deformation constant. Clearly, the resulting attenuation α_s is zero for a free-electron model ($K_y = 0$), so that we can regard the residual attenuation for shear waves as a "real-metal" effect (Leibowitz, 1954b), which is characteristic of a Fermi surface with a nonvanishing shear deformation contribution.

When the condition $a \gg 1$ is not fulfilled, the second term in Eq. (92) leads to a significant contribution to the residual shear-wave attenuation in the superconducting state. This is the so-called collision-drag term discussed originally by Holstein (1959) and later by Claiborne and Morse (1964). For a spherical Fermi surface, Leibowitz (1964b) has shown that the second term of Eq. (92) in this case reduces to their result, namely, the collision-drag attenuation α_C is given by

$$\alpha_C = g\alpha_{nt} \tag{96}$$

where $g = (3/2a^2)\{[(1 + a^2)/a] \tan^{-1} a - 1\}$. This contribution vanishes in the limit of $a \gg 1$. Thus, if, in general, we write

$$\alpha_R = \alpha_D + \alpha_C \tag{97}$$

and if g is assumed to be independent of the Fermi-surface topology, we have

$$\alpha_R = \alpha_D + g\alpha_{nt} \tag{98}$$

Hence, a plot of α_R/α_{nt} versus g should be a straight line, with intercept equal to the value of α_D/α_{nt}.

As noted previously, the rapid fall of the shear-wave attenuation at the transition temperature is due to the loss of electromagnetic coupling between the electrons and the sound wave in the superconducting state. The temperature dependence of this rapid-fall region provides information about the penetration depth in the superconductor. Thus, in general, the screening is describable in terms of an effective permeability

$$1/\mu = 1 - (4\pi i/c^2 q^2)\omega\sigma(\omega, q) \tag{99}$$

where $\sigma(\omega, q)$ is the frequency- and wave number-dependent conductivity. For frequencies very much below the gap value, it is readily shown that the permeability is given by the simple static approximation

$$1/\mu = 1 + [1/q\lambda(T)]^2 \tag{100}$$

where the penetration depth $\lambda(T)$ has the local value near T_c, namely,

$$\frac{1}{\lambda(T)} = \frac{1}{\lambda(0)}\left[2\left(1 - \frac{T}{T_c}\right)\right]^{1/2} \tag{101}$$

$\lambda(0)$ being the London penetration depth. The corresponding attenuation is $\sigma_N \mu^2$, where σ_N is the static normal-state conductivity. Fossheim (1967) has shown that $\lambda(0)$ can be determined from the temperature range ΔT, corresponding to the region of rapid fall in the superconducting shear-wave attenuation at low frequencies.

For higher frequencies, which are still small compared to the gap parameter, there are nonlocal deviations from this temperature dependence, and the ultrasonic measurements provide a measurement of $\lambda(q)$. Cullen and Ferrell (1966) have shown that, in this limit, the attenuation is given by the product $\sigma_1 \mu^2$, where σ_1, the real part of the conductivity, can be written

$$(\sigma_1/\sigma_N) - 1 \approx Ax \log Bx \tag{102}$$

In this equation, the parameter x is given by Δ/Δ_q, where Δ_q is the energy gap at a temperature for which the permeability has the value $1/2$. The dimensionless parameters A and B are expressible in terms of Δ_q,

$$A = \frac{\Delta_q}{2k_B T_c}; \qquad B = \frac{8\Delta_q}{e\hbar\omega} = 1.15 \left(\frac{\lambda(0)}{\xi_0}\right)\left(\frac{v_F}{v_s}\right) \qquad (103)$$

where e is the natural base of logarithms and ξ_0 is the coherence distance. Clearly, B is a constant for a given material, independent of frequency and temperature. At a sufficiently high frequency, the rise in σ_1 below the transition temperature can predominate over the onset of screening. A corresponding "rapid-rise" behavior of the attenuation, similar to the rise in nuclear spin relaxation rate, would then occur just below the transition temperature. For lead, Cullen and Ferrell estimate that $A \log B = 0.73$ and that there is a maximum in α_s/α_n for $x \approx 1/3$, corresponding to a frequency of 2.7 GHz. The predicted form of the shear-wave attenuation for this case is shown in Fig. 5. Graphs giving the frequency dependence of the shear-wave acoustic attenuation in aluminum, indium, tin, and lead have recently been published by Fossheim and Torvatn (1969).

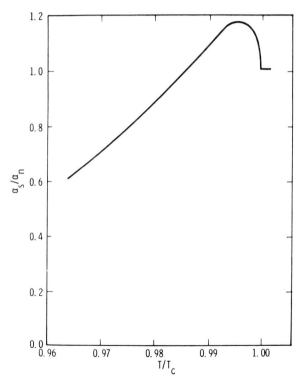

FIG. 5. Predicted transverse ultrasonic attenuation versus reduced temperature for lead at 2.7 GHz (after Cullen and Ferrel, 1966).

The quantum-mechanical arguments leading to Eq. (83) involve only the scattering of longitudinal ultrasonic phonons by thermally excited quasiparticles, so that the theory as originally proposed is for the regime $ql \gg 1$. More recently, it has been shown by Tsuneto (1961) and Kadanoff and Pippard (1966) that Eq. (83) is also valid for arbitrary values of ql in the case of longitudinal waves. The semiclassical treatment by the latter authors shows that this result only holds if the mean free path remains constant through the superconducting–normal transition. Anisotropy of the energy gap might cause this constancy to break down, leading to departures from the BCS relation when $ql \ll 1$. For transverse waves, the BCS relation holds for the residual attenuation provided that $\hbar\omega \ll k_B T_c$.

In superconductors containing localized magnetic impurities or in which other mechanisms break time-reversal symmetry, the behavior of α_s/α_n is materially changed from that characteristic of pure superconductors (Kadanoff and Falko, 1964; Griffin and Ambegaokar, 1965; Nam, 1967; Snow, 1968). It may be shown that, for this case, Eq. (81) is replaced by

$$\alpha_s/\alpha_n = (1/\bar{\omega}) \int_{-\infty}^{\infty} [f(\omega) - f(\omega + \bar{\omega})] N(\omega) N(\omega + \bar{\omega}) C(\omega, \bar{\omega}) \, d\omega \quad (104)$$

where $\bar{\omega}$ is now written for the phonon energy (i.e., the system of units is chosen so that $\hbar = 1$), and ω denotes the electron energy referred to the Fermi level. As before, $N(\omega)$ is the quasiparticle density of states given by

$$N(\omega) = \text{Im}\{u(\omega)/[1 - u^2(\omega)]^{1/2}\} \quad (105)$$

where $u(\omega)$ is the fundamental parameter occurring in the Abrikosov–Gor'kov model of superconductors containing paramagnetic impurities. If Γ_s is the level broadening due to these centers, then $u(z)$ is the solution of the equation

$$\frac{z}{\Delta} = u(z) \left[1 - \frac{i\Gamma_s/\Delta}{[u^2(z) - 1]^{1/2}} \right] \quad (106)$$

Finally, $C(\omega, \bar{\omega})$ is a direct generalization of the BCS coherence factor introduced previously, and can be written

$$C(\omega, \bar{\omega}) = 1 - \frac{\omega(\omega + \bar{\omega})}{\Delta^2 |u(\omega)|^2 |u(\omega + \bar{\omega})|^2} \quad (107)$$

For the limit of a pure superconductor, $\Gamma_s/\Delta \to 0$, $u \to \omega/\Delta$, and Eq. (104) reverts once more (except for changes in units and notation) to Eq. (81). In general, it may be solved numerically as a function of the ratio of the impurity concentration n_i to n_{cr}, the critical concentration which destroys superconductivity at absolute zero. At low frequencies, where the phonon energy is very much less than the gap parameter Δ, the behavior is as shown in Fig. 6. Clearly, there is a pronounced deviation from the BCS prediction; of particular interest is the finite slope of α_s/α_n at the critical temperature for finite n_i. For higher frequencies, when the phonon energy is comparable to the gap parameter, the addition of magnetic impurities smears out the absorption edge, as shown in Fig. 7. No experimental verification of either

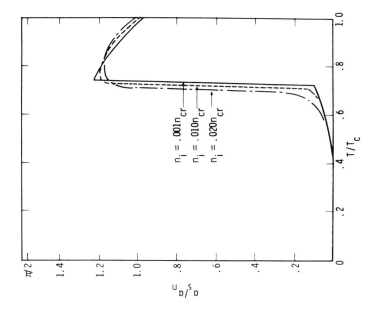

FIG. 7. Variation of normalized attenuation α_s/α_n with reduced temperature as a function of n_i for $\hbar\omega \approx \Delta$ (after Griffin and Ambegaokar, 1965).

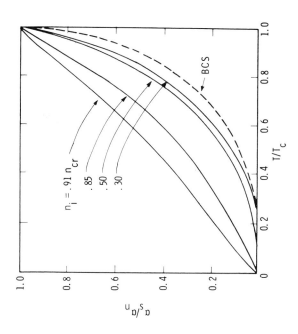

FIG. 6. Variation of normalized attenuation α_s/α_n with reduced temperature as a function of magnetic impurity concentration n_i for $\hbar\omega \ll \Delta$ (after Griffin and Ambegaokar, 1965).

of these predictions has thus far been obtained. Nam (1967) has derived a formula similar to (104) for the ratio α_s/α_n of a strong coupling superconductor. Strong coupling effects on the superconducting longitudinal attenuation in the low-frequency limit have been treated by Woo (1967, 1968).

IV. Measuring Techniques

Most measurements of the electronic contribution to the attenuation in metals have been made using the pulse technique, although a CW method has sometimes been used for semimetals (Reneker, 1959). A typical set up is shown schematically in Fig. 8. Radiofrequency pulses on the order of 1 μsec in duration and several hundred volts in amplitude are generated by a suitable oscillator, which is triggered from some form of synchronizing unit. The

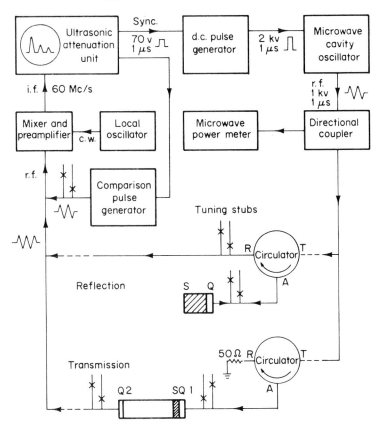

Fig. 8. Schematic arrangement of electronic equipment for ultrasonic attenuation measurements in the frequency range 250 MHz to 1.3 GHz: Q, quartz transducer; Q1, transmitter; Q2, receiver quartz; S, metal sample (after Perz and Dobbs, 1967).

pulses pass through suitable matching stubs to a transducer bonded to the sample. At frequencies below 500 MHz, the transducer is usually a suitably oriented quartz crystal operating in an overtone mode. Nonaq[2] or a high-viscosity Dow-Corning silicone fluid are commonly used as bonding agents. Above 1 GHz, evaporated cadmium sulfide transducers operating in their fundamental mode, such as developed by de Klerk and Kelly (1965), have been used with considerable success. For very high frequencies (~ 10 GHz), the usual form of transducer is an X-cut rod, excited in a nonresonant surface mode. Nonaq, as well as indium films, have also been used as bonds in this frequency range.

Measurements can be made either in reflection or transmission. In the former case, the same transducer serves for both transmitting and receiving the acoustic pulses. With this arrangement, it is necessary to use some form of circulator to protect the receiver from overload due to the transmitter pulse. Even with the transmission technique, however, receiver blanking is desirable to prevent its paralysis by unavoidable leakage signal from the transmitter. The effects of leakage can also be minimized by the use of a buffer rod of Z-cut quartz between the sample and transmitting crystal, so as to provide an acoustic delay for the signal without introducing appreciable additional attenuation.

Since the specimen attenuation is usually very large except at the very lowest frequencies, the acoustic echo train consists generally of a single pulse. Thus, the use of a calibrated exponential matching waveform as discussed by Morse (1959) is not practicable in most cases. The usual method of measurement involves a pulse-comparison technique to measure the relative change in the specimen attenuation as some external variable such as temperature or magnetic field is varied. In this technique, a pulse of the same frequency as the acoustic signal is injected into the input of the receiver through a suitable attenuator and isolator. The comparison oscillator and main transmitter are triggered on alternate synchronizing cycles, so that, by a suitable delay, the acoustic and comparison pulses may be brought into close proximity on the display oscilloscope. If the attenuator is adjusted so that the two pulses are always of equal amplitude, it is clear that changes in the attenuator setting will be identical to changes in the specimen attenuation, independent of the receiver linearity. Attenuation changes may be recorded automatically by the use of a boxcar integrator or some form of pulse sampler, such as employed by Toxen and Tansal (1965). These devices may be conveniently calibrated by the attenuator in the comparison circuit. It should be noted that attenuation changes have also been measured by an insertion technique, using an attenuator in the main receiver line.

At frequencies below 500 MHz, the design of sample holder is not critical, a typical unit employed by Perz and Dobbs (1966) for transmission measurements being shown in Fig. 9. This assembly is contained in a vacuum

---

[2] A water-soluble stopcock grease manufactured by Fisher Sci. Corp., Pittsburgh, Pennsylvania.

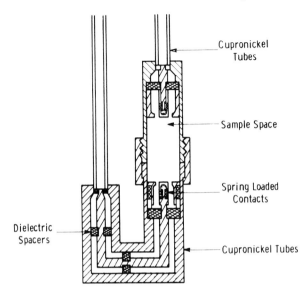

Cupronickel
Tubes

Sample Space

Spring Loaded
Contacts

Dielectric
Spacers

Cupronickel Tubes

FIG. 9. Sample holder for transmission measurements below 1.5 GHz (after Perz and Dobbs, 1967).

can, which is filled with a suitable pressure of exchange gas for thermal contact with the surrounding cryogenic fluid. By means of a heater and electronic controller, the temperature of the sample holder can be varied over a wide range. Matching stubs external to the cryostat are normally employed, since signal to noise is rarely a problem at these frequencies. Typical sample thicknesses are on the order of several millimeters for most metals. For frequencies above 1 GHz, the sample holder is some form of resonant cavity with provision for remote tuning in the cryostat itself. Figure 10 shows an ultrasonic probe for transmission measurements at 9 GHz down to 0.3°K, as used by Fagen (1967). At this frequency, the sample thickness is only a few mils, since the electronic attenuation is extremely large. The usual specimen configuration consists of the sample cemented between two X-cut rods, which act as the transmitting and receiving transducers, respectively, and which are excited by surface modes.

A typical form of the resulting data for lead, obtained by Love and Shaw (1964), is shown in Fig. 11. Above about 12°K, the acoustic phonons interact with the thermal phonons and dislocations in the metal, giving a monotonic increase of relative attenuation with temperature. Below 12°K, the attenuation again rises, as the electron mean free path becomes appreciable compared with the sound wavelength. This increase continues until the condition $ql \gg 1$ is satisfied, at which point there is saturation in the normal-state attenuation. For the superconducting state, the attenuation decreases rapidly below the transition temperature and reaches a constant limit at absolute zero. If it is assumed that this residual value is independent of any

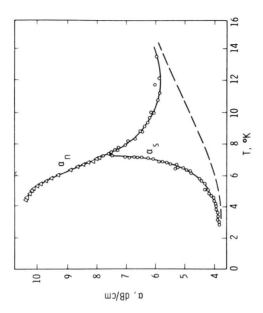

FIG. 11. Ultrasonic attenuation in lead for longitudinal waves propagating along [100] at 50 MHz. The dashed line is the background attenuation excluding the electronic component (after Love and Shaw, 1964).

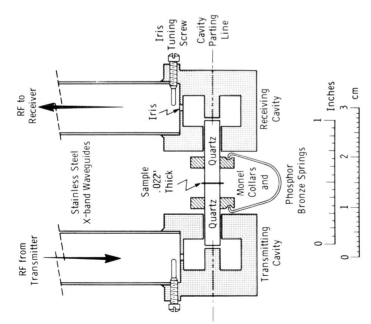

FIG. 10. Basic arrangement for hypersonic transmission measurements at 9.3 GHz (after Fagen, 1967).

electronic effects, it follows that the value of $\alpha_n - \alpha_s$ extrapolated to absolute zero gives directly the electronic contribution to the sample attenuation.

In practice, this method of determining the electronic attenuation is subject to a number of difficulties. The first concerns the existence of an amplitude dependence of both α_n and α_s, owing to an interaction between the acoustic waves and the dislocations in the metal (e.g., Tittman and Bömmel 1966). Under the influence of the sound wave, these dislocation networks vibrate like elastic strings between their pinning points and are subject to viscous forces due to the conduction electrons.[3] The resulting energy dissipation manifests itself as an amplitude-dependent attenuation, as can be seen from Fig. 12. Although these effects are not fully understood quantitatively, they can, for the most part, be eliminated by making measurements as a function of transmitter pulse height and extrapolating to zero amplitude. It has been pointed out by Mason (1966), however, that there is still a dislocation contribution to α_s even at zero amplitude. The magnitude of this effect is open to some question, and there is evidence that, at least in some

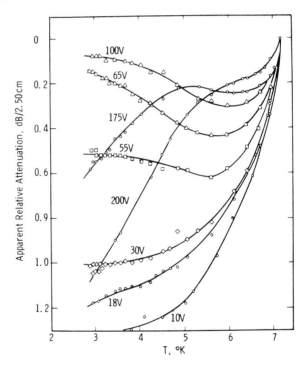

Fig. 12. Apparent relative attenuation for lead in superconducting state as a function of temperature and voltage across transducer (after Love and Shaw, 1964).

[3] This viewpoint is, strictly speaking, only valid at low frequencies. For a more rigorous discussion of the effect, see Tittman and Bömmel (1966).

cases, it is fairly small compared to the total electronic attenuation. Thus, in Fig. 11, the extrapolated high-temperature behavior shown by the dashed line, which represents the expected background attenuation in the absence of electronic effects, agrees well with the limiting attenuation in the super-conducting state. This agreement suggests that, at least in this case, the zero-amplitude dislocation contribution to α_s is quite small.

The remaining difficulty in this technique results from the use of the quenching magnetic field used to obtain α_n. For the high-purity samples normally employed, there is a change in attenuation due to the magneto-acoustic effect discussed by Peverly (1966) and Roberts (1968) in an earlier volume in this series. In most cases, however, this change is not very large and can be corrected by measuring $\alpha_n(H)$ and extrapolating to zero field. Figure 13 shows typical data obtained by this method for indium samples of several different purities. For the highest-purity material, in which the condition $ql \gg 1$ is satisfied for all frequencies of measurement, the electronic attenuation clearly exhibits a linear dependence on frequency, in agreement with Eq. (58). As the specimen purity decreases, the region of linearity also decreases, although, clearly, the limiting slope $d\alpha/df$ is unaffected by the mean free path. Finally, for the lowest-purity material, in which $ql \ll 1$, there is a quadratic dependence of α on frequency. Thus, in all cases, the general predictions of the theory are verified.

For materials which are not superconductors or in which the transition temperature is too low to be readily accessible by conventional cryogenic techniques, the electronic contribution to the attenuation cannot be measured

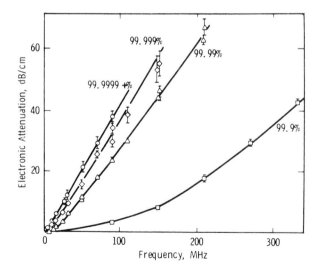

FIG. 13. Electronic component of normal state attenuation in indium as a function of frequency for different specimen purities. The propagation direction is normal to (111) (after Bliss and Rayne, 1969).

in the way just described. It is still possible, however, to measure the relative changes in the total attenuation of the sample as a function of temperature. The electronic contribution to the attenuation is given by the difference between the total sample attenuation extrapolated from high temperatures and that measured at absolute zero. In the case of superconductors, the value of electronic attenuation obtained in this way (cf. Fig. 11) agrees well with that computed from the difference $\alpha_n - \alpha_s$, so that there is reasonable certainty as to the validity of the procedure in general. It should be noted, however, that the extrapolation of the background attenuation does present difficulties if there is a significant amplitude-dependent attenuation due to dislocations. Empirical correction procedures have been devised in such cases [see, for example, Macfarlane and Rayne (1967)], but there seems little doubt that the use of neutron or γ-ray irradiation to pin the dislocation loops, and hence to reduce the amplitude effect, is a better solution. Figure 14 shows the effects of γ-irradiation in the measurements on copper by Wang *et al.* (1968). When buffer rods are used with this technique, allowance must be made for the temperature dependence of the attenuation in the quartz. The correction procedure is quite simple, however, since the attenuation is small and the temperature dependence is accurately known.

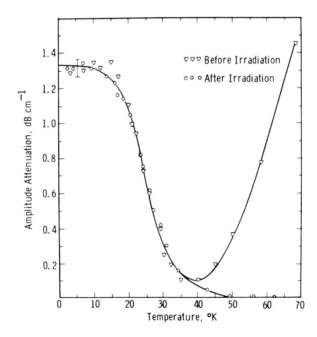

Fig. 14. Effect of γ-irradiation on the temperature dependence of the attenuation of longitudinal waves propagating along [111] in copper at 30 MHz (after Wang *et al.*, 1968).

V. Results

A. Normal Metals and Normal State of Superconductors

Ideally, the alkali metals, in particular potassium, should provide the best test of the theory developed previously. Shoenberg and Stiles (1964) have shown that the Fermi surface of potassium deviates only very slightly from sphericity, and hence it would be expected that the electronic attenuation would be predicted accurately by the free-electron equations (17). Unfortunately, there are no experimental data for this metal, even though it can be prepared in high-purity single crystals, which are not too difficult to handle. The noble metals are the next most favorable possibilities, since their Fermi surfaces have been determined with great accuracy by Shoenberg and Roaf (1962). Moreover, they are monovalent, with conduction bands lying in a single zone, so there is no possibility of complications due to interband transitions. In the following section, a detailed comparison of theory and experiment for these metals will be given. This will be followed by a similar comparison for more complicated cases; namely, polyvalent and transition metals with Fermi surfaces lying in several zones, and, finally, the semimetal bismuth.

1. *Noble Metals*

The electronic contribution to the longitudinal-wave acoustic attenuation in the noble metals has been measured up to 300 mHz, corresponding to values of ql in excess of thirty for the case of copper, by Macfarlane and Rayne (1967). Low-frequency data (Wang *et al.*, 1968) for copper have also been obtained covering the range $ql \lesssim 1$. All the latter measurements appear to be consistent with the free-electron equation (13), the values of mean free path l required to fit this expression being independent of propagation direction for a given crystal; the fit to an f^2 dependence is quite good. The temperature dependence of the mean free path has also been deduced from these data, using Eq. (13) and assuming Matthiessen's rule

$$1/l = (1/l_0) + [1/l(T)] \tag{108}$$

where l_0 is the mean free path associated with impurity scattering. It is found that $l(T)$ for phonon scattering varies as T^{-5} up to 40°K, in agreement with the Bloch–Grüneisen law. There is, however, considerable discrepancy between these values of $l(T)$ and the corresponding values obtained from electrical conductivity data. However, this discrepancy is not surprising, since, by Eq. (56), the attenuation depends on an integral over the Fermi surface of the product $\mathscr{D}^2 l$. On the other hand, the electrical conductivity involves the integral of l alone. Thus, the disagreement is simply a measure of the anisotropy of the deformation parameter $\mathscr{D}$, which, as will be shown shortly, is quite large.

The frequency dependence of the electronic acoustic attenuation for longitudinal-wave propagation along the principal directions in copper is shown in Fig. 15. The data show the expected limiting linear frequency dependence, in agreement with Eq. (58). There is, however, a large anisotropy in the limiting slopes as a function of propagation direction, which cannot be explained in terms of elastic anisotropy. This can be seen from Table III, which shows the ratios of the experimental limiting slopes to those computed for a free-electron metal with one electron per atom and the actual sound velocity for the various propagation directions. Also shown are the results for silver and gold. The large variations in this ratio from unity represent deviations from the free-electron theory. Since this comparison compensates for the different acoustic properties of the three metals along the three principal propagation directions, the observed behavior must be due to real-metal, electronic-structure effects, namely, departures of the actual Fermi surface from spherical shape and anisotropies in the deformation properties for the three metals.

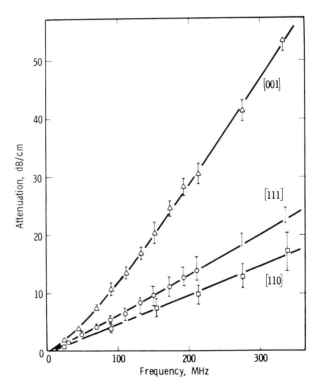

FIG. 15. Frequency dependence of the electronic acoustic attenuation in copper for longitudinal-wave propagation along the principal directions (after Macfarlane and Rayne, 1967).

TABLE III

COMPARISON OF LIMITING SLOPES FOR LONGITUDINAL-WAVE
PROPAGATION ALONG THE PRINCIPAL DIRECTIONS IN THE
NOBLE METALS WITH PREDICTIONS OF THE FREE-ELECTRON
(FE) MODEL[a]

Metal	Propagation Direction	$\lim_{a \to \infty} d\alpha/df$ (dB cm^{-1} MHz^{-1})		Ratio expt/FE
		Expt	FE	
Copper	[001]	0.186	0.0985	1.88
	[110]	0.072	0.0691	1.03
	[111]	0.051	0.0747	0.68
Silver	[001]	0.095	0.0811	1.17
	[110]	0.062	0.0603	1.02
	[111]	0.050	0.0649	0.77
Gold	[001]	0.122	0.0531	2.29
	[110]	0.093	0.0445	2.09
	[111]	0.031	0.0464	0.67

[a] Computed for one electron per atom and relevant sound velocity.

The observed deviations from free-electron behavior can be understood qualitatively in terms of the effective zones for the actual Fermi-surface geometry of the noble metals. From Eq. (55), it is clear that the most effective electrons contributing to attenuation are those for which $|\cos \varphi| < 1/ql$. Figure 16 shows a stereographic projection of the curves of constant $\cos \varphi$ for [001], [110], and [111] propagation in copper[4] computed from the analytical form for its Fermi surface due to Shoenberg and Roaf (1962), together with the corresponding projection for the free-electron model. In all cases, the effective zones for $ql = 10$ have been cross-hatched. It is immediately evident that there are more effective electrons for [100] propagation than in the free-electron model. A similar situation also obtains in silver and gold. Thus, if the deformation parameter is not too anisotropic, one would expect the attenuation along [001] should have a limiting slope larger than the free-electron value. This is just what is observed experimentally, suggesting that much of the deviation from the free-electron model is due to the nonspherical Fermi surfaces of the noble metals. The same conclusion can be drawn from Figs. 16c and 16d, which show the effective zones in copper for propagation along [110] and [111]. For example, the low observed attenuation for [110] propagation is explained by the fact that the

---

[4] Note that φ now denotes the angle between the electron velocity and the propagation direction, which in this case is along Oz.

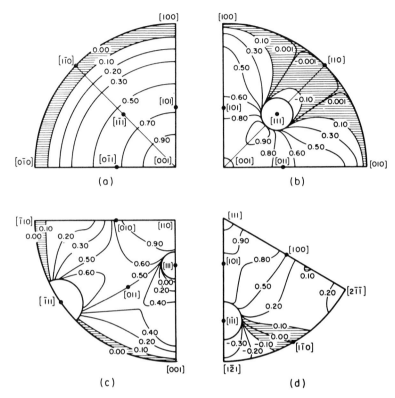

Fig. 16. Stereographic projection of lines of constant cos φ in (a) free-electron model, (b) [001] propagation, (c) [110] propagation, and (d) [111] propagation in copper (after Macfarlane and Rayne, 1967).

neck interrupts the effective zone in such a way that there are fewer electrons available for the attenuation process. In this direction, the limiting slopes for all three metals should be smaller than the free-electron values, in agreement with experiment.

Whether the Fermi-surface geometry alone is sufficient to explain the observed deviations from the free-electron model can be determined by detailed calculations of the attenuation using Eq. (55), assuming an isotropic deformation tensor. From the charge-conservation condition (38), it is easily shown that, in this case,

$$K_{ij} = (1/S)[-V_{\mathrm{F}} + (8L_3/\sqrt{3})A_{\mathrm{N}}]\,\delta_{ij} \tag{109}$$

where S is the area of the free Fermi surface, V_F is the volume enclosed by the surface, A_N is the neck area of a contact region on the {111} zone faces, and L_3 is one component of the vector to the center of the neck at [111]. The magnitudes of the diagonal components of this tensor are $-1.53, -1.60,$

and -1.51 for copper, silver, and gold, respectively, in units of the inverse lattice parameter expressed in angstroms. For comparison, the free-electron deformation parameter is -1.637 in the same units.

The resulting curve of α/f versus ql for [001] propagation in copper is shown as curve A in Fig. 17; curve C is the prediction of the free-electron model. The data points corresponding to Fig. 15 are plotted for $l = 5.9 \times 10^{-3}$ cm, which is the value of mean free path calculated from the residual resistance ratio. It is seen that the measured attenuation is much larger than the free-electron result and that, although the correction supplied by using the actual Fermi surface with an isotropic deformation tensor is in the right direction, it is too large. Similar results are obtained for [001] propagation in gold, except that, in this case, the isotropic curve lies below the data, rather than above it as in the case of copper. For silver, an isotropic deformation tensor gives good agreement with the experimental data. Table IV summarizes the results of the calculated limiting slopes for the principal propagation directions in all three metals. The limiting slopes for [001] propagation in copper and gold are large or infinite due to a nearly flat spot on the effective zones. This region, which can be seen in Fig. 16b as a sharp bend in the curve for $\cos \varphi = 0.001$, gives rise to a singularity in the integrand of the first term of Eq. (55). The singularity is responsible

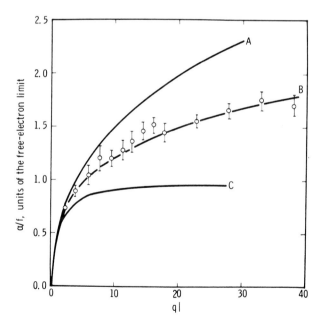

FIG. 17. Variation of α/f for [100] propagation in copper as a function of ql. Curve C is the free-electron prediction, while curves A and B are the predictions for an isotropic and nonisotropic deformation parameter (after Macfarlane and Rayne, 1967).

for the delay or absence of saturation in the attenuation behavior. Comparison of Tables III and IV shows that the assumption of an isotropic deformation tensor is generally insufficient to explain the observed attenuation behavior.

TABLE IV

RATIOS OF THE LIMITING SLOPE WITH AN
ISOTROPIC DEFORMATION PARAMETER TO
THE FREE-ELECTRON LIMITING SLOPE FOR
LONGITUDINAL-WAVE PROPAGATION
ALONG THE PRINCIPAL DIRECTIONS IN
COPPER, SILVER, AND GOLD

Metal	Propagation direction		
	[001]	[111]	[110]
Copper	>6.1	0.502	0.288
Silver	1.08	1.15	0.663
Gold	>6.0	0.805	0.409

A better fit to the experimental data can be obtained by appropriate expansions of the deformation-tensor components in spherical harmonics. For example, the component $K_{zz} \equiv K_z$, which is contained in the attenuation integral for longitudinal-wave propagation along [001], must have the symmetry of the point group D_{4h}. It may be shown that only those harmonics which form basis functions for the representations A_g and $E_g{}^1$ of the cubic point group, in the notation of Altmann and Cracknell (1965), are involved in its expansion. If these harmonics denoted by X_l and Y_l, respectively, then we have

$$K_{zz} = A_0 X_0 + A_4 X_4 + A_6 X_6 + A_8 X_8 + \cdots + B_2 Y_2$$
$$+ B_4 Y_4 + B_6 Y_6 + B_{8,1} Y_{8,1} + B_{8,2} Y_{8,2} + \cdots \qquad (110)$$

The coefficients A_l may readily be determined from the results of de Haas–van Alphen (dHvA) measurements on the noble metals under hydrostatic pressure (Templeton, 1966). For this case, Eqs. (29) and (110) give

$$\Delta k_n = K_{ij} \varpi_{ij} = (\Delta V / V)(\textstyle\sum A_l X_l) \qquad (111)$$

where the sums involving Y_l vanish by symmetry. Applying this equation to the measured area changes of the neck and [111] belly orbits, and again using the charge-conservation condition, we have three simultaneous linear equations in the coefficients A_l. Truncation of the expansion (110) at terms of order six then gives the coefficients shown in Table V.

TABLE V

EXPANSION COEFFICIENTS A_l FOR THE DIAGONAL COMPONENTS OF
THE DEFORMATION TENSOR K_{ij} FOR THE NOBLE METALS
OBTAINED FROM dHvA DATA[a]

	Metal		
Coefficient	Copper	Silver	Gold
A_0	-1.52 ± 0.01	-1.60 ± 0.01	$-1.51 + 0.04$
A_4	-0.09 ± 0.04	-0.22 ± 0.06	-0.2 ± 0.1
A_6	0.02 ± 0.02	$-0.2 + 0.1$	-0.2 ± 0.2

[a] Distances in **k**-space measured in units of inverse lattice parameter expressed in angstroms. The free-electron value of A_0 is -1.637.

For uniaxial tension along [001], it may be shown that

$$\Delta k_n = \frac{\Delta V}{V}\left(\sum_l A_l X_l + \frac{1+\sigma}{1-2\sigma}\sum_l B_l Y_l\right) \tag{112}$$

where σ is the relevant Poisson ratio. Application of this equation to the area changes of the [001] belly orbit and the so-called rosette orbit, obtained from dHvA measurements under tension by Shoenberg and Watts (1965), gives two more simultaneous equations involving the coefficients B_l. If the expansion is again truncated at order six, there is only one free parameter B_6, which can be used to match the acoustic attenuation data. The curve B in Fig. 17 represents the best such fit for copper with $B_2 = 0.01$, $B_4 = 0.33$, and $B_6 = 0.40$, in units of the inverse lattice parameter. A similar, but less satisfactory, fit is obtained for gold. For copper, the resulting deformation coefficient K_{zz} varies between -1 and -2 over most of the Fermi surface, increasing to 0.27 at [001] and decreasing to -2.74 at [100]. It is of interest that the corresponding parameter calculated from a single OPW model by Shoenberg and Watts (1965) has a similar anisotropy, so that the behavior inferred from the attenuation data is at least physically reasonable. Just as in the case of the isotropic deformation parameter, however, the presence of a flat spot on the effective zones for copper and gold again causes a lack of saturation in the plot of α/f versus ql. The experimental data do not extend to sufficiently high values of this parameter to determine whether this behavior is, in fact, observed.

For [111] and [110] propagation, the attenuation of longitudinal waves depends on the off-diagonal components of the static deformation tensor K_{ij}. These can also be expressed in terms of appropriate linear combinations of spherical harmonics, which form the basis functions for the irreducible representations A_{1g}, E_g^1, and T_g^3 of the full cubic point group. In the case of [111] propagation, a reasonable fit to the data for copper and gold is obtained using expansions up to $l = 6$ for both the diagonal and nondiagonal components of K_{ij}. Nevertheless, this fit is not unique, since there are more disposable coefficients than independent experimental data to fit. For [110], this situation is even more pronounced, so that the form of the off-diagonal components of K_{ij} is very uncertain.

In principle, a combination of dHvA and acoustic attenuation data could be used in the manner just outlined to obtain an accurate harmonic expansion of the components of K_{ij}. The results thus far obtained indicate that much more accurate measurements would be necessary to determine the expansion coefficients with any certainty. It is clear that the deformation tensor is quite anisotropic, thus necessitating an expansion up to quite large values of l. Under these conditions, the physical meaning of the procedure becomes rather obscure. Nevertheless, the partial success in finding low-order harmonic expansions, which simultaneously explain both sets of data, strongly suggests that the general theory of electronic acoustic attenuation is valid and that the electron–phonon interaction at low wave-numbers can, in fact, be obtained from a static deformation tensor.

A severe limitation of the above analysis is the implicit assumption of an isotropic electron mean free path. Recent longitudinal magnetoresistance data of Powell (1966) indicate that, in fact, the mean free path is anisotropic, being somewhat smaller near [111] than [100]. This anisotropy would have only small effects on the limiting slopes, but it would materially affect the shapes of the α/f curves near $ql \approx 1$. It is possible that, if the deformation tensor could be determined independently, the form of the acoustic attenuation data for low ql could give significant information about the anisotropy of the mean free path.

2. *Polyvalent Metals*

a. Aluminum. The normal-state acoustic attenuation in aluminum has been studied by a number of workers. Longitudinal-wave measurements by Lax (1959) below 1 MHz on polycrystalline material, using a resonant technique, are consistent with Eqs. (14) and (56) for the limit $ql \ll 1$. The attenuation obeys an f^2 dependence quite closely and has the same temperature dependence as the electrical conductivity. The ratio of the observed attenuation to the free-electron value, expressed in terms of the conductivity σ, namely,

$$\alpha_1 = \frac{4}{15} \frac{f^2}{\rho v_s{}^3} \frac{h^2}{e^2} (3\pi^2 N)^{2/3} \sigma \tag{113}$$

is approximately 1.25 and is independent of frequency. Just as in the noble metals, the deviation of this ratio from unity represents band-structure effects, which modify the integral of $\mathscr{D}^2 l$ over the Fermi surface. As discussed previously, it is also possible that interband relaxation effects are important, since the conduction electrons are contained in more than one band. Similar results have recently been obtained by Wang and McCarthy (1969) for both longitudinal and shear waves propagating in single-crystal aluminum at frequencies such that $ql \lesssim 1$.

The attenuation of both longitudinal and shear waves has been measured in both the normal and superconducting states by David *et al.* (1962) for propagation along [100] and [110]. These measurements have been made on zone-refined single crystals with residual resistance ratios in excess of several thousand, at frequencies from 5 to 35 MHz. The corresponding maximum value of ql certainly corresponds to $ql > 1$. Although the data do not extend to sufficiently high frequencies to ensure saturation behavior in the plots of α/f versus f, they indicate deviation from free-electron behavior for longitudinal wave propagation along [110]. For propagation along [100], the ratio α_t/α_1 in the same crystal at both 5 and 15 MHz deviates by less than 10% from the corresponding ratio for the free-electron model given by Eq. (17), namely,

$$\alpha_t/\alpha_1 = (8/\pi^2)(v_{s1}/v_{st})^2 \tag{114}$$

However, for [110] propagation, the deviation of α_t/α_1 from the free-electron formula is much greater, so that band-structure effects again appear to be significant.

Attenuation data over a much wider frequency interval, with ql ranging from 0.1 to 20, for longitudinal waves propagating in high-purity aluminum have recently been obtained by Hepfer and Rayne (1968). Figure 18 summarizes the results for [110] propagation on samples of 99.999% purity and also on zone-refined material of purity in excess of 99.9999%. The frequency values for the high-purity sample have been scaled

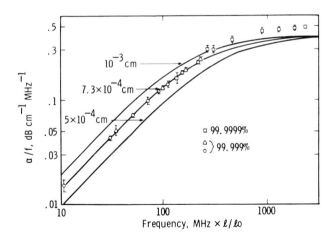

Fig. 18. Variation of α/f for [110] propagation in aluminum as a function of frequency for specimens of different purities. The full lines are the free-electron curves for various values of l_0, the mean free path for the lower-purity sample (after Hepfer and Rayne, 1968).

by a constant factor, so that all the data lie on a common smooth curve. This procedure is equivalent to multiplying the frequency by l/l_0, the ratio of the mean free path to the value l_0 for the lower-purity sample. The curves in Fig. 18 are the predictions of the free-electron theory (cf. Eq. 13) for three values of l_0, assuming three conduction electrons per atom and the relevant sound velocity. At low frequencies ($ql \ll 1$), the data are fitted quite well by the theory for $l_0 = 7.3 \pm 0.7 \times 10^{-4}$, which is larger than the value $l_0 = 5.8 \pm 0.6 \times 10^{-4}$ obtained from resistivity data. The ratios of these two estimates compares favorably with those obtained by Lax (1959), considering the uncertainty in obtaining the mean free path from residual resistance measurements. At high frequencies ($ql \gg 1$), the limiting behavior for [110] propagation does not agree with the free-electron theory, the saturation value of α/f being about 30% too high. Table VI gives the limiting values of α/f for the three principal directions. The observed anisotropy is in agreement with the results of other recent attenuation experiments by Fil' *et al.* (1968), Timms and Dobbs (1968), and Berre and Olsen (1965). Of particular interest is the very low attenuation, relative to the free-electron prediction, observed for propagation along [111].

TABLE VI

COMPARISON OF LIMITING VALUES OF α/f FOR LONGITUDINAL-WAVE
PROPAGATION ALONG THE PRINCIPAL DIRECTIONS IN ALUMINUM WITH
THE PREDICTIONS OF THE FREE-ELECTRON MODEL AND A PSEUDO-
POTENTIAL MODEL WITH AN ISOTROPIC DEFORMATION PARAMETER

| Propagation Direction | $\lim_{a \to \infty} \alpha/f$ (dB cm^{-1} MHz^{-1}) | | | Ratio Expt/FE |
	Expt.[a]	FE[b]	Pseudopotential[c]	
[001]	0.38	0.43	>2.75 (0.23)	0.89
[110]	0.52	0.41	0.35	1.28
[111]	0.29	0.40	0.36	0.73

[a] For [001] propagation, the quoted α/f may not correspond to the correct limiting value owing to nonsaturation behavior.

[b] Computed assuming three electrons/atom and relevant sound velocity.

[c] The number in parentheses is the value of α/f for $ql = 20$, which is the estimated maximum experimental limit for [001] propagation.

Since interband relaxation effects can be neglected in the high-ql regime, the attenuation for longitudinal waves is given by Eq. (60). Preliminary calculations by Hepfer (1968) based on this expression, using a pseudopotential model for the Fermi surface of aluminum developed by

Ashcroft (1963) and assuming an isotropic deformation parameter, give the limiting values of α/f shown in Table VI. Just as in the case of copper and gold, the presence of a flat spot on the effective zone for [001] propagation causes a singularity in the integrand of Eq. (55) and leads to a lack of saturation in the plot of α/f versus f. For the other propagation directions, the ordering of the attenuation is in agreement with experiment, although, clearly, the absolute value for [110] propagation is too low. Thus, it is clear that the deformation tensor must be anisotropic and that, in the region of the effective zone for [110] propagation, the diagonal components must be larger than the free-electron value. This conclusion is supported by the results of recent dHvA measurements by Melz (1966) on aluminum under pressure. These experiments show that the pressure derivative for the normal cross sections of the third-zone arms is much larger than that predicted by the free-electron theory. It would thus be expected that the corresponding derivatives for the second-zone cross-sectional area would behave similarly, although the deviations from free-electron predictions would be less. Since the effective electrons for [110] propagation are almost exclusively in the second zone, the relevant deformation parameter, and hence the attenuation, would be larger than in the simple isotropic model, as required. It is clear, however, that further experimental and theoretical investigation of these conclusions is required.[5]

b. Indium. The normal-state longitudinal-wave attenuation for indium has been extensively investigated by Bliss and Rayne (1968), Fossheim and Leibowitz (1966), Sinclair (1967), and Fil' *et al.* (1967). Table VII gives the experimental data for the principal propagation directions, as well as the predictions of the free-electron model, assuming three electrons per atom and the relevant sound velocity. Of particular interest are low values of attenuation along [100] and [001], where the limiting α/f is approximately one half the predicted value.

Since indium is a trivalent face-centered-tetragonal metal with a c/a ratio of 1.08, it has a free-electron Fermi surface similar to that of aluminum. Owing to the tetragonal distortion, however, the third-zone arms α and β along $\langle 011 \rangle$ and $\langle 110 \rangle$ are no longer equivalent. Pseudopotential calculations by Ashcroft and Lawrence (1968) and Anderson (1968) indicate that the crystal potential removes the α arms, leaving a third-zone surface consisting of an interconnected ring of β arms in the (001) plane. For [100] propagation, approximately one half of the effective zone follows the length of the α arms in the free-electron model, so that their elimination would reduce the perimeter of the effective zone by the same amount. Assuming that the deformation constant is unchanged, it can be seen from Eq. (60) that

[5] Note added in proof: Hepfer and Rayne (1969) have calculated the deformation tensor for aluminum using a pseudopotential model for the Fermi surface. Good agreement is obtained with both the dHvA results of Melz (1966) and the values of α/f quoted in Table VI. Considerable anisotropy in the deformation parameter is found for all three directions of sound propagation.

TABLE VII

COMPARISON OF LIMITING VALUES OF α/f FOR
LONGITUDINAL-WAVE PROPAGATION ALONG
PRINCIPAL DIRECTIONS IN INDIUM WITH
PREDICTIONS OF FREE-ELECTRON MODEL[a]

Propagation Direction	$\lim_{a \to \infty} \alpha/f$ (dB cm^{-1} MHz^{-1})		Ratio Expt/FE
	Expt	FE	
[110]	0.23	0.497	0.47
[001]	0.27	0.519	0.52
[110]	0.40	0.429	0.93
(011)	0.43	0.471	0.91
(111)	0.42	0.440	0.96

[a] Computed for three electrons/atom and relevant sound velocity.

the attenuation would be reduced to one half the free-electron value, as required. For [001] propagation, where the effective zone follows the length of the β arms, the effect at the crystal potential would be to reduce their contribution to the attenuation, because of the change in the reciprocal Gaussian curvature. The second-zone contribution would also be modified, both as a result of the reduction in its perimeter as well as the changes in its curvature. A quantitative calculation of these effects using the actual Fermi surface of indium is not available. However, rough estimates indicate that the predicted attenuation behavior is approximately in agreement with experiment, assuming an isotropic deformation parameter. It should be noted that the apparent ineffectiveness of the β arms has also been explained by Sinclair (1967) by the effects of small-angle phonon scattering in decreasing the effective mean free path of those electrons on an effective zone running close to and parallel to a zone edge. In a number of the above experiments, attenuation measurements have been made at very low temperatures, where phonon scattering is completely negligible. Thus, the reduction in the mean free path is correspondingly small, and hence this explanation of the observed attenuation behavior is probably incorrect.

Attenuation measurements by Bliss and Rayne (1968) for quasilongitudinal modes propagating in the (011) plane show that the attenuation follows the form of Eqs. (90) and (94). The discontinuity in the difference $(\alpha_n - \alpha_s)$ at the transition temperature, due to the term J_2, is illustrated in Fig. 19. For propagation directions approximately 20° from [001] and [100], this term becomes appreciable, indicating that the integral L_1 in Eq. (76) is quite large. Approximate calculations indicate that the large value of L_1 is due to the Fermi-surface topology of indium and not to an

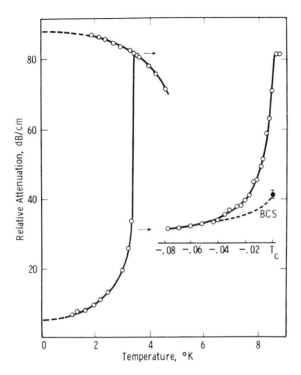

FIG. 19. Electronic attenuation in normal and superconducting indium for 150 MHz quasilongitudinal waves propagating in the (010) plane along a direction 19° from [001]. The dashed line in the inset is the BCS prediction for $\Delta(0) = 1.75k_B T_c$ (after Bliss and Rayne, 1968).

anomaly in the deformation parameter. It is of interest that the residual part of α_n associated with the term J_1 in Eq. (90) exceeds the free-electron value for propagation near [001]. This behavior can only be explained if the deformation parameter is anisotropic.

The attenuation of shear waves propagating along the principle directions in indium has also been measured by Fossheim and Leibowitz (1966) for the regime $ql \gg 1$. Table VIII gives the total attenuation α_t measured at 90 MHz. From Eqs. (66) and (67), it is clear that α_t depends in a rather complex way on both the Fermi-surface geometry and the shear-deformation parameter. No quantitative analysis of the data has this far been attempted.

 c. *Zinc, Cadmium, and Thallium.* The normal-state electronic attenuation for propagation along the directions [10$\bar{1}$0], [11$\bar{2}$0], and [0001] has been measured in zinc by Gonz and Neighbours (1965) and by Lea and Dobbs (1968). Data on cadmium have been reported by Lea *et al.* (1968), while normal-state measurements on thallium have been made by Weil and Lawson (1966) and Willard (1968). From Table IX, it can be seen that, for

TABLE VIII

COMPARISON OF SHEAR-WAVE ATTENUATION IN
INDIUM WITH PREDICTIONS OF FREE-ELECTRON MODEL

Propagation direction	Polarization direction	Frequency (MHz)	Attenuation (dB cm^{-1})			
			α_t	α_D	α_t^E	α_t^f
[001]	$\perp$[001]	90	159	54	105	111
[100]	[001]	90	158	53	106	111
[110]	[1$\bar{1}$0]	90	172	60	112	118
[110]	[001]	90	145	47	98	111

zinc, the limiting attenuation for propagation along the hexagonal axis is very large in comparison to that predicted by the free-electron model. This behavior is thought to be due to the large deformation tensor associated with the second-zone hole surface and also to the existence of flat regions on the effective zone, with very large values of the reciprocal Gaussian curvature. No dHvA measurements under pressure or pseudopotential calculations have been made to check the former hypothesis (see note added in proof, p. 215).

Cadmium behaves in a similar way, but reliable attenuation data along [0001] have not been reported. For thallium, on the other hand, the attenuation for [0001] propagation is anomalously small. No explanation of this behavior has yet been advanced. The dependence of α/f versus ql for

TABLE IX

COMPARISON OF LIMITING VALUES OF α/f FOR LONGITUDINAL-WAVE
PROPAGATION ALONG PRINCIPAL DIRECTIONS IN ZINC, CADMIUM, AND
THALLIUM WITH PREDICTIONS OF FREE-ELECTRON MODEL[a]

	$\lim_{a \to \infty} \alpha/f$ (dB cm^{-1} MHz^{-1})					
	[11$\bar{2}$0]		[10$\bar{1}$0]		[0001]	
Metal	Expt	FE	Expt	FE	Expt	FE
Zinc	0.067	0.182	0.128	0.182	1.29	0.475
Cadmium	0.048	0.156	0.165	0.156	1.08[b]	0.356
Thallium	0.3	0.520	0.6	0.520	0.08	0.380

[a] Assuming the relevant sound velocity and two electrons/ atoms in the case of zinc and cadmium, and three electrons/atoms in the case of thallium.

[b] Tentative value estimated from difference between normal-state attenuation at 4.2°K and 77°K.

this metal does not fit the free-electron theory in any of the principal propagation directions. It is believed that this behavior is due, in part, to anisotropy of the electron mean free path. The temperature dependence of phonon-limited mean free path, inferred from the attenuation data, follows a $T^{-3.6}$ law, in contrast to the T^{-5} dependence obtained from electrical resistance measurements. Attenuation data do not extend to sufficiently large values of ql to enable the limiting values of α/f to be determined with any accuracy.

d. Lead. Normal-state attenuation measurements on high-purity lead from 10 to 210 MHz for propagation along [001] have been made by Fate (1968). The dependence of α/f on ql, normalized to the limiting value at $ql = \infty$, is fitted quite well by Eq. (13) for the free-electron model. However, the limiting value of 0.18 dB cm^{-1} mHz^{-1} is much lower than the free-electron prediction of 0.58 dB cm^{-1} mHz^{-1}. From dHvA measurements under pressure by Anderson *et al.* (1969), it is known that the observed changes in extremal areas are much larger than the predictions of the free-electron deformation parameter. Since the Fermi-surface geometry is not appreciably altered by the crystal potential, it is difficult to understand why the effective zone integral, and, hence, the attenuation, should be lower than the free-electron results. Clearly, further experimental and theoretical work are necessary to clarify this situation.

e. Tin. There is relatively little information about the normal-state electronic attenuation in white tin. The most extensive data of Shepelev and Filimonov (1965) are for quasilongitudinal modes propagating along nonprincipal crystallographic directions. As noted previously, the value of α_n for such modes contains a contribution involving the integral J_2, which can be quite large. There is insufficient detail in the published results to enable the contribution from the effective zone integral J_1 to be calculated. Even if the anisotropy of J_1 were known, however, there is little hope of relating it to the detailed Fermi-surface topology, since the band structure of tin has been shown by Gold and Priestley (1960) to be relatively complex.

3. Transition Metals

a. Tungsten and Molybdenum. The normal-state longitudinal-wave attenuation of high-purity tungsten and molybdenum has been measured up to 1 GHz by Jones and Rayne (1964) for propagation along the principal crystallographic directions. Considerable anisotropy is observed in the limiting value of α/f, as shown by Table X, although both metals are very nearly isotropic in their elastic properties. Thus, the effective zone integral in Eq. (60) must depend strongly on crystal orientation, a not very surprising result, in view of the complex Fermi-surface topology (Mattheiss 1965). It is nevertheless interesting to note that the values of α_l differ markedly from the predictions of the free-electron theory. Thus, if we adopt a simple two-band spherical model to represent the electron and hole surfaces, we may rewrite Eq. (59) in the form

$$\alpha_l = (\hbar f/\rho v_s^2)k_1^2(K_1^2 + K_2^2) \tag{115}$$

TABLE X

LIMITING VALUES OF α/f FOR LONGITUDINAL-WAVE PROPAGATION IN
TUNGSTEN AND MOLYBDENUM ALONG PRINCIPAL CRYSTALLOGRAPHIC
DIRECTIONS

Direction	$\lim_{a \to \infty} \alpha/f$ (dB cm^{-1} MHz^{-1})		$(\alpha/f)_W/(\alpha/f)_{Mo}$	$(\rho v_s^2)_{Mo}/(\rho v_s^2)_W$
	W	Mo		
[100]	0.057	0.063	0.92	0.91
[110]	0.071	0.089	0.90	0.81
[111]	0.034	0.050	0.78	0.78

where $k_1 = k_2$, since the metals are compensated. If, further, we assume that the deformation coefficients are given by the free-electron value, i.e., $K_1 = K_2 = -\frac{1}{3}k_1$, we then have

$$\alpha_1 = \frac{1}{3}\pi^2(\hbar k_1 N/\rho v_s^2)f \qquad (116)$$

where N is now the total number of carriers in the two bands. Anomalous skin-effect measurements (Fawcett and Griffiths, 1962) give $N \approx 0.4$ per atom, so that for tungsten α/f has a predicted limiting value of 6.5×10^{-3} dB cm^{-1} MHz^{-1}. This figure is at least an order of magnitude lower than the experimental result, and hence the deformation parameter must be much larger than that predicted by the free-electron model. Unfortunately, there are no dHvA data to check this prediction.

Table X shows that the dependence of the limiting α/f on orientation is qualitatively the same in both metals. In view of the similarity of their band structure, it is not unreasonable to suppose that they have similar deformation properties, and hence that the effective zone integrals in Eq. (60) are the same. With this hypothesis, we have, for a given orientation,

$$(\alpha_1)_W/(\alpha_1)_{Mo} = (\rho v_s^2)_{Mo}/(\rho v_s^2)_W \qquad (117)$$

A comparison of these ratios, computed from the attenuation data and the available elastic constants measured by Bolef and de Klerk (1962), is given in the table. The agreement is rather better than might be expected, in view of the rather crude initial assumptions used in deriving Eq. (117).

From Eq. (60), it can be seen that the limiting value of α/f for an assumed isotropic deformation parameter should depend on the integral of the reciprocal Gaussian curvature around the effective zone for the propagation direction under consideration. Reference to the Fermi surface in Fig. 20 shows that the attenuation in tungsten and molybdenum for propagation along [100] and [111] should be determined principally by the electron jack at Γ, since the radius of curvature on the hole surface at H, at right angles to the effective zone, is small. An approximate calculation of the relevant line

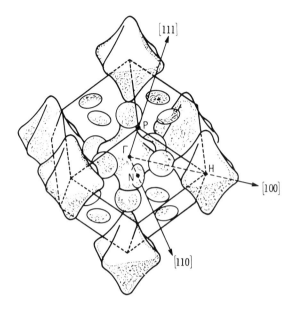

F IG. 20. Fermi surface of molybdenum showing the electron jack at Γ and the hole octahedron at N (after Mattheiss, 1966).

integrals shows that the expected ratios of the attenuation in these directions are in agreement with experiment. For propagation along [110], the effective zone passes over the flat faces of the hole octahedron. There is thus an additional contribution to the integral in Eq. (60), resulting in a larger attenuation for this direction.

b. Vanadium, Niobium, and Tantalum. Measurements of the normal-state longitudinal-wave attenuation have been made in niobium by Weber (1964) and Perz and Dobbs (1967), and in vanadium, niobium, and tantalum by Levy *et al.* (1963) for both longitudinal and shear waves. Only in the case of niobium, however, has the specimen purity been sufficient to attain the condition $ql > 1$. The measurements of Dobbs and Perz give limiting α/f values of 0.19, 0.30, and 0.27 dB cm^{-1} MHz^{-1} for propagation along [100], [111], and [110], respectively. Since the Fermi surface of niobium is rather complex, it is not unexpected that the attenuation behavior does not conform to the predictions of the free-electron model. For propagation along [100], the latter model gives a limiting α/f equal to 0.011 dB cm^{-1} MHz^{-1}, assuming a spherical Fermi surface with an area 18% of that for a sphere containing five electrons per atom, as suggested by Swenson (1962). It thus would appear that the deformation characteristics of niobium are also anomalous.

For $a \ll 1$, Eq. (14) gives for the ratio of α_t/α_l in the free-electron model

$$\alpha_t/\alpha_l = \tfrac{4}{3}(v_{sl}/v_{st})^3 \tag{118}$$

The data of Levy *et al.* (1963) on low-purity vanadium, niobium, and tantalum for propagation along [110] are consistent with this relation. The agreement is surprising, in view of the known complex Fermi surfaces for these metals (Mattheiss, 1964).

c. *Rhenium.* The limiting electronic attenuation of longitudinal waves propagating along the principal directions in high-purity rhenium, with a residual resistivity ratio in excess of 10,000, has been measured in the frequency range 250 MHz to 1.25 GHz by Jones and Rayne (1966). There is considerable anisotropy in the attenuation behavior, the limiting values of $d\alpha/df$ being 0.035, 0.077, and 0.134 dB cm^{-1} MHz^{-1} for propagation along [0001], [10$\bar{1}$0], and [11$\bar{2}$0], respectively. For the latter direction, the limiting behavior does not occur until quite high values of frequency. Since the measurements all relate to material of the same purity, it is clear that the electron mean free path on the effective zone for [11$\bar{2}$0] propagation must be unusually small. No explanation of this result has thus far been proposed, although it seems likely that it is in some way connected with the extremely complicated Fermi-surface topology of rhenium (Mattheiss, 1966).

4. Semimetals

Although bismuth has only about 10^{-5} carriers per atom, there is an appreciable electronic attenuation (Reneker, 1959) as a result of the large components of its deformation tensor. Detailed calculations by Inoue and Tsuji (1967) exist for the orientation dependence of the electronic attenuation in bismuth, but no *direct* experiments have been made to check the predicted behavior and to obtain information about its deformation tensor. The components of the latter have, however, been obtained from related measurements of giant quantum oscillations in the attenuation as a function of magnetic field by Mase *et al.* (1966) and Walther (1968).

It can be shown that, as the field H is varied, sharp peaks $\alpha_p(H)$ occur in the attenuation, the magnitude of which are given by

$$\alpha_p(H)/H = (\hat{\mathbf{e}} \cdot \mathbf{C} \cdot \hat{\mathbf{q}})^2 (m_b/v_s^2 \cos \alpha) F(\omega, T) K^{-1}(q_b l_b) \tag{119}$$

where

$$F(\omega, T) = (em_0/16\pi\hbar^2\rho)\omega/k_B T \tag{120}$$

In the former equation, $\mathbf{C}$ is the deformation tensor defined as $\Delta E = C_{ij}\varpi_{ij}$, which is clearly related to K_{ij} by the equation $K_{ij} = (1/\hbar v)C_{ij}$. The unit vector $\hat{\mathbf{e}}$ defines the polarization of the ultrasonic wave, m_b is the relevant component of the effective mass along the magnetic field direction, α is the angle between the magnetic field and the propagation direction, and $K^{-1}(q_b l_b)$ is a correction factor which describes the reduction of the peak absorption due to collisions. For $q_b l_b \to \infty$, this correction factor approaches unity.

To evaluate the components of the deformation tensor $\mathbf{C}$, the anisotropy of $\alpha_p(H)/H$ has been measured at fixed frequency and temperature with different $\hat{\mathbf{q}}$ and $\hat{\mathbf{e}}$ vectors. By suitable choice of these latter quantities, selective coupling with the electron and hole ellipsoids in bismuth can be obtained. From Eq. (119), the ratio of the contributions of the rth and sth ellipsoids can be written

$$\pm \left\{\frac{[\alpha_p(H)/H]_r}{[\alpha_p(H)/H]_s}\right\}^{1/2} = \frac{(\hat{e} \cdot C \cdot \hat{q})_r}{(\hat{e} \cdot C \cdot \hat{q})_s} \qquad (121)$$

where the right-hand side can be expressed in terms of the components of the relevant deformation tensors. Using the experimental data, the above equation gives a set of simultaneous equations which can be solved for the ratios of the deformation-tensor components for the electrons and holes. Only one assignment of the signs for these components gives a consistent solution to these equations. Thus, if one component for the electron and hole bands is determined absolutely from Eq. (119), the whole deformation tensor can be evaluated to within one sign ambiguity. The resulting components for the principal electron ellipsoid are: $C_{11}/C_{22} = -0.37, C_{33}/C_{22} = -0.29, C_{23}/C_{22} = +0.25$, and $C_{22} = \pm 5.9$ eV, while, for the hole ellipsoid, the deformation components are $C_{33}/C_{22} = -1.03, C_{33} = -1.2$ eV. The diagonal components agree well with those obtained from piezoresistance measurements made by Jain and Jaggi (1964). No independent determination exists for the shear constant C_{23}.

B. Superconductors

Superconductivity has been found in many polyvalent metals, in the majority of the transition elements, and in an extremely large number of intermetallic compounds and alloys. The range of superconducting transition temperatures for these materials extends from $21°K$ for the niobium–aluminum–germanium compound $Nb_3Al_{0.8}Ge_{0.2}$ to well below $1°K$, the lowest value recorded so far for a pure metal being that of tungsten at approximately 15 m°K. For most of these materials, the electron mean free path is extremely short, either because of impurity or defect scattering or as the result of a disordered structure. The preparation of single crystals of these superconductors has been attempted for some of the elements and for a few alloys and compounds. Of these, only about ten elements have been produced so far with sufficiently high purity to permit ultrasonic studies with present techniques in the $ql > 1$ regime, as is required for the study of anisotropic effects associated with the Fermi-surface topology. Measurements have been confined mainly to those polyvalent metals with fairly low melting points, although data on reasonably pure single crystals of a few transition elements have also be obtained. Experimental difficulties have tended to limit both the temperature and frequency range in these experiments to the regions above $0.3°K$ and below 500 MHz, respectively.

Attenuation measurements of both longitudinal and shear waves have been carried out, principally to test the validity of Eq. (83) in as wide a range of materials as possible. Surprisingly good agreement with the predictions of the BCS theory is obtained. If Eq. (83) is written in the form

$$\ln[(2\alpha_n/\alpha_s) - 1] = \Delta/k_B T \tag{122}$$

it can be seen that a plot of $\ln[(2\alpha_n/\alpha_s) - 1]$ versus $1/T$ should give a straight line at sufficiently low temperatures.[6] The slope of this line is proportional to $\Delta(0)$, the gap at absolute zero. A typical plot of the data for tin (Perz, and Dobbs, 1967) is shown in Fig. 21. For most materials, the values of gap parameter $2A = 2\Delta(0)/k_B T_c$ obtained in this way are within 20% of the theoretical value of 3.56 and are in fair agreement with the values obtained by other experimental techniques. Similar results are obtained for both longitudinal and transverse waves when the appropriate allowances are made for the difference in behavior near T_c. Anisotropies in the effective value of the gap parameter for a specific material, inferred from attenuation measurements for sound propagation along different crystal directions (see Fig. 21), are generally considered to be real and have been tentatively accounted for in terms of anisotropy in the electronic structure and/or phonon spectrum. These effects appear to be quite small for most materials, particularly for those with free-electron-like band structures. The biggest deviations from the BCS value have been reported for the strong coupling superconductors, lead and mercury, where the temperature dependence of the attenuation in the superconducting state is extremely anomalous. This

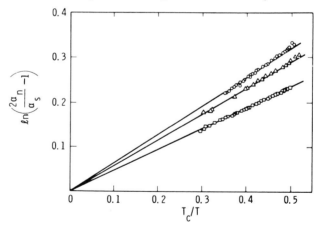

FIG. 21. Plot of $\ln[(2\alpha_n/\alpha_s) - 1]$ versus T_c/T in superconducting tin to determine the energy gap $\Delta(0)$. Circles: Propagation along [001] in pure tin, giving $2A = 3.15$; squares: propagation along [001] in impure tin, giving $2A = 3.44$; triangles: propagation along [310] in pure tin, giving $2A = 4.25$ (after Perz and Dobbs, 1967).

[6] For $T \ll T_c$, the ratio $\alpha_n/\alpha_s \gg 1$, and the equation can be written in the approximate form $\ln(\alpha_s/\alpha_n) = -\Delta/k_B T + $ const. The limiting slope of the plot of $\ln(\alpha_s/\alpha_n)$ versus $1/T$ again gives $\Delta(0)$. This method of analysis has been used frequently [e.g. Morse *et al.*, (1959)]. An evaluation of the various methods of extracting the gap parameter from ultrasonic data has recently been given by Perz (1970).

behavior has not yet received adequate theoretical explanation, but it appears to be associated with the complex nature of the electronic structure of these materials.

All of the above experiments have been carried out under the conditions $ql > 1$ and $\hbar\omega_{phonon} \ll \Delta(0)$. The results obtained in impure materials and dilute alloys, where $ql \ll 1$, are in good agreement with theory for the case of an isotropic superconductor where the energy-gap anisotropy has been removed by impurity scattering. The extremely sparse information available from the few experiments carried out at high frequencies, where $\hbar\omega_{phonon} \gtrsim \Delta$, appears to be in qualitative agreement with theory.

1. *Monovalent and Noble Metals*

Superconductivity has not yet been detected in any of the pure alkali metals or in the noble metals, although investigations have been carried out to extremely low temperatures in the millidegree range. Therefore, in contrast to the situation in the normal state, there are no ultrasonic measurements in the superconducting state available for discussion at the present time.

2. *Polyvalent Metals*

Ultrasonic experiments have been carried out in pure single-crystal samples of many of the polyvalent metals. Sample purities in excess of 99.999% are readily obtainable in these materials, enabling ql values exceeding twenty in the 100 MHz range to be achieved routinely in the superconductors aluminum, tin, indium, zinc, cadmium, lead, mercury, and thallium. A considerable amount of detailed experimental information has been accumulated on most of these metals. The principal features of the normal-state electronic structures of the majority of these metals have been well established, and it is in these materials that a detailed understanding of the ultrasonic properties of the superconducting state will probably first be achieved. They have therefore been the subject of numerous investigations in recent years, the results of which will be discussed in the following sections.

a. Aluminum. Attenuation measurements in the superconducting state of aluminum were carried out initially by Morse and Bohm (1959) for both longitudinal and transverse waves. The temperature in these experiments extended only to $1.06°K$ ($T_c = 1.17°K$), so that no gap-parameter estimates were possible. Subsequent measurements to lower temperatures have been carried out by Morse and Claiborne (1964) employing an adiabatic demagnetization cryostat, but the experiments are still in the $ql \lesssim 1$ regime due to limited sample purity and the low frequencies used. Later work by David (1964) and co-workers (1962, 1963), Fil' *et al.* (1968), and Timms (1968) on substantially purer specimens has enabled the range of ql to be significantly increased. The use of ^3He cryostats in this work has provided a considerable improvement in temperature stability and control, facilitating much more precise measurements than in the earlier investigations. The results of all of these experiments are shown in Table XI. In general, there is agreement

TABLE XI

Values of Gap Parameter $2A = 2\Delta(0)/k_B T_c$ Obtained from Ultrasonic Measurements[a]

Metal	$2A$	Propagation direction	Polarization direction	Remarks	Reference
Al	3.4 ± 0.3	13° from [100]	—	$\hbar\omega \approx 2\Delta(T_d)$	Fagen and Garfunkel (1967)
	3.7 ± 0.2	[100]	—	—	Fil' et al. (1968)
	(3.5 ± 0.2)				
	3.8 ± 0.2	[110]	—	—	
	(3.6 ± 0.2)				
	3.8 ± 0.2	⊥(111)	—	—	
	(3.25 ± 0.2)				
	3.7 ± 0.2	⊥(122)	—	—	
	(3.3 ± 0.2)				
	3.6 ± 0.2	[100]	—	—	David et al. (1962)
	3.6 ± 0.2	[110]	—	—	
	3.5 ± 0.2	[100]	—	—	Timms (1968)
	3.5 ± 0.2	[110]	—	—	
	3.5 ± 0.2	[111]	—	—	
Tl	3.90 ± 0.1	[0001]	—	—	Willard et al. (1968)
	3.52 ± 0.15	[11$\bar{2}$0]	—	—	
	3.70 ± 0.20	[10$\bar{1}$0]	—	—	
Zn	3.41	[0001]	—	—	Lea and Dobbs (1968)
	3.79	[10$\bar{1}$0]	—	—	
	3.64	[11$\bar{2}$0]	—	—	
	3.4 ± 0.2	[10$\bar{1}$0]	—	—	Bohm and Horowitz (1963)
	3.8 ± 0.2	[11$\bar{2}$0]	—	—	

Sn	3.4 ± 0.2	[001]	[100]	—	Leibowitz (1964a)
	3.7 ± 0.2	[100]	[011]	—	
	3.3 ± 0.2	[100]	[001]	—	
	3.14 ± 0.04	[001]	—	No allowance	Perz and Dobbs (1967)
	4.24 ± 0.04	[310]	—	for possible	
	3.55 ± 0.04	[100]	—	discontinuity	
	3.84 ± 0.07	[110]	—	near T_c	
	3.9 ± 0.2	⊥(101)	—	As above	Shepelev (1963)
	4.1 ± 0.2	⊥(301)	—		
	4.8 ± 0.3	⊥(111)	—	As above	Shepelev and Filimonov (1965)
	4.4	⊥(112)	—		
	3.9	⊥(211)	—		
	4.0	⊥(113)	—		
	4.3	⊥(311)	—		
	3.1 ± 0.1	C_4	—		Bezuglyi et al. (1960)
	3.5 ± 0.2	C_2	—		
	3.2 ± 0.1	[001]	—		
	4.3 ± 0.2	[100]	—		Morse, et al. (1959)
	3.8 ± 0.1	[110]	—		
Nb	3.52 ± 0.04	[100]	—		Perz and Dobbs (1966)
	3.61 ± 0.04	[111]	—		
	3.52 ± 0.05	[110]	—		
Mo	3.3 ± 0.2	[100]	—	0.4 GHz	Jones and Rayne (1967)
	3.5 ± 0.2	[110]	—		
	3.1 ± 0.2	[111]	—		
	3.1 ± 0.2	[111]	—	9 GHz	Pike (1969)

TABLE XI—(cont.)

Metal	A	Propagation direction	Polarization direction	Remarks	Reference
Re	2.9 ± 0.1	[0001]	—	—	Jones and Rayne (1966)
	3.0 ± 0.1	[10$\bar{1}$0]	—	—	
	3.5 ± 0.1	[11$\bar{2}$0]	—	—	
Hg	4.0 ± 0.3	Polycrystalline longitudinal wave	—	9 GHz	Ferguson and Burgess (1967)
Pb	No reliable analysis for $ql \geqslant 1$				
In	3.2 ± 0.2	⊥(100)	—	—	Fil' et al. (1967)
	3.1 ± 0.2	⊥(001)	—	—	
	3.1 ± 0.2	⊥(110)	—	—	
	3.18	[100]	—	—	Sinclair (1967)
	3.04	[001]	—	—	
	3.40	[110]	—	—	
	2.88	[100]	[010]	—	
	3.32	[100]	[001]	—	
	3.15	[001]	[100]	—	
	3.00	[11C]	[001]	—	

[a] The data are for longitudinal waves except where indicated.

among these relatively low-frequency ultrasonic measurements, the values of the gap parameter being fairly close to the BCS value, but somewhat larger. No indications of any amplitude-dependent effects have been reported in these investigations.

The gap values are significantly higher than those obtained from specific-heat measurements (Zavaritskii, 1958) and microwave-absorption studies (Biondi *et al.*, 1964). Fil' *et al.* (1968) argue that this discrepancy can only be explained on the basis of a relatively large gap anisotropy. From Eq. (87), and neglecting the temperature dependence of $\delta\Delta$, we have, to within a constant,

$$\ln \alpha_s - \tfrac{1}{2} \ln t \sim -\Delta_{\min}(0) f(t)/t \tag{123}$$

The resulting $\Delta_{\min}$ obtained from a fit of their data to this formula are given by the entries in parentheses in Table XI. It is claimed that the correct gap values lie between the latter and those obtained by the usual fit to the BCS form. From Table II, it can be seen that the anisotropy parameter $\langle a^2 \rangle_{\text{av}}$ for aluminum indicates that there should be at least a 10% variation in the gap parameter. The ultrasonic data do not appear to be consistent with this prediction.

Fagen and Garfunkel (1967) have made attenuation measurements near 9 GHz on essentially single-crystal aluminum for propagation near [100]. Their data give a value for $2\Delta(0)$ of $(3.4 \pm 0.3)k_B T_c$, which is in agreement with the results from infrared and microwave absorption measurements within the quoted probable error. As can be seen from Fig. 22, there is a discontinuity in the attenuation near T_c of about $0.2\alpha_n$, and the acoustically determined value of T_c appears to be 0.005°K below the accepted value for pure Al of 1.175°K. This behavior is consistent with the previously discussed calculations of Privorotskii (1962) and Bobetic (1964), which show that a discontinuity in the attenuation should occur at a temperature T_d such that $\hbar\omega = 2\Delta(T_d)$. It should be noted, however, that the discontinuity could also be due to the deviation of the propagation direction from a pure-mode axis [cf. Eq. (90)]. Experiments over a wider frequency interval on accurately oriented samples are clearly required to eliminate this possibility.

From a plot of α_R/α_t versus g (see Eq. 98), Leibowitz (1964b) has re-analyzed the data of David *et al.* (1962) to determine the deformation contribution to the shear-wave attenuation in aluminum. The results are sensitive to the assumed value of electron mean free path and do not give unambiguous results. Nevertheless, the analysis indicates that, even in aluminium, there are substantial real-metal effects, No attempt was made to estimate a gap-parameter value in this analysis.

b. Indium. The early measurements of the attenuation of longitudinal waves in the superconducting state of a polycrystalline sample of indium by Morse and his collaborators (1956, 1957) were of great importance in providing a direct experimental verification of one of the fundamental predictions on ultrasonic attenuation of the BCS theory. Since that time, indium has been the subject of many ultrasonic investigations with both longitudinal

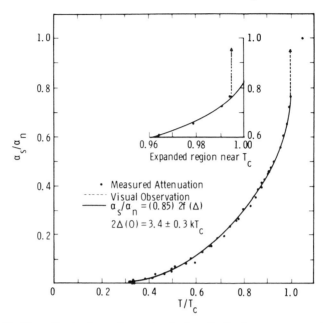

Fig. 22. Normalized attenuation α_s/α_n for longitudinal-wave propagation in aluminum at 9.3 GHz. The propagation direction is approximately 15° from [100] (after Fagen and Garfunkel, 1967).

and transverse waves. The extremely convenient value of the transition temperature of 3.45°K, the ready availability of very high-purity material, and the relatively simple techniques required for preparing oriented single crystals have greatly facilitated experiments on this metal.

As in the case of several other free-electron-like metals with low melting points, large contributions to the ultrasonic attenuation in the megahertz range due to interactions with dislocations have considerably complicated the experimental situation. A number of recent studies Leibowitz and Fossheim, 1966; Sinclair, 1967, Bliss and Rayne, 1968) have been conducted in both the normal and superconducting states for a wide range of strain amplitude. These measurements have enabled amplitude effects to be corrected for or eliminated, and the electronic contribution to the attenuation has been successfully isolated. The results obtained are in general agreement with the BCS theory, although the anisotropic nature of the Fermi surface, particularly with respect to those regions near Brillouin-zone edges, has a considerably greater effect than anticipated. The gap-parameter values, which are considered to be reliable, are presented in Table XI. They are seen to be quite close to the value appropriate for a weakly coupled superconductor.

For quasilongitudinal-wave propagation along directions with effective zones on the Fermi surface of low symmetry, the attenuation falls discontinuously near T_c, although the elastic properties of the material indicate that the shear-wave component present in the sound is extremely small.

The measurements of Bliss and Rayne (1968) for sound propagation in the (010) plane have shown that the contribution from the J_2 term in Eq. (90) is significant, discontinuities of up to 60% of α_n being observed. It is evident that estimates of the gap parameter for the effective zone, based upon only the J_1 term in the normal and superconducting states, can be subject to large error. In view of the anomalously large values of gap parameter which have been reported for several materials, this complication may be of a rather general nature.

The transverse-wave measurements of Leibowitz and Fossheim (1966) provide further evidence for the importance of the shear components of the deformation tensor in indium. As can be seen from Fig. 23, there is a considerable residual attenuation below T_c, which is associated with the shear-deformation properties of the Fermi surface given by Eq. (95). The experimental condition $ql \gg 1$ ensures that the collision-drag contribution to the shear-wave attenuation is negligible. Table VIII shows the deformation and electromagnetic contributions, α_D and α_t^E respectively, to the total shear-wave attenuation α_t. There is good agreement between α_t^E and α_t^f, the calculated free-electron value for the shear-wave attenuation. No independent estimate is available for comparison with the deformation contribution α_D derived from these measurements. Fossheim (1967) has analyzed the temperature dependence of the electromagnetic contribution to the superconducting shear-wave attenuation in indium to determine the London penetration depth $\lambda(0)$.

  c. Zinc, Cadmium, and Thallium. Measurements in the superconducting state of zinc ($T_c = 0.84°K$) have been performed by Bohm and Horowitz (1963), Gonz and Neighbours (1965), and, more recently, in considerably greater detail, by Lea and Dobbs (1968). In the later work, experiments were

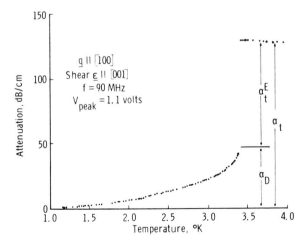

FIG. 23. Temperature dependence of shear-wave attenuation in indium, showing the deformation contribution α_D and the electromagnetic contribution α_t^E (after Fossheim and Leibowitz, 1966).

carried out on single crystals of 6N purity, with a residual resistivity ratio of about 2000, for longitudinal-wave propagation along the [0001], [10$\bar{1}$0], and [11$\bar{2}$0] directions at frequencies from 20 to 160 MHz. No evidence of any amplitude dependence of the attenuation is reported. Analysis of the low-temperature measurements yields values of the gap parameter quite close to the BCS value for all propagation directions studied, as shown in Table XI. There is no apparent correlation between the relatively small anisotropy observed in the gap parameter and the very large variation in the normal-state attenuation coefficient with propagation direction. Each measured gap parameter in this work is interpreted as a weighted average of the gap variation over the appropriate effective zone. The observed anisotropy of $\pm 5\%$ is considerably less than the root-mean-square aniso-tropy of $\pm 22\%$ inferred from the value of $\langle a^2 \rangle_{\mathrm{av}}$ given in Table II.

The superconducting transition temperature of cadmium is 0.56°K for pure material. At present, data are available from measurements in the normal state, which show a pronounced similarity in behavior for cadmium and zinc. Only very preliminary experiments in the superconducting regime of cadmium have been carried out, and no estimated gap parameter values are available.

Ultrasonic attenuation studies in the superconducting state of thallium ($T_{\mathrm{c}} = 2.38$°K) have been made by Saunders and Lawson (1964) and Weil and Lawson (1966) for longitudinal waves, and by Weil and Lawson (1966) for transverse waves. These measurements extend only to 1.2°K, and considerable extrapolation is required to estimate the limiting attenuation, thereby introducing a large uncertainty in the inferred values of the gap parameter. Willard *et al.* (1968) have carried out measurements using a ^{3}He cryostat in which considerably lower temperatures could be achieved, the range of experiment being extended down to 0.46°K. The attenuation of longitudinal waves propagating along the [0001], [$\bar{1}$2$\bar{1}$0], and [10$\bar{1}$0] directions was measured over the frequency range from 10 to 130 MHz in samples with residual resistivity ratio of 7400. No indication of any amplitude-dependent effects was found. By employing the usual analytical procedure for the inversion of Eq. (83), the values of the gap parameter shown in Table XI are obtained from the low-temperature data. Substantial deviations from BCS-like behavior are observed close to T_{c}, the attenuation falling off considerably more rapidly than predicted by theory, particularly for propa-gation along [10$\bar{1}$0] at low frequencies. The anisotropy observed in the gap parameter for the low-temperature data is relatively small, all values being within 10% of the BCS prediction and in substantial agreement with the results obtained by other experimental techniques. The discrepancies be-tween the results of Saunders and Lawson and the later studies are ascribed to the greater probable error in the earlier experiments due to the restricted temperature range available to these workers. No efforts have been made to correlate the gap-parameter values obtained in these experiments with the electronic structure of the material.

d. Tin. Ever since the early work of Mackinnon (1954) and Bömmel

(1955), the superconducting state of tin has been under investigation continuously. This situation is primarily due to the convenient transition temperature, $3.76°K$, and the relative ease with which pure single-crystal samples can be prepared from this material. Tin appears to be the most anisotropic in its ultrasonic properties of all the nontransition metals, in both the normal and superconducting states. Due to the extreme complexity of the electronic structure of this material, the large body of data obtained in these experiments has not so far been subjected to the detailed analysis possible in the simpler metals. However, Perz and Dobbs (1967) have attempted to assign the various observed effective gap parameters to specific sheets of the Fermi surface in a consistent way. Although the complications due to strain-amplitude-dependent effects appear to be absent in this material, the very complicated Fermi-surface topology has given rise to some difficulties in interpretation of the data. Shepelev (1963) has reported surprisingly large gap-parameter values for certain propagation directions. These large values have been obtained under experimental conditions where an exceptionally rapid decrease in α_s near T_c is observed. Use of the normal fitting procedure to Eq. (83) invariably results in an anomalously large value for the gap parameter, even though tin is a weakly coupled superconductor. The probable origin of this difficulty was pointed out by Dobbs and Thomas (1966), and also independently by Perz (1966). The rapid drop in α_s near T_c is considered to be due to a significant contribution to α_n from the J_2 term in Eq. (90), which would vanish near T_c. When allowance is made for this contribution, the value of the gap parameter is close to those obtained for other effective zones derived from measurements where no anomalous behavior is observed. In general, the results obtained for longitudinal and shear waves by many workers are internally consistent and in general agreement with the results of other experiments. The values of gap parameter which are shown in Table XI give an anisotropy which is consistent with the value of $\langle a^2 \rangle_{av}$ in Table II.

Measurements on dilute tin–indium alloys by Claiborne and Einspruch (1965) and Perz and Dobbs (1967) have established that there is a smooth transition from the anisotropic behavior in the $ql \gg 1$ limit to isotropic behavior as ql decreases to small values. The energy-gap anisotropy decreases with the electron mean free path on alloying, and it has been shown in the later work of Claiborne and Einspruch (1966) that this behavior is independent of the nature of the doping agent. Essentially identical results are obtained when tin is doped with indium, cadmium, antimony, and bismuth. In all cases, the anisotropy in energy gap, which amounts to about 20% between the [001] and [110] effective zones for the pure material, disappears and the BCS value for an isotropic, weakly coupling superconductor results for all propagation directions studied in this system. This value of 3.56 represents the average gap over all relevant sheets of the Fermi surface, and is considered by Perz and Dobbs (1967) to be due to the removal of the focusing effect obtained at large ql values, rather than a reduction of the electron mean free path into the Anderson limit of $l < \xi_0$.

Phillips (1969) has recently measured the longitudinal ultrasonic attenuation along [100] and [001] in tin single crystals at 500 kHz. The resistance ratio of the samples varied from 800 to 30,000, and in all cases $ql \ll 1$, while satisfying the condition $l \gg \xi_0$ necessary to maintain the anisotropy of the energy gap. No difference is found between propagation along [100] and [001] in the most heavily doped samples, in contrast to the suggestion of Kadanoff and Pippard (1966). As shown in Fig. 24, there is a systematic decrease in attenuation with increasing purity, the attenuation curve lying above and below the BCS form for low- and high-purity specimens, respectively. The effect is more marked for [001] propagation. Using a simple model to calculate the details of elastic scattering in an anisotropic superconductor, Phillips has shown that scattering across the Fermi surface can account for the lack of orientational differences in the attenuation of the doped samples. Phonon scattering across the Fermi surface is shown to explain the variation of attenuation with purity; the anisotropy of this effect is consistent with the gap anisotropy.

The attenuation of hypersonic shear waves up to frequencies in excess of 1 GHz has been measured near the transition temperature in superconducting tin by Page and Leibowitz (1969). There is good agreement with the theoretical predictions of Cullen and Ferrell (1966).

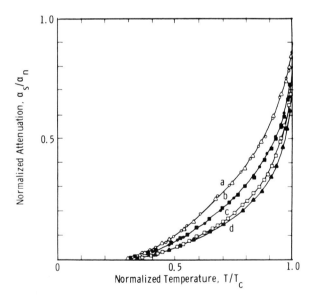

FIG. 24. Temperature variation of normalized attenuation α_s/α_n in tin samples of different purities for [100] and [001] longitudinal-wave propagation at 500 kHz: curve *a*, heavily doped; curve *b*, lightly doped; curve *c*, undoped, [001] propagation; curve *d*, undoped [100] propagation (after Phillips, 1969).

e. Lead. Ever since the pioneering experiments of Bömmel (1954), ultrasonic studies in lead have yielded remarkable results, the anomalous behavior depending on both sample purity and sound frequency. In the recent publication by Fate *et al.* (1968), which contains the most comprehensive set of measurements so far, it was concluded that the ultrasonic technique appears to be incapable of providing data from which reliable gap-parameter values can be extracted. By suitable selection of the temperature range and background attenuation, values for the gap parameter as disparate as 2.12 and 3.72 can be inferred from the same set of data. Earlier work by Deaton (1966) and others report similar behavior. Measurements of the gap parameter in the regime $ql \lesssim 1$ by Tittman and Bömmel (1966) yield a value for $2\varDelta$ of approximately 5.0 ± 0.3, independent of frequency from 50 to 1080 MHz, for longitudinal waves propagating along [111]. A very weak dependence on strain amplitude is reported. This value is substantially higher than the gap parameter derived from tunneling measurements, which is an unexpected result for an isotropic, strongly coupling, dirty superconductor.

In general, the situation is as summarized in Fig. 25, where it is seen that the attenuation near T_c falls more rapidly than theory predicts, and, at low temperatures, is significantly higher in value. When the mean free path is phonon-limited, the attenuation ratio α_s/α_n is frequently dependent and the

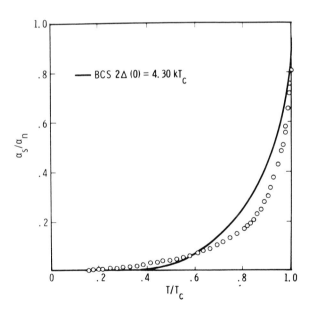

Fig. 25. Typical dependence of the reduced attenuation α_s/α_n as a function of the reduced temperature $t = T/T_c$ for propagation along [100] in lead (after Fate *et al.*, 1968).

maximum departure from the BCS theory occurs at low frequency.[7] On the other hand, the ratio α_s/α_n is nearly frequency independent when the mean free path is impurity-limited. In this case, the ratio depends on the concentration and type of impurities, but agrees more closely with the BCS theory.

A possible explanation of these results is that the phonon-limited mean free path is different in the superconducting and normal states, in contrast to the situation for impurity scattering. This difference reflects the lowering of the mean free path due to its energy dependence in the normal state (Ambegaokar and Woo, 1965), as well as any intrinsic effect. From Eq. (80), we can then write

$$\alpha_s/\alpha_n = [F(ql_s)/F(ql)]2f(\Delta) \qquad (124)$$

where $F(ql)$ is written for the attenuation integral occurring in the Pippard theory, and l_s is the electron mean free path in the superconducting state. Fate *et al.* (1968) have analyzed their data using this formula, assuming the form of $F(ql)$ given by the free-electron theory [cf. Eq. (13)]. Their analysis approximately explains the main features of the experimental data, but cannot even qualitatively explain the low-temperature behavior (cf. Fig. 25). Since the above equation reduces to the BCS form for l equal to l_s, it cannot explain the dependence of α_s/α_n on impurity concentration for heavy doping. Indeed, it consistently underestimates the deviations from the BCS theory in the impurity-scattering regime. The more general strong-coupling theory of Woo (1967, 1968) is claimed to give a better fit to the data.

Deaton (1969) has measured the superconducting shear-wave attenuation in high-purity lead single crystals for [001] and [110] propagation. As in the case of longitudinal waves, there is a pronounced frequency dependence of the variation of α_s/α_n with reduced temperature, making estimates of the gap parameter rather unreliable. The data indicate that the ratio α_D/α_{nt} is larger in lead than in either aluminum or indium. The values of $\lambda(0)$, computed from the temperature dependence of α_t^E using the analysis of Fossheim (1967), appear to agree with those obtained from thin-film data.

f. Mercury. The first measurements of ultrasonic attenuation in superconducting mercury ($T_c = 4.1°K$) were carried out by Mackinnon and Myers (1959) on a polycrystalline sample of 99.999% purity at 10 MHz. These measurements were extended in frequency to 60 MHz by Chopra and Hutchinson (1959). In both of these investigations, deviations from BCS-like behavior were reported. The later, more sophisticated investigations of Thomas *et al.* (1966) and Newcomb and Shaw (1968) on oriented single-crystal samples have confirmed the anomalous behavior, which exhibits a pronounced dependence upon frequency, in the range 10–250 MHz, for pure

---

[7] This behavior has been verified by Sebastian and Liebowitz (1969), who have measured the attenuation of longitudinal waves in very pure lead near the transition temperature at frequencies up to 450 MHz. Their data indicate that the departure from BSC behavior becomes less significant at higher frequencies in this temperature range.

samples in which $ql > 1$. In general, the attenuation at the higher frequencies near T_c falls considerably more rapidly than theoretical predictions for all propagation directions studied. Although strong coupling effects undoubtedly are present, it is possible that at least part of the anomalous behavior of α_s in pure samples may arise from a contribution to α_n associated with the J_2 term in Eq. (90), as discussed previously. At low frequencies, or at small ql values produced by alloying, the deviations from theory are minimal, and a fit to the BCS relationship yields an approximate gap parameter of 3.5.

The extremely high-frequency measurements of Ferguson and Burgess (1967) on polycrystalline samples at 9 GHz show no evidence of anomalous behavior, and reasonably good agreement with theory is obtained for a gap parameter of approximately 4.0 ± 0.3. No indications of any amplitude-dependent effects have been observed in these investigations. Their data do not exhibit a discontinuity near the critical temperature, either because the phonon energy is much smaller than the energy gap, or because of an averaging over the gaps in a polycrystalline sample.

3. *Transition Metals*

Pure single crystals have been prepared from only a few transition elements, primarily because of difficulties arising from the high melting points of the majority of these materials, typically in excess of $1500°C$, and their tendency to form solid solutions with gases such as nitrogen. Of the super-conducting transition metals, only three—niobium, molybdenum, and rhenium—have been the subjects of ultrasonic investigations, where the product ql was significantly greater than unity. Only one other super-conducting transition element, tungsten, is available in very pure form, but its transition temperature of $0.015°K$ places it well outside the range of present experimental techniques.

a. Niobium, Vanadium, and Tantalum. Measurements of the attenuation of longitudinal waves propagating along the principal directions [100], [110], and [111] in the normal and superconducting states of niobium of moderate purity (residual resistivity ratio of approximately 300) have been carried out by Perz and Dobbs (1967) over the frequency range from 290 MHz to 1.28 GHz. The attenuation in the vicinity of the transition temperature is found to decrease considerably more rapidly than the BCS theory predicts, although fitting the data obtained at lowest temperatures by the usual procedure yields values for the gap parameter quite close to the BCS value, as can be seen from Table XI. Niobium is a strong superconductor, with $T_c/\theta_D = 0.038$, so that deviations from BCS-like behavior, particularly near T_c, are not unexpected. Perz and Dobbs have shown that the observed behavior of the attenuation near T_c is related to the high value of the coupling constant and is consistent with the specific-heat anomaly of T_c for this material. A small anisotropy in gap parameter is observed at the highest frequencies employed ($ql \approx 5$), with the minimum value occurring for the effective zone for the propagation direction of highest crystalline symmetry.

Transverse attenuation measurements for $ql < 1$ have been made in impure niobium, vanadium, and tantalum by Levy *et al.* (1963). Their data can be fitted quite well to the BCS theory with the usual gap parameter.

b. Molybdenum. Initial measurements by Horowitz and Bohm (1962) on relatively impure molybdenum in the $ql < 1$ regime yielded an average value for the gap parameter in agreement with the BCS theory. Subsequently, the superconducting-state longitudinal-wave attenuation has been measured up to 1 GHz by Jones and Rayne (1967) for propagation along the principal crystallographic directions [100], [110], and [111]. Data were obtained down to $0.40°\text{K}$ (or reduced temperature $t = T/T_c = 0.44$) on electron-beam zone-refined specimens having a resistance ratio of approximately 6000. For each propagation direction, the results obtained in the temperature range $0.4 < t < 0.6$ were fitted by the usual least-squares analysis to Eq. (83). No evidence of any amplitude-dependent effects was detected, thereby excluding complications due to dislocation interactions. A small anisotropy in the gap parameter is observed, as indicated by the values in Table XI. It is found that the ordering of the energy-gap parameter is similar to that of the normal-state attenuation coefficient.

Measurements near 9 GHz by Pike (1969) for longitudinal sound propagation along [111] give results for both α_n and the gap parameter in good agreement with the low-frequency values. In contrast to the case of Al, no evidence of a discontinuity in the attenuation near T_c is observed, although there is still an apparent shift in the transition temperature for the reasons discussed previously. The absence of a discontinuity is explained in terms of the existence of multiple energy gaps in this metal.

c. Rhenium. The attenuation of longitudinal waves propagating along the principal directions [0001], [10$\bar{1}$0], and [11$\bar{2}$0] has been measured in the normal and superconducting states in the frequency range 250 MHz to 1.25 GHz by Jones and Rayne (1966). The normal-state measurements have been discussed in an earlier section of this chapter. Observations in the superconducting state ($T_c = 1.70°\text{K}$) extend to a temperature of $0.45°\text{K}$ (a reduced temperature $t = T/T_c$ of 0.26). As in the case of molybdenum, no evidence of any amplitude-dependent effects has been observed. Use of the usual fitting procedure to Eq. (83) gives values of the superconducting energy-gap parameter shown in Table XI. Again, the observed small anisotropy in gap parameter appears to be similar in symmetry to that of the normal-state attenuation coefficient. The value of gap parameter which appears appropriate to the effective zone for [11$\bar{2}$0] propagation differs markedly from the values obtaining for the other principal directions. This effect seems to be quite selective, since additional measurements for propagation along a direction in the basal plane midway between [10$\bar{1}$0] and [11$\bar{2}$0] yield results essentially the same as for [10$\bar{1}$0] propagation. Since the basal plane in a crystal of hexagonal symmetry is degenerate for longitudinal-mode propagation, the effect is presumably electronic in origin. It is probably associated with a pronounced variation of the electron mean free path over the Fermi surface, the value on the effective zone for [11$\bar{2}$0] propagation being substantially shorter than the average value.

Note added in proof: Approximate calculations by Lea (1968) indicate that the observed anisotropy of the electronic attenuation for zinc and cadmium may be due to the absence of the third-zone butterfly in their Fermi surfaces. For magnesium, in which the butterfly is present, the electronic attenuation is essentially isotropic and is not too different from the free electron value (Llewellyn, 1969).

ACKNOWLEDGMENTS

One of the authors (JAR) would like to acknowledge the support of the National Science Foundation in preparing the manuscript for this article.

REFERENCES

Akhiezer, A. I., Kaganov, M. I., and Liubarskii, G. Ia. (1957). *Zh. Eksperim. i Teor. Fiz.* [*Soviet Phys. JETP (English Transl.*) **5**, 685].
Altmann, S. L., and Cracknell, A. P. (1965). *Rev. Mod. Phys.* **37**, 19.
Ambegaokar, V., and Woo, J. (1965). *Phys. Rev.* **139**, A1818.
Anderson, J. R. (1968) (Private communication).
Anderson, J. R., O'Sullivan, W. J., and Schirber, J. E. (1969). *Phys. Rev.* **153**, 721.
Ashcroft, N. W. (1963). *Phil. Mag.* **8**, 2055.
Ashcroft, N. W., and Lawrence, W. E. (1968). *Phys. Rev.* **175**, 938.
Bardeen, J., Cooper, L. N., and Schrieffer, J. R. (1957). *Phys. Rev.* **108**, 1175.
Berre, K., and Olsen, T. (1965). *Phys. Status Solidi* **11**, 657.
Bezuglyi, P. A., Galkin, A. A., and Korolyuk, A. P. (1960). *Zh. Eksperim i Teor. Fiz.* **39**, 7 [*Soviet Phys. JETP (English Transl.*) **12**, 4 (1961)].
Biondi, M. A., Garfunkel, M. P., and Thompson, W. A. (1964). *Phys. Rev.* **136**, A1471.
Bliss, E. S., and Rayne, J. A. (1968). *Phys. Letters* **26A**, 278.
Bliss, E. S., and Rayne, J. A. (1969). *Phys. Rev.* **177**, 673.
Blount, E. I. (1959). *Phys. Rev.* **114**, 418.
Bobetic, V. M. (1964). *Phys. Rev.* **136**, A1535.
Bohm, H. V. and Horowitz, N. H. (1963). *Proc. Intern. Conf. Low Temp. Phys., 8th, London, 1962*, p. 198. Butterworths, London.
Bolef, D. I., and de Klerk, J. (1962). *J. Appl. Phys.* **33**, 2311.
Bömmel, H. E. (1954). *Phys. Rev.* **96**, 220.
Bömmel, H. E. (1955). *Phys. Rev.* **100**, 758.
Chambers, R. G., (1952). *Proc. Phys. Soc. (London)* **A65**, 458.
Chambers, R. G. (1961). *Proc. Intern. Conf. Low Temp. Phys., 7th, Toronto, Ont., 1960*, p. 11. Univ. of Toronto Press, Toronto.
Chopra, K. L., and Hutchinson, T. S. (1959). *Can. J. Phys.* **37**, 1100.
Claiborne, L. T., and Einspruch, N. G. (1965). *Phys. Rev. Letters* **15**, 862.
Claiborne, L. T., and Einspruch, N. G. (1966). Phys. Rev. **151**, 229.
Claiborne, L. T., and Morse, R. W. (1964). *Phys. Rev.* **136**, A893.
Clem, J. R., (1966). *Ann. Phys.* (*N.Y.*) **40**, 268.
Cullen, J. R., and Ferrell, R. A. (1966). *Phys. Rev.* **146**, 282.
David, R. (1964). *Philips Res. Rept.* **19**, 524.
David, R., van der Laar, H. R., and Poulis, N. J. (1962). *Physica* **28**, 330.
David, R., van der Laar, H. R., and Poulis, N. J. (1963). *Physica* **29**, 357.
Deaton, B. C. (1966). *Phys. Rev. Letters* **16**, 577.
Deaton, B. C. (1969). *Phys. Rev.* **177**, 688.
de Klerk, J., and Kelly, E. F. (1965). *Rev. Sci. Instr.* **36**, 506.

Dobbs, E. R., and Thomas, G. P. (1966). *Proc. Intern. Conf. Low Temp. Phys.*, *10th*, *Moscow*, *1966*. **II B**, 260.

Fagen, E. A. (1967). Ph.D. Thesis, Univ. of Pittsburgh, Pittsburgh, Pennsylvania.

Fagen, E. A., and Garfunkel, M. P. (1967). *Phys. Rev. Letters* **18**, 897.

Farrell, D., Park, J. G., and Coles, B. R. (1964). *Phys. Rev. Letters* **13**, 328.

Fate, W. A. (1968). *Phys. Rev.* **172**, 402 (1968).

Fate, W. A., Shaw, R. W., and Salinger, G. L. (1968). *Phys. Rev.* **172**, 413.

Fawcett, E., and Griffiths, D. (1962). *J. Phys. Chem. Solids* **23**, 1631.

Ferguson, R. B., and Burgess, J. H. (1967). *Phys. Rev. Letters* **19**, 494.

Fil', V. D., Shevchenko, O. A., and Bezuglyi, P. A. (1967). *Zh. Eksperim. i Teor. Fiz.* **52**, 891 [*Soviet Phys JEPT (English Transl.*), **25**, 587].

Fil', V. D., Shevchenko. O. A., and Bezuglyi, P. A. (1968). *Zh. Eksperim. i Teor. Fiz.* **54**, 413 [*Soviet Phys. JETP (English Transl.)* **27**, 223].

Fossheim, K. (1967). *Phys. Rev. Letters* **19**, 81.

Fossheim, K., and Leibowitz, J. R. (1966). *Phys. Letters* **22**, 140.

Fossheim, K., and Torvatn, B. (1969), *J. Low Temp. Phys.* **1**, 341.

Gold, A. V., and Priestley, M. G. (1960). *Phil. Mag.* **5**, 1089.

Gonz, J. D., and Neighbours, J. R., (1965). *Bull. Am. Phys. Soc.* **10**, 319.

Gorter, C. J., and Casimir, H. G. B. (1934). *Physik. Z.* **35**, 963.

Griffin, A., and Ambegaokar, V. (1965). *Proc. Intern. Conf. Low Temp. Phys. 9th.*, *Columbus*, *Ohio*, *1964*, p. 524. Plenum Press, New York.

Hepfer, K. (1968). Private communication.

Hepfer, K., and Rayne, J. A. (1968). *Phys. Letters* **28A**, 163.

Hepfer, K., and Rayne, J. A. (1969). *Phys. Letters* **30A**, 281.

Holstein, T. (1956). *Res. Memo* 60-94698-3-M17. Westinghouse Res. Lab., Pittsburgh, Pennsylvania, unpublished.

Holstein, T. (1959). *Phys. Rev.* **113**, 479.

Horowitz, N. H., and Bohm, H. V. (1962). *Phys. Rev. Letters* **9**, 313.

Inoue, S., and Tsuji, M. (1967). *J. Phys. Soc. Japan* **22**, 192.

Jain, L., and Jaggi, R. (1964). *Phys. Rev.* **135**, A708.

Jones, C. K., and Rayne, J. A. (1964). *Phys. Letters* **13**, 282.

Jones, C. K., and Rayne, J. A. (1966). *Phys. Letters* **21**, 510.

Jones, C. K., and Rayne, J. A. (1967). *Phys. Letters* **26A**, 75.

Kadanoff, L. P., and Falko, I. I. (1964). *Phys. Rev.* **136**, A1170.

Kadanoff, L. P., and Pippard, A. B. (1966). *Proc. Roy. Soc.* **A292**, 299.

Kittel, C. (1955). *Acta Met.* **3**, 295.

Lax, E. (1959). *Phys. Rev.* **115**, 1591.

Lea, M. J. (1968). Ph.D. Thesis, Univ. of Lancaster, Lancaster, England.

Lea, M. J., and Dobbs, E. R. (1968). *Phys. Letters* **27A**, 556.

Lea, M. J., Llewellyn, J. D., Peck, D. R., and Dobbs, E. R. (1968). *Proc. Intern. Conf. Low Temp. Phys.*, *11th*, *St. Andrews*, 1968, p. 733.

Leibowitz, J. R. (1964a). *Phys. Rev.* **133**, A84.

Leibowitz, J. R. (1964b). *Phys. Rev.* **136**, A22.

Leibowitz, J. R., and Fossheim, K. (1966). *Phys. Rev. Letters* **17**, 636.

Levy, M., Kagiwada, R., and Rudnick, I. (1963). *Phys. Rev.* **132**, 2039.

Llewellyn, J. D. (1969). Ph.D. Thesis, Univ. of Lancaster, Lancaster, England.

Love, R. E., and Shaw, R. W. (1964). *Rev. Mod. Phys.* **36**, 260.

Macfarlane, R. E., and Rayne, J. A. (1967). *Phys. Rev.* **162**, 532.

Mackinnon, L. (1954). *Phys. Rev.* **98**, 1181, 1210.

Mackinnon, L., and Myers, A. (1959). *Proc. Phys. Soc. (London)* **73**, 291.

Markowitz, D., and Kadanoff, L. P. (1963). *Phys. Rev.* **131**, 563.
Mase, S., Fujimori, Y., and Mori, H. (1966) *J. Phys. Soc. Japan* **21**, 1744.
Mason, W. P. (1966). *Phys. Rev.* **143**, 299.
Mattheiss, L. F. (1960). *Phys. Rev.* **151**, 450.
Mattheiss, L. F. (1964). *Phys. Rev.* **134**, A970.
Mattheiss, L. F. (1965). *Phys. Rev.* **139**, A1893.
Mattheiss, L. F. (1966). *Phys. Rev.* **151**, 1176.
Melz, P. J. (1966). *Phys. Rev.* **152**, 540.
Morse, R. W. (1959). *Progr. Cyrog.* **1**, 219.
Morse, R. W., and Bohm, H. V. (1957). *Phys. Rev.* **108**, 1094.
Morse, R. W., and Bohm, H. V. (1959). *J. Acoust. Soc. Am.* **31**, 1523.
Morse, R. W., Olsen, T., and Gavenda, J. D., (1959) *Phys. Rev. Letters* **3**, 15.
Morse, R. W., Tomarkin, P., and Bohm, H. V. (1956). *Phys. Rev.* **101**, 1610.
Mühlschlegel, B. (1959). *Z. Physik* **155**, 313.
Nam, S. B. (1967). *Phys. Rev.* **156**, 470.
Newcomb, C. P., and Shaw, R. W. (1968). *Phys. Rev.* **173**, 509.
Page, E. A., and Leibowitz, J. R. (1969). *Bull. Am. Phys. Soc.* **14**, 380.
Perz, J. M. (1966). *Can. J. Phys.* **44**, 1765.
Perz, J. M. (1970). *J. Phys.* **3**, C347.
Perz, J. M., and Dobbs, E. R. (1966). *Proc. Roy. Soc.* **A296**, 113.
Perz, J. M., and Dobbs, E. R. (1967). *Proc. Roy. Soc.* **A297**, 403.
Peverly, J. R. (1966) *In* "Physical Acoustics" (W. P. Mason, ed.), Vol. IVA. Academic Press, New York.
Phillips, W. A. (1969). *Proc. Roy. Soc.* **A309**, 259.
Pike, G. E. (1969). Ph.D. Thesis, Univ. of Pittsburgh, Pittsburgh, Pennsylvania.
Pippard, A. B. (1955). *Phil. Mag.* **46**, 1104.
Pippard, A. B. (1960). *Proc. Roy. Soc.* **A257**, 165.
Pokrovskii, V. A. (1961). *Zh. Eskperim. i Teor. Fiz.* **40**, 898 [*Soviet Phys. JEPT* (*English Transl.*) **13**, 628].
Powell, R. L. (1966). NBS Rept. No. 9176, unpublished.
Privorostskii, I. A. (1962). *Zh. Eksperim. i. Teor. Fiz.* **42**, 450 [*Soviet Phys. JETP* (*English Transl.*) **15**, 315].
Privorotskii, I. A. (1962). *Zh. Eksperim. i Teor. Fiz..* **43**, 1331 [*Soviet Phys. JETP* (*English Transl.*) **16**, 945 (1963)].
Radebaugh, R., and Keesom, P. H. (1966). *Phys. Rev.* **149**, 209.
Reneker, D. (1959). *Phys. Rev.* **115**, 303.
Roberts, B. W. (1968). *In* "Physical Acoustics" (W. P. Mason, ed.), Vol. IVB. Academic Press, New York.
Saunders, G. A., and Lawson, A. W. (1964). *Phys. Rev.* **135**, A1161.
Sebastian, R. L., and Leibowitz, J. R. (1969). *Bull. Am. Phys. Soc.* **14**, 380.
Seraphim, D. P., Novick, D. T., and Budnick, J. I. (1961). *Acta Met.* **9**, 446.
Shepelev, A. G. (1963). *Zh. Eksperim. i. Teor. Fiz.* **45**, 2076 [*Soviet Phys. JETP* (*English Transl.*) **18**, 1423 (1964)].
Shepelev, A. G., and Filimonov, G. D. (1965). *Zh. Eksperim. i Teor. Fiz.* **48**, 1054 [*Soviet Phys. JETP* (*English Transl.*) **21**, 704].
Shoenberg, D., and Roaf, D. J. (1962). *Phil. Trans. Roy. Soc. London Ser A***225**, 85.
Shoenberg, D., and Stiles, P. J. (1964). *Proc. Roy. Soc.* **A281**, 62.
Shoenberg, D., and Watts, B. R. (1965). *Proc. Intern. Conf. Low Temp. Phys., 9th, Columbus, Ohio, 1964*, p. 831. Plenum Press, New York.
Sinclair, A. C. E. (1967). *Proc. Phys. Soc.* (*London*) **92**, 962.

Smilowitz, B. (1965). Thesis, Univ. of Pittsburgh, Pittsburgh, Pennsylvania.

Snow, J. A. (1968). *Phys. Rev.* **172**, 455.

Swenson, C. A. (1962). *Phys. Rev. Letters* **9**, 370.

Templeton, I. M. (1966). *Proc. Roy. Soc.* **A292**, 413.

Thomas, R. L., Wu, H. C., and Tepley, N. (1966). *Phys. Rev. Letters* **17**, 22.

Timms, W. E. (1968). Thesis, Univ. of Lancaster, unpublished.

Timms, W. E., and Dobbs, E. R. (1968). *Proc. Intern. Conf. Low Temp. Phys., 11th, St. Andrews, (1968)*, p. 748.

Tittman, B. R., and Bömmel, H. E. (1966). *Phys. Rev.* **151**, 189.

Toxen, A. M., and Tansal, S. (1965). *Phys. Rev.* **137**, A211.

Tsuneto, T. (1961). *Phys. Rev.* **121**, 402.

Walther, K. (1968). *Phys. Rev.* **174**, 782.

Wang, E. Y., and McCarthy, K. A. (1969). *Phys. Rev.* **183**, 653.

Wang, E. Y., Kolouch, R. J., and McCarthy, K. A. (1968). *Phys. Rev.* **175**, 723.

Weber, R. (1964). *Phys. Rev.* **133**, A1487.

Weil, R. B., and Lawson, A. W. (1966). *Phys. Rev.* **141**, 452.

Willard, Jr., H. J. (1968). *Phys. Rev.* **175**, 367.

Willard, Jr., H. J., Shaw, R. W., and Salinger, G. L. (1968). *Phys. Rev.* **175**, 362.

Woo, J. W. F. (1967). *Phys. Rev.* **155**, 429.

Woo, J. W. (1968). *Phys. Rev.* **172**, 423.

Zavaritskii, N. V. (1958). *Zh. Eksperim. i Teor. Fiz.* **34**, 1116 [*Soviet Phys. JETP (English Transl.)* **7**, 773].

—4—

Excitation, Detection, and Attenuation of High-Frequency Elastic Surface Waves

K. DRANSFELD* and E. SALZMANN†

Physik-Department der Technischen Hochschule
München, Germany

* During the preparation of this article, one of the authors (K.D.) spent a semester at the Physics Department of Cornell University.

† Present address: Rohde and Schwarz, München, Germany.

I. Introduction

The existence of elastic waves traveling along the free surface of an elastic half-space was first predicted by Lord Rayleigh (1885). The particle motion of these so-called Rayleigh waves is confined to a layer approximately one wavelength thick at the surface of the elastic solid and its amplitude decays exponentially with depth. Rayleigh waves have a certain resemblance to waves on fluid surfaces: In both cases the particle motion occurs in elliptical orbits in a plane perpendicular to the surface and parallel to the direction of propagation. The restoring forces are, of course, very different: elastic forces in solids, gravity and surface tension in fluids (see, e.g., Landau and Lifshitz 1959).

Originally, Rayleigh waves were of specific interest only to seismologists: As surface waves propagate into only two dimensions, their energy decays more slowly with increasing distance from their point of origin than that of bulk waves, namely, proportional to $1/r$ as compared with $1/r^2$ for bulk waves. Therefore, at large distances from the epicenter of an earthquake, surface shocks are caused mainly by surface waves. (The seismological aspects of elastic surface waves are discussed in detail, e.g., by Ewing et $al.$, 1957.)

Rayleigh waves became interesting for technical applications when Firestone and Frederick (1946) had succeeded in their piezoelectric generation. At frequencies of about 5 MHz, the wavelength and hence the penetration depth is of the order of 1 mm, and this fact was utilized for materials testing: It became possible to detect very tiny surface cracks, since Rayleigh waves are reflected by any discontinuities in the surface.

It was only in recent years that transduction methods for Rayleigh waves have been sufficiently improved for their excitation up to frequencies in the gigahertz range. At these frequencies a variety of new applications for Rayleigh waves becomes apparent, most of which are based on their following properties: (1) Rayleigh waves have a small penetration depth into the elastic solid; consequently, (2) the acoustic energy is accessible at any point along the path of a Rayleigh wave; and (3) the elastic properties of the surface can be easily influenced, so that it is possible to guide Rayleigh waves on the surface in a manner similar to the way electromagnetic waves are guided by waveguides.

Because of these properties, Rayleigh waves turn out to be a valuable research tool in thin-film physics and surface physics, and they are of particular interest in conjunction with microelectronic circuits, as discussed in Section V.

The purpose of this chapter is to describe the experimental aspects of Rayleigh waves at high frequencies, their excitation, detection, and absorption as well as some preliminary applications. The theory of Rayleigh wave propagation on crystal surfaces is mentioned only as far as necessary for an understanding of the experiments. (A more detailed discussion of these problems is given by Farnell (1970) in another volume of this series. See also Lim and Farnell, 1968.)

II. General Properties of Rayleigh Waves[1]

In this section the general formalism for the calculation of the phase velocity and the particle displacement amplitude is given for Rayleigh waves. First, Rayleigh waves in isotropic solids will be discussed, then crystals and piezo-electric materials will be considered.

Let the position of a particle inside an elastic solid be determined by three cartesian coordinates x_1, x_2, x_3. If u_k denotes the excursion of a particle from its equilibrium position into the direction x_k, the elastic strain tensor is defined as

$$S_{kl} = \tfrac{1}{2}(u_{k,l} + u_{l,k}) \tag{1}$$

where an index preceded by a comma denotes differentiation with respect to a space coordinate.

The elastic stress tensor T_{ij} is related to the strain tensor by the elastic stiffness constants c_{ijkl} as

$$T_{ij} = c_{ijkl}S_{kl} \tag{2}$$

where the summation convention for repeated tensor indices is employed.

The elastic force F_j exerted on a volume element along the x_j direction is given by

$$F_j = T_{ij,i} \tag{3}$$

from which the three equations of motion follow,

$$\rho\ddot{u}_j = T_{ij,i} = c_{ijkl}u_{k,li} \tag{4}$$

where ρ is the mass density and the dot denotes differentiation with respect to time.

For the existence of Rayleigh waves, an elastic half-space with a free surface has to be considered. Let this surface be determined by the coordinate $x_3 = 0$ with positive values x_3 inside the solid. Solutions of Eq. (4) corresponding to surface waves must have an amplitude decaying with depth into the solid. Consequently, a solution for waves traveling in the x_1 direction is of the form

$$u_i = \beta_i \exp[-k\alpha_i x_3 + ik(x_1 - vt)] \tag{5}$$

where β_i is an amplitude factor, α_i is a positive constant expressing the decay of the amplitude with increasing distance from the surface, v is the velocity of the surface wave, and $k = 2\pi/\lambda$ is its wave vector.

---

[1] This section is based on books and papers by Landau and Lifshitz (1959), Victorov (1967), Stonely (1955), Coquin and Tiersten (1967), and Campbell and Jones (1968).

Amplitude and phase velocity of the Rayleigh wave are calculated by substituting Eq. (5) into Eq. (4) and using the boundary condition of vanishing normal stress at the free surface, i.e.,

$$T_{3j} = 0, \quad \text{for} \quad x_3 = 0 \tag{6}$$

A. Isotropic Solids

For isotropic solids (having only two independent elastic constants) the phase velocity v of a Rayleigh wave can be expressed in terms of the transverse sound velocity v_t by the equation

$$\gamma^6 - 8\gamma^4 + 8\gamma^2[3 - 2(v_t^2/v_1^2)] - 16[1 - (v_t^2/v_1^2)] = 0 \tag{7}$$

where $\gamma = v/v_t$ and v_1 is the longitudinal velocity of sound. This so-called Rayleigh equation has six roots for v depending only on the ratio v_t/v_1, which in turn depends only on Poisson's ratio σ of the elastic solid,

$$\gamma = v_t/v_1 = [(1 - 2\sigma)/(2 - 2\sigma)]^{1/2} \tag{8}$$

The solution for γ corresponding to the Rayleigh wave is approximated by

$$\gamma = (0.87 + 1.12\sigma)/(1 + \sigma) \tag{9}$$

(See Bergmann, 1954; Victorov, 1967.)

For $0 < \sigma < 0.5$, the phase velocity of Rayleigh waves varies between $0.87v_t$ and $0.96v_t$ (Fig. 1), which shows that Rayleigh waves are the slowest elastic waves ($v < v_t < v_1$).

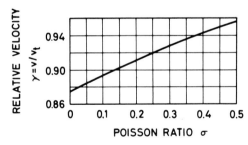

Fig. 1. Phase velocity v of a Rayleigh wave compared to the velocity v_t of transverse waves as a function of Poisson's ratio σ. (See Landau and Lifshitz, 1959; or Mason, 1958.)

The particle motion in a Rayleigh wave occurs in elliptic orbits in a plane through the direction of propagation and normal to the surface. Figure 2 shows the horizontal amplitude u_0 and the vertical amplitude w_0 as functions

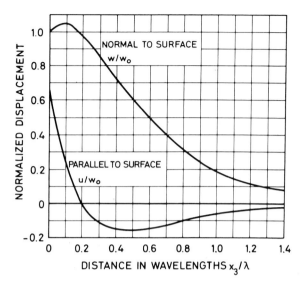

FIG. 2. Horizontal (u) and vertical (w) amplitude of the particle motion in a Rayleigh wave (normalized to the vertical amplitude at the surface w_0) as function of the distance x_2 (expressed in wavelengths λ) from the surface. (See Mason, 1958; or Victorov, 1967.)

of distance from the surface. It can be seen that the sense of rotation along the elliptic orbit is reversed at a depth of approximately 0.2λ. From the magnitude of the respective amplitudes it is obvious that the major part of the acoustic energy is transported in a layer of thickness λ adjacent to the surface.

B. CRYSTALS

For crystals there exist at least three independent elastic constants (e.g., in cubic crystals). Therefore, aside from some particular symmetry directions, Rayleigh waves exist in most cases in a "generalized" form. This means that there may also be a component of the particle motion along the x_2 direction. The decay of the amplitude (see Eq. 5) is described by complex constants α_i having a positive real part, which leads to an oscillatory decay.

For the design of experiments with Rayleigh waves it is important to know that in crystals the directions of phase and group velocity are not necessarily collinear. From a polar diagram of the phase velocity v as a function of the direction of propagation φ, the direction of energy transport is easily found (Fig. 3). Directions of collinear phase and group velocity are determined by the condition

$$dv/d\varphi = 0 \tag{10}$$

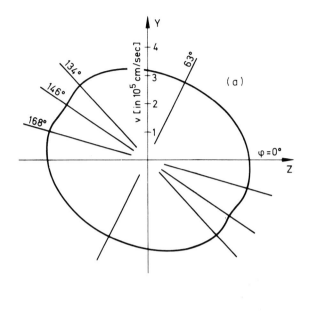

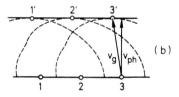

FIG. 3. (a) Polar diagram of the phase velocity v_{ph} of Rayleigh waves on the surface of X-cut quartz and directions of collinear phase and group velocity, $v_{\mathrm{ph}} \parallel v_{\mathrm{g}}$. (After data of Engan et al. 1967; and Coquin and Tiersten 1967.) (b) Construction of the direction of energy flow (group velocity) of a Rayleigh wave with phase velocity parallel to the Y-axis, using Huygens' principle. The wavefront 1′, 2′, 3′, is the envelope of the Huygens wavelets (dashed lines) emitted from the points 1, 2, 3 on the initial wavefront.

C. PIEZOELECTRIC SOLIDS

For piezoelectric solids the equations describing the particle motion in a Rayleigh wave have to be supplemented by terms expressing the piezoelectric behavior of the solid (see, e.g., Coquin and Tiersten, 1967; Bleustein, 1969[2]).

If E_k and D_i are components of the electric field and the dielectric displacement, respectively, e_{kij} and ε_{ik} are the peizoelectric and dielectric

---

[2] The Bleustein wave is a face shear surface wave in piezoelectric solids which has no counterpart in nonpiezoelectric solids of the same elastic symmetry.

tensor, respectively, and φ is the electric potential, the new relations for the piezoelectric solid are

$$T_{ij} = c_{ijkl} S_{kl} - e_{kij} E_k \tag{11}$$

$$D_i = e_{ikl} S_{kl} + \varepsilon_{ik} E_k \tag{12}$$

Using

$$E_k = -\varphi_{,k} \tag{13}$$

and

$$D_{i,i} = 0 \tag{14}$$

the equations of motion and an additional equation for φ are

$$\rho \ddot{u}_j = c_{ijkl} u_{k,li} + e_{kij} \varphi_{,ki} \tag{15}$$

$$0 = e_{ikl} u_{k,li} - \varepsilon_{ik} \varphi_{,ki} \tag{16}$$

Solutions of this set of equations corresponding to surface waves are of the same form as Eq. (5).

In addition to the boundary condition of Eq. (6) (vanishing normal stress at the surface), continuity of the electric potential φ and the normal component of the dielectric displacement must be provided across the surface.

In contrast to Rayleigh waves in nonpiezoelectric materials, electric fields are associated with Rayleigh waves in piezoelectric solids. Inside the solid the amplitude decay of these fields is similar to that of the particle motion (Fig. 4). Outside the solid the electric fields normal and parallel to the surface decay exponentially with increasing distance from the surface.

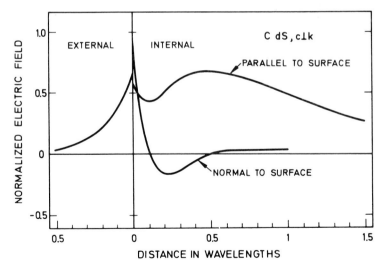

FIG. 4. Electric fields associated with a Rayleigh wave propagating on the surface of a piezoelectric crystal (on a basal plane of cadmium sulfide) (after White, 1967).

The existence of these piezoelectric fields at the surface makes it possible to detect—and, conversely, to generate—Rayleigh waves by means of electrodes deposited on the surface (see Section III,B).

III. Excitation and Detection of Rayleigh Waves

Rayleigh waves are excited by applying a space- and time-periodic mechanical stress to the surface of an elastic solid. For effective excitation at a given frequency the spatial period has to be chosen equal to the corresponding Rayleigh wavelength. This spatial periodicity of the applied stress is obtained, for example, by placing a suitable mechanical structure on the surface, or, in piezoelectric solids, by exposing the surface to a spatially periodic electrical rf field.

Most of these methods for excitation described in the following three sections A–C can also be used for the detection of Rayleigh waves. Other methods (see Sections D and E) are suitable only for either excitation or detection.

A. MECHANICAL TRANSDUCERS

The mechanical methods—which have been discussed in detail by Victorov (1967)—are characterized by the fact that a spatially periodic vibration of a transducer solid is mechanically coupled to the surface of another solid on which the Rayleigh wave is to be excited. The acoustic contact between both surfaces is often improved by means of an oil film. The different transducers are distinguished only by the methods by which the spatial variation of the driving stress is obtained.

Firestone and Frederick (1946) used a rectangular Y-cut quartz plate as transducer (Fig. 5a). Its thickness-shear vibration (shear motion along the X-direction) was coupled by a thin film of oil to the surface of an elastic solid, where it excited Rayleigh waves with a wavelength equal to twice its X dimension, $\lambda = 2a$. Most effective excitation of Rayleigh waves was found for a quartz plate having a ratio of length a to thickness d (see Fig. 5a) equal to 7:1. However, in this transduction process for Rayleigh waves the greater part of the mechanical energy is radiated as a shear wave into the depth of the solid.

An analogous transduction method is obtained when the Y-cut quartz plate is replaced by an X-cut crystal. Its thickness vibration exerts a periodic stress on the surface and again excites Rayleigh waves with a wavelength $\lambda = 2a$.

One of the most commonly used mechanical methods for the excitation of Rayleigh waves is the so-called "wedge method" sketched in Fig. 5(b) (Cook and Valkenburg, 1954; Minton, 1954): An X-cut quartz (thickness vibrator) launches a plane longitudinal wave into a prismatic block bonded to the surface of a solid by means of an oil film. The longitudinal wave in the wedge (wavelength λ_w) impinging on the interface at an angle ϑ causes

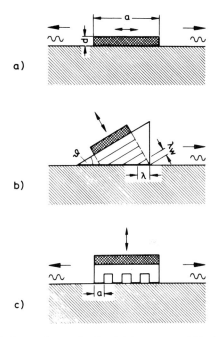

FIG. 5. Mechanical transducers for Rayleigh waves. (a) Excitation by means of a Y-cut quartz (thickness-shear vibrator). Rayleigh wavelength $\lambda = 2a$. (b) Excitation by means of a longitudinal wave launched in a plastic wedge; $\sin \vartheta = \lambda_w/\lambda = v_w/v$; the wedge angle ϑ is determined by the respective wavelengths or velocities of the longitudinal wave in the wedge and the Rayleigh wave in the adjacent solid. (c) Excitation by means of a longitudinal transducer (e.g., an X-cut quartz) on top of an aluminum comb. Rayleigh wavelength $\lambda = 2a$.

a periodic perturbation of the surface with a spatial period equal to $\lambda_w/\sin \vartheta$. Maximum conversion into a Rayleigh wave occurs when ϑ is chosen so that this period equals the Rayleigh wavelength λ, i.e., $\sin \vartheta = \lambda_w/\lambda = v_w/v$. Here v_w is the longitudinal sound velocity in the material of the wedge and v is the velocity of the Rayleigh wave on the surface of the adjacent solid. Since in a given isotropic solid the velocities of longitudinal and transverse waves are always greater than the velocity of Rayleigh waves ($v_l > v_t > v$), it follows from the above condition for ϑ that v_w in the wedge has to be smaller than v in the solid. Therefore, longitudinal and transverse waves cannot penetrate from the wedge into the adjacent solid; they are totally reflected at the interface. Instead, an inhomogeneous wave traveling parallel to the interface and decaying exponentially with depth excites a Rayleigh wave.

Advantages of the wedge method are that for all frequencies the wedge angle ϑ depends only on the velocities of the longitudinal wave in the wedge and the Rayleigh wave in the solid, and that the Rayleigh wave is radiated in the "forward direction." A disadvantage follows from the fact that v_l

has to be smaller than v: this makes it necessary to use plastic wedges which have a sufficiently low longitudinal sound velocity, but on the other hand also exhibit a high attenuation for ultrasonic waves, especially at high frequencies.

A method similar to the wedge method is obtained when the surface of the solid and a longitudinal sound radiator are immersed together in a tank of liquid (Rollins et al., 1968). Since the longitudinal sound velocity in liquids is usually smaller than the Rayleigh-wave velocity in solids, it is possible to direct the sound beam to the surface at such an angle ϑ that a Rayleigh wave is excited. However, the frequency has to be sufficiently low or the distance over which the Rayleigh wave propagates sufficiently small in order to avoid a fast reconversion of the Rayleigh wave into a longitudinal wave in the ambient liquid (see Section IV,B).

A very effective method for the excitation of Rayleigh waves (first proposed by Sokolinskii, 1958) is described in more detail by Victorov (1967); a comb structure consisting of an aluminum plate with a periodic array of parallel grooves is pressed against the surface of a solid (Fig. 5c). When this plate is driven by an X-cut quartz from the other side the uniform motion of the set of "teeth" leads to the excitation of a Rayleigh wave if the distance between neighboring teeth equals the Rayleigh wavelength to be generated ($2a = \lambda$).

No restrictions exist concerning the materials of the comb and the adjacent solid. Furthermore, the transduction efficiency can be enhanced by increasing the number of teeth. On the other hand, since the Rayleigh wavelength is determined by the spacing of the teeth, increased transduction efficiency reduces the bandwidth of the transducer: as the bandwidth is inversely proportional to the number of teeth (see Section III,B5), effective transduction is limited to a small range around a center frequency or harmonics of this fundamental frequency. In addition to the Rayleigh wave, an appreciable amount of energy is radiated as a longitudinal bulk wave into the solid.

By these mechanical methods, Rayleigh waves have been successfully excited up to a maximum frequency of about 50 MHz, corresponding to a minimum wavelength of the order of 0.1 mm (for a typical Rayleigh-wave velocity of 3×10^5 cm/sec, $\lambda = 0.1$ mm at 30 MHz). At higher frequencies, e.g., 1 GHz, the method of Firestone and Frederick (1946) would require quartzes with very inconveniently small dimensions; the wedge method cannot be used, since the longitudinal wave would already be reabsorbed in the wedge before reaching the surface; the comb method would require such a narrow spacing of the teeth that manufacturing would become very difficult.

Further transduction methods which extend the frequency range up to gigahertz frequencies will be described in the following sections.

For cryogenic applications these mechanical methods which rely on a good mechanical contact between two surfaces are even less useful, since the quality of the acoustic contact varies with temperature (see Arzt and Dransfeld, 1965).

B. Transducers for Direct Piezoelectric Coupling

Instead of creating a space- and time-dependent stress by purely mechanical means (Section III,A), it is possible in piezoelectric solids to excite it by exposing a suitably cut piezoelectric-crystal surface to a spatially periodic electric rf field. By piezoelectric coupling, this electric field produces

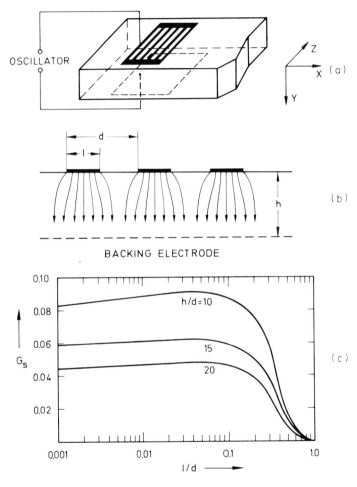

Fig. 6. Excitation of Rayleigh waves in quartz by direct piezoelectric coupling to an electric field: *Single-phase* electrode array. (a) Electrode configuration. (b) Schematic electric field near the surface. At the frequency ω_0 of maximum efficiency for Rayleigh wave excitation the Rayleigh wavelength equals the spatial period d of the electric field. (c) Geometric efficiency factor G_s (see Eq. 17) of one grating line as a function of the ratio l/d (for $\omega = \omega_0$). The magnitude of G_s is normalized to unity length in the Z direction and refers to the crystal orientation shown in (a). (Coquin and Tiersten, 1967.)

a corresponding periodic mechanical stress which gives rise to a Rayleigh wave. At present this transduction method is the one most widely used for the excitation of Rayleigh waves at high frequencies (up to approximately 3 GHz; Slobodnik, 1969).

The periodic electric field is produced by connecting an rf generator to a set of parallel electrodes on the surface. Two electrode configurations are of particular interest:

1. All electrodes have the same electric potential and the rf source is connected between these electrodes on the crystal surface and a backing

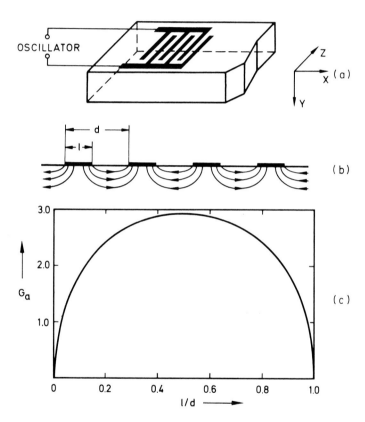

FIG. 7. Excitation of Rayleigh waves in quartz by direct piezoelectric coupling to an electric field: *Alternate-phase* electrode array. (a) Electrode configuration. (b) Schematic electric field near the surface. The spatial period of the electric field, and consequently the Rayleigh wavelength at the frequency ω_0 of maximum transduction efficiency, is $2d$. (c) Geometric efficiency factor G_a (see Eq. 17) of one electrode pair as a function of the ratio l/d (for $\omega = \omega_0$). The magnitude of G_a is normalized to unity width in the Z direction and refers to the crystal orientation shown in (a). (Coquin and Tiersten, 1967.)

electrode below the crystal ("single-phase electrode array,"[3] Fig. 6a). and
2. The electric potential alternates between neighboring electrodes ("alternate phase array,"[3] or "interdigital transducer," Fig. 7a).

Both methods for direct piezoelectric excitation of Rayleigh waves can, conversely, be used for their detection. When the transducers are electrically matched to the impedance of the rf generator and receiver, respectively, their efficiency is the same for both applications (see, e.g., Coquin and Tiersten, 1967).

According to Coquin and Tiersten (1967), the efficiency $\eta_{\rm T}$ of a transducer for direct piezoelectric transduction can be defined as the ratio of the (electric) power extracted from the surface wave by the transducer to the acoustic power of the incident surface wave. This ratio is expressed by the product of three dimensionless factors: the efficiency G of the electrode geometry ($G_{\rm s}$ for the single-phase array and $G_{\rm a}$ for the alternate phase array, respectively), the electric quality factor Q of the transducer, and the efficiency of the piezoelectric material, which depends on the orientation of the crystal surface and the direction of Rayleigh wave propagation on this surface,

$$\eta_{\rm T} = GQ\eta_{\rm m} \tag{17}$$

The magnitudes of these three factors will be evaluated in the next sections.

1. Single-Phase Array

A single-phase electrode array for the excitation of Rayleigh waves is shown in Fig. 6a (see Arzt *et al.*, 1967). A metallic grating with all its lines connected at their ends is placed on the surface of a thin quartz plate and an electric rf voltage is applied between this electrode and a backing electrode below the quartz. The spatially periodic distribution of the electric field is shown in Fig. 6(b). At the fundamental frequency of operation the spatial period d, i.e., the grating constant, equals the Rayleigh wavelength to be excited.

The efficiency of this electrode configuration depends strongly on the magnitude and inhomogeneity of the resulting electric field. Since the field strength is a function of the crystal thickness, it is obvious that highest transduction efficiency is obtained with thin plates. However, the minimum thickness is limited by the requirement that for Rayleigh-wave propagation the plate has to behave like an elastic half-space. This condition is satisfied only for plates with a minimum thickness of about ten Rayleigh wavelengths.

Another factor which determines the strength as well as the inhomogeneity of the electric field is the length l of the electrodes (measured in the direction of Rayleigh wave propagation) as compared to their spacing d (see Fig. 6b). For a quartz crystal the geometric efficiency factor $G_{\rm s}$ of the single-phase array was calculated by Coquin and Tiersten (1967). Their

---

[3] The terms "single-phase" and "alternate-phase" array were first used by Coquin and Tiersten (1967).

calculation is based on infinitely thin and ideally conducting electrodes and weak piezoelectric coupling, so that the electric power extracted from the Rayleigh wave per transducer electrode is small compared with the total acoustic power. Hence, the efficiency of the entire transducer can be approximated by the product of the efficiency per electrode multiplied by the number of electrodes. The mathematical expression for G_s is

$$G_s = F_1{}^2 C_1/(\varepsilon_{11}\varepsilon_{22})^{1/2} \tag{18}$$

with

$$F_1 = \cos^2(\pi l/2d) \tag{19}$$

and

$$C_1 = \frac{\varepsilon_{11}d/h}{1 - (\varepsilon_{22}/\varepsilon_{11})(d/\pi h)\sin(\pi l/2d)} \tag{20}$$

where F_1 is the ratio of the electric potential at the electrode to the electric surface potential with no electrode present; C_1 denotes the capacitance per unit width (measured perpendicular to the direction of Rayleigh-wave propagation) of one single electrode—it can be seen that for the extreme cases $l = d$ (i.e., a totally metallized surface) and $h \gg d$ (very thick quartz plate) the denominator of C_1 becomes unity and C_1 approaches the value for a parallel-plate capacitor; ε_{11} and ε_{22} are the dielectric constants in the X and Y directions, respectively; and it can be seen that the magnitude of G_s also depends on the crystal and its orientation through the relevant dielectric constants.

In Fig. 6c, G_s is plotted as a function of l/d for different ratios of the quartz thickness h to the Rayleigh wavelength ($\lambda = d$), with the magnitude of G_s referring to the orientation of the quartz crystal given in Fig. 6a. The efficiency is best for ratios $l/d \leqslant 0.1$. (The slight deterioration as l decreases should become more pronounced for a finite electrode conductivity.)

2. Alternate-Phase Array

Figures 7(a) and 7(b) show the electrode configuration and the schematic electric field distribution for an alternate-phase electrode array (see, e.g., White and Voltmer, 1965). An electric rf voltage is applied between two interleaved comb-shaped electrodes on the crystal surface. This avoids the necessity of a backing electrode below the crystal, and hence the efficiency of this interdigital transducer is independent of the crystal thickness. If d is the spacing of neighboring electrodes, the spatial period of the electric field equals $2d$ and the fundamental frequency of operation has to be chosen so that the Rayleigh wavelength is twice the spacing ($\lambda = 2d$). For this case, the geometric efficiency factor G_a per electrode pair as calculated by Coquin and Tiersten is plotted in Fig. 7(c) as a function of the ratio l/d. The magnitude of G_a again refers to a Y-cut quartz crystal with the Rayleigh wave propagating in the X direction (Fig. 7a). Optimal efficiency is obtained with a ratio $l/d = 0.5$.

3. *Comparison of the Two Configurations*

From the magnitude of G_s and G_a it is easily seen that in general the alternate-phase array is far more effective than the single-phase array. The optimal values for both configurations (see Figs. 6c and 7c) are $G_a = 2.8$ for $l/d = 0.5$ and $G_s = 0.09$ for $l/d = 0.1$ and for the minimum crystal thickness $h = 10\lambda = 10d$. This shows that compared even with the optimal case for the single-phase array, the alternate-phase array is about 30 times more efficient.

However, for application at high frequencies—e.g., in the gigahertz range, where λ is of the order of 1μ—the single-phase array may also have some advantages: since for this configuration the spatial period of the electric field equals the spacing d of neighboring electrodes, the fundamental frequency reached with a single-phase array is twice as high as that of an interdigital transducer having the same grating spacing. Furthermore, a short occurring between neighboring grating lines, particularly likely at small grating spacings, i.e., at high frequencies, does not inactivate the entire single-phase transducer, since the electric potential is the same for all its lines. The drawback of smaller efficiency has to be compensated for by applying a higher electric power.

4. *Transducer Impedance*

Figure 8 shows the real and imaginary parts of the electrical admittance of an interdigital transducer on a piezoelectric ceramic sample (Thomann, 1969). The corresponding equivalent circuit of the transducer is inserted into Fig. 8 (see also Joshi and White, 1969).

For the resonance frequency ω_0 of the transducer (approximately 3.14 MHz in the example above) the impedance of the series circuit consisting of C_s and L_s is zero. Consequently, at ω_0 the impedance of the transducer is determined only by its static capacitance C_0 and the radiation resistance R_t. (The static capacitance C_0 can be tuned out by means of an external parallel inductor L_p so that $\omega_0 L_p = 1/\omega_0 C_0$).

It follows from the equivalent circuit that the radiation resistance R_t is inversely proportional to the square of the number N of transducer periods.

If all transducer fingers (or electrode pairs, respectively) are connected to the same electric rf voltage, the mechanical displacement amplitude a of the resulting Rayleigh wave is proportional to N. Hence, it follows for the acoustic power P_a of the Rayleigh wave

$$P_a \propto a^2 \propto N^2 \tag{21}$$

This mechanical power equals the electric power consumed by the transducer, which for an rf voltage V is $P_{el} = V^2/R_t$ (where V is the root-mean-square value of the voltage). Thus, from $P_a = P_{el}$ it follows that

$$R_t \propto 1/N^2 \tag{22}$$

For the example shown in Fig. 8, R_t is about 10 Ohms.

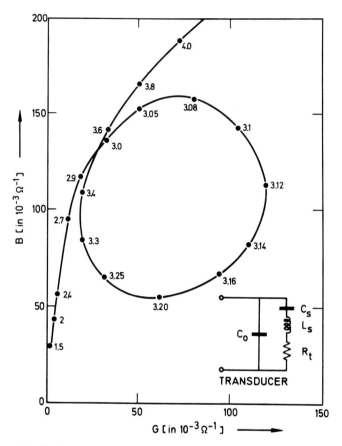

FIG. 8. Real (G) and imaginary (B) part of the admittance of an interdigital Rayleigh-wave transducer on a piezoelectric ceramic

[Pb(Zr 0.54, Ti 0.46)O_3 + 1% Nd_2O_3].

Parameter along the curve is the frequency (in MHz). Insert: Equivalent circuit representing the transducer (Thomann, 1969).

The electromechanical coupling constant k (for definition, see, e.g., Cady, 1964; Mason, 1950, 1958) calculated from the equivalent circuit is

$$k^2 = C_s/(C_s + C_0) \qquad (23)$$

(see, e.g., Thomann, 1969), where C_s can be determined from Fig. 8 as follows.

The electric quality factor Q of the equivalent series circuit representing the transducer is given by

$$Q = 1/\omega_0 C_s R_t \qquad (24)$$

Since Q also represents the -3 dB fractional bandwidth of the transducer, two frequencies ω_1 and ω_2 can be defined where the acoustic power generated by the transducer drops by 3 dB relative to the power at the resonance frequency ω_0. In terms of the equivalent circuit, these frequencies are defined by those points on the B–G plot where the real part of the radiation admittance is half its value at ω_0. From these frequencies C_s is calculated to be

$$C_s = \frac{1}{Q\omega_0 R_t} = \frac{|\omega_1 - \omega_2|}{\omega_0{}^2 R_t} \tag{25}$$

This determines the coupling constant k for the ceramic sample above to be $k = 0.22$.

For a different equivalent circuit representing an interdigitial transducer the radiation resistance has been calculated by Collins *et al.* (1968b) (see also Broers *et al.*, 1969; Collins *et al.*, 1968c, d).

5. *Coupling Efficiency*

In general, the coupling efficiency between a Rayleigh wave and surface electrodes is of course largest for materials with high piezoelectric coupling constants. Thus, the conversion efficiency from electric energy to elastic energy is considerably higher in lithium niobate and most piezoelectric ceramics than it is for crystalline quartz. It is, however, important to know for a given piezoelectric crystal the surface orientations and propagation directions for which most efficient conversion is possible.

Coquin and Tiersten (1967) have calculated the coupling efficiency between Rayleigh waves and electrodes deposited appropriately on the surface for a few specific surface orientations and propagation directions in crystalline quartz. Figures 9a and b, respectively, show their result for the angular dependence of the material efficiency factors η_m (as defined in Section III,B, Eq. 17) for X- and rotated Y-cut quartz surfaces. It can be seen that for rotated Y-cuts the efficiency for Rayleigh-wave excitation is best for an angle of rotation of about $-30°$. On an X-cut surface, excitation is most effective for Rayleigh waves propagating along the Y axis. However, the direction of the group velocity is not collinear with the direction of the phase velocity along the Y axis (see Fig. 3), whereas phase and group velocity are collinear for Rayleigh waves propagating along the X axis on all rotated Y-cut surfaces (Coquin and Tiersten, 1967).

While Coquin and Tiersten (1967) calculated the elastic energy flow in the surface wave and the electric energy that can be extracted from it by means of surface electrodes in order to derive the coupling efficiency, a simpler method sufficiently accurate for the evaluation of optimal crystal cuts was proposed by Campbell and Jones (1968): In piezoelectric crystals the effective elastic constants pertaining to a particular Rayleigh wave are "stiffened" due to the piezoelectric fields associated with elastic deformations. This results in a higher sound velocity than without piezoelectric fields. If the piezoelectric fields at the surface are short-circuited, e.g., by metallizing the surface (assuming an infinitesimally thin, perfect conductor), the sound

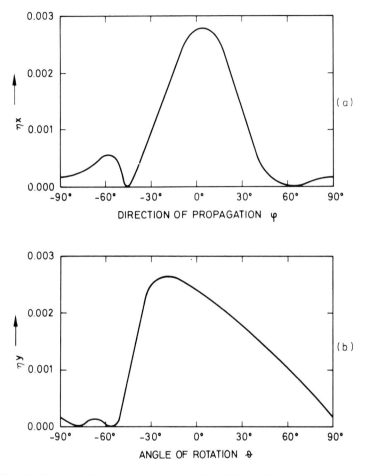

FIG. 9. Material efficiency factor η_m (see Eq. 17) for Rayleigh waves in quartz
(Coquin and Tiersten, 1967.) (a) Efficiency factor η_x for Rayleigh waves propagating
on an X-cut surface. Here φ is the angle between the Y axis and the direction of the phase
velocity of the Rayleigh wave. (b) Efficiency factor η_y for Rayleigh waves propagating
along the X-axis on a rotated Y-cut surface. (The Y surface is rotated about the X axis,
and ϑ is the angle between the crystallographic Y axis and the Y' axis normal to the
rotated surface.)

velocity is reduced by an amount Δv which can be regarded as a measure for
the coupling strength between the Rayleigh wave and the corresponding
surface electrodes: if no velocity change occurs, there are no tangential
piezoelectric fields at the unmetallized surface, and hence, there is no coupling
to an electrode grating transducer, which responds mainly to these tangential
fields. A large velocity change, on the other hand, indicates strong electric
fields existing at the free surface, and thus, a high coupling efficiency.

Mathematically, the problem of calculating the stiffened and unstiffened velocities is much easier than calculating the amount of electric energy extracted from the Rayleigh wave. It just means introducing new boundary conditions on the electric potential φ outside the crystal (see Eqs. 15 and 16): for the calculation of the stiffened velocity, i.e., for a free surface, $\varphi = 0$ at $x_3 = -\infty$, and for the (smaller) unstiffened velocity $\varphi = 0$ at $x_3 = 0$.

Some representative velocities of Rayleigh waves in LiNbO$_3$ as calculated by Campbell and Jones (1968) are plotted in Fig. 10 where h denotes the dis-

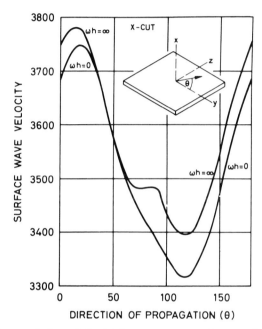

FIG. 10. Phase velocity of Rayleigh waves propagating on the free ($\omega h = \infty$) and metallized ($\omega h = 0$) surface, respectively, of X-cut lithium niobate (h denotes the distance at which an infinitesimally thin, perfect conductor is located above the crystal surface.) Coupling to surface electrodes is most effective for directions exhibiting a large velocity change. (Campbell and Jones, 1968.)

tance of the perfect conductor from the crystal surface, so that the curve $\omega h = \infty$ represents the stiffened velocity on a free surface and the curve $\omega h = 0$ represents the unstiffened velocity on the metallized surface. It can be seen that there should be no coupling between a transducer grating and a Rayleigh wave propagating at an angle of $\theta = 50°$, whereas high coupling should exist for an angle θ of about 100°.

6. Frequency Response

To a first approximation, the frequency response of a grating transducer can be represented by that of a linear array of antennas.

In the case of a single-phase transducer each grating line simply corresponds to one radiator, so that the number N of radiators equals the number of lines. In an alternate-phase transducer one antenna is represented by two neighboring grating lines, so that n grating lines correspond to $(n-1)$ antennas. Each transducer period contains one pair of antennas radiating with a phase difference of π. If this pair of antennas is regarded as one radiator, the number N of radiators equals the number of transducer periods. $N = (n-1)/2$. (For a sufficiently large number n the number of radiators is half the number of grating lines.)

Let ω_0 denote the fundamental transducer frequency and $\Delta\omega$ the deviation from this frequency. Then the amplitude of the resulting Rayleigh wave as a function of the frequency $\omega = \omega_0 + \Delta\omega$ is

$$|a(\omega)| = \left| \frac{\sin(N\pi \, \Delta\omega/\omega_0)}{N \sin(\pi \, \Delta\omega/2)} \right| \tag{26}$$

or, for large N,

$$|a(\omega)| = \left| \frac{\sin x}{x} \right| \tag{27}$$

where $x = N\pi \, \Delta\omega/\omega_0$. The amplitude at the fundamental frequency $a(\omega_0)$ is normalized to unity in both equations. (Compare, e.g., diffraction by an optical grating; Born and Wolf, 1959.)

From Fig. 11 it can be seen that the fractional bandwidth $\Delta\omega_{1,2}/\omega_0$

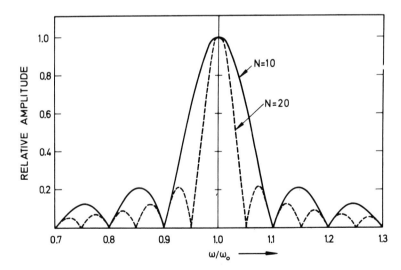

FIG. 11. Frequency response of a grating transducer for Rayleigh waves. Here N is the number of grating elements, and ω_0 is the fundamental frequency of the transducer which is determined by the spacing of the grating lines.

between the first minima on either side of the fundamental maximum is

$$\Delta\omega_{1,2}/\omega_0 = 2/N \qquad (28)$$

for both the single-phase and the alternate-phase array, with N as defined above.

For large values of $\Delta\omega = \omega - \omega_0$ deviations from the above behavior are to be expected due to the approximation that the radiators should have zero extension in the direction of Rayleigh-wave propagation. This is a very crude approximation, as is seen, e.g., from a calculation by Coquin and Tiersten (1967): Figure 12 shows their result for the frequency response of

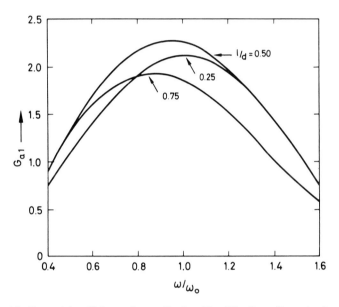

FIG. 12. Geometric efficiency factor G_{a1} (see Eq. 17) of an alternate-phase transducer consisting of a single electrode pair for several values of the ratio l/d (see Fig. 7b). (Coquin and Tiersten, 1967.)

an alternate-phase transducer consisting of only two electrodes. If it were possible to represent this transducer by a line source, it should exhibit a flat frequency response. However, although qualitatively a 3 dB fractional bandwidth of roughly 100% is obtained, it is obvious that the ratio of electrode length l to electrode spacing d is also relevant.

The ratio l/d plays a particularly important role when grating transducers are to be operated at harmonic frequencies. Suitable choice of this ratio can improve transduction at selected harmonics (Tseng, 1968).

Further deviations from the calculated frequency response are, of course, to be expected when the transducers are connected to frequency-dependent electrical matching networks for maximum power transfer (see, e.g., Collins *et al.*, 1969).

C. THIN-FILM TRANSDUCERS

The mechanical methods described in Section III,A have a reasonable efficiency only at low frequencies. The methods of Section III,B allow the excitation of Rayleigh waves at high frequencies, but only on piezoelectric materials (single crystals or ceramics). In this section some techniques will be discussed which also appear promising for the excitation of high-frequency Rayleigh waves on nonpiezoelectric materials. These methods utilize evaporated films of piezoelectric materials, e.g., CdS, ZnO, or AlN (aluminum nitride) films (see, e.g., Foster, 1965; de Klerk and Kelly, 1965; Wauk and Winslow, 1968).

Although evaporated CdS films have a polycrystalline structure, they can be used as ultrasonic transducers, since the piezoelectric C-axes of all the crystallites can be oriented toward the evaporation source, e.g., perpendicular to the substrate surface. For the excitation of Rayleigh waves such a piezoelectric film may be used in conjunction with an electrode grating: An interdigital grating (see Section III,B2) is placed on top of the CdS film, or—if the substrate is a dielectric—evaporated between substrate and film. With metallic substrates a single-phase grating (see Section III,B1) may be placed on top of the film and the driving voltage is connected between the grating and the substrate below the film.

Evaporated thin-film transducers exhibit a broadband frequency response (see, e.g., Foster, 1965), which means that the film thickness is not very critical. Therefore, in conjunction with a grating electrode, the bandwidth of the Rayleigh-wave transducer is determined mainly by the number of grating lines (see Section III,B6) and the film thickness may be chosen to be of the order of 1μ for frequencies between 100 MHz and 1 GHz.

It is favorable to keep the film as thin as possible, since it changes the elastic properties at the surface of the substrate, and, e.g., Rayleigh waves in layered media (substrate and film) are no longer nondispersive: For films thick compared to the Rayleigh wavelength (i.e., also to the penetration depth of the Rayleigh wave) the velocity approaches that of Rayleigh waves in the film material, and only for sufficiently thin films does the velocity approach that of Rayleigh waves on the substrate. In addition to this fundamental Rayleigh mode, higher Rayleigh modes and Love modes (horizontal shear waves) occur (see, e.g., Ewing *et al.*, 1957; Schnitzler, 1967). Since these modes exist only in the transducer area covered by the film, they would have to be transformed into a pure nondispersive Rayleigh mode on the free surface. This problem is avoided by using very thin films.

Humphryes and Ash (1969) describe a method to excite Rayleigh waves using bulk compressional waves impinging on the surface from *within* the solid (in contrast to the wedge method, where the bulk wave impinges on the surface from *outside* of the specimen through a wedge consisting of a different material—see Section III,A). For an undisturbed surface the mechanical boundary conditions do not allow the excitation of a Rayleigh wave for this geometry, but partial transformation of the incident bulk wave into a Rayleigh wave is possible if a periodic array of parallel grooves is ruled into the

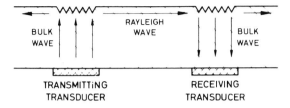

Fig. 13. Conversion of bulk (longitudinal or transverse) waves into Rayleigh waves by means of parallel grooves in the surface. For normal incidence of the bulk wave the spacing of the slots equals the Rayleigh wavelength. (After Humphryes and Ash, 1969.)

surface (Fig. 13). For normal incidence of the bulk wave the spacing of the grooves equals the Rayleigh wavelength. The reverse transformation effect, radiation of bulk waves from periodic surface perturbations (Victorov, 1967; Rischbieter, 1965/66), is used to detect the Rayleigh wave by means of a bulk-wave transducer, e.g., a piezoelectric film on the opposite side of the substrate.

In contrast to depositing an electrode grating on top of a piezoelectric film, ruling or etching grooves into a surface is a much easier production process. Therefore, this method may prove especially useful in "micro-sound" circuits in conjunction with electronic integrated circuits on a common silicon substrate, with the electronic circuits (including the CdS transducers) on one side and the acoustic circuits on the other side (Humphryes and Ash, 1969; compare also Section V).

Preliminary experiments conducted at 2 MHz with triangular grooves (their depth being approximately equal to the Rayleigh wavelength) showed a minimum insertion loss of 7.5 dB for the two Rayleigh-wave transducers of Fig. 13, where a loss of 6 dB is due to the bidirectivity of the transducers. (The insertion loss of the Rayleigh-wave transducers was determined by measuring the total insertion loss of the arrangement of Fig. 13 and subtracting the insertion loss of the two bulk-wave transducers, which was measured separately.) Preferential Rayleigh-wave radiation into one direction is obtained with asymmetrical grooves for which a forward-to-backward ratio between 5 dB and 10 dB over a 20% bandwidth has been observed. Equally encouraging results may possibly be obtained at much higher frequencies (Humphryes and Ash, 1969).

D. Excitation by Pulsed Surface Heating

If the surface of a crystal is heated for a short time above the ambient temperature, an excess number of thermal bulk and surface phonons propagating away from the heated surface region can be detected at some distance as a delayed heat pulse, by means of a bolometer. This heat-pulse technique has proved to be very useful at low temperatures for the study of bulk phonons having a frequency of the order of 30 GHz $= kT/h$ (von Gutfeld and Nethercot, 1964; see also Weis, 1969). It appears possible to adopt the same techniques for generating and detecting surface phonons of the same energy. At present, however, no experiments have been reported.

Lee and White (1968) pointed out that a transient surface heating also leads to the excitation of surface waves by the effect of thermal expansion (which need not be present in the heat-pulse method mentioned above). They used the beam of a Q-switched laser focused on the polished surface of a fused silica block. In order to achieve a reasonable absorption of thermal energy, a thin aluminum film was deposited on the surface. The heat pulses cause high temperature gradients along the surface, and, due to the mechanical inertia of the substrate, these result in strong mechanical stresses, which in turn result in elastic waves. A part of the elastic energy propagates along the surface as a Rayleigh wave, the frequency of which is determined by the main component of the Fourier spectrum of the laser pulse.

It was demonstrated by a simple test that it really is surface heating which accounts for the excitation of the acoustic wave, and not perhaps transfer of photon momentum: If the surface is an ideal optical reflector, the momentum transfered to it is twice the photon momentum, whereas the photon momentum is transferred only once to an ideally absorbing surface. As the resulting acoustic wave had a maximum amplitude when the surface was covered with a black layer, it is obvious that absorption of thermal energy causes the excitation of the elastic wave (see White, 1963). Excitation of Rayleigh waves is enhanced when the laser beam is incident through a stripe mask: this leads to a *space-periodic* surface heating and increases the amplitude of a Rayleigh wave with a wavelength equal to the spacing of the stripes.

Although this method requires high power densities (up to 2.2 MW/cm² were used) it has a number of advantages: (1) it can be used to excite high mechanical amplitudes; (2) the source of transduction can be easily moved along the surface; (3) excitation of Rayleigh waves is possible on any kind of elastic solid, at high or low temperatures and without direct mechanical access to the surface, e.g., in vacuum. The latter property is particularly advantageous if the Rayleigh wave is also detected without direct contact with the surface, e.g., by the optical methods described in Section III,E below.

The highest frequency of Rayleigh waves that can be excited with this method is determined by the shortest laser pulses available (the experiments by Lee and White (1968) have been conducted at frequencies between 5 MHz and 95 MHz.)

E. OPTICAL DETECTION

Two properties of a Rayleigh wave make its detection by optical means possible: (1) the surface along which the Rayleigh wave travels becomes sinusoidally corrugated, and (2) due to the elastooptical effect, the index of refraction changes periodically within a layer equal to the penetration depth of the Rayleigh wave.

If the surface excursion and the Rayleigh wavelength are sufficiently large, the first of the above properties can be used to deflect a beam of light having a diameter small compared to the Rayleigh wavelength (compare Fig. 31 below; Adler *et al.*, 1968).

The reflected beam swings periodically past a knife edge in front of a photomultiplier, thus producing an electric signal of the same frequency as that of the Rayleigh wave. In order to reduce multiplier noise, it is favorable to position the knife edge so that the undeflected beam is just blocked from the multiplier tube.

Since for acoustic waves carrying the same amount of energy the amplitudes are inversely proportional to the square of their respective frequencies, detection of Rayleigh waves by deflecting a beam of light is limited to low frequencies. At higher frequencies the angular excursion of the beam of light is very small, and therefore, positioning of the knife edge becomes very critical. As far as focusing of the beam of light is concerned, 1 GHz ($\lambda \approx 3\mu$) appears to be a practical limit.

At high frequencies it is more appropriate to consider a Rayleigh wave as a moving optical diffracting grating, both because of its surface corrugation (see, e.g., Korpel *et al.*, 1967; Mayer *et al.*, 1967; Auth and Mayer, 1967), as well as the periodically changing refractive index (see Ippen, 1967). This grating modulates the phase of reflected and transmitted light. Thus, for opaque solids diffraction orders can be detected in the reflected light, and for transparent solids diffraction orders can also be detected in transmission.

For normal incidence of the beam of light the directions ϑ_1 of the first diffraction orders depend on the Rayleigh wavelength and the wavelength of the light λ_L as

$$\sin \vartheta_1 = \lambda_L / \lambda \tag{29}$$

The intensity of the diffraction orders can be calculated from the Raman–Nath theory (1935): When a plane wavefront of light (intensity I_0) is *reflected* from a sinusoidally corrugated surface the reflected wavefront becomes corrugated with twice the amplitude of the surface corrugation (Fig. 14).

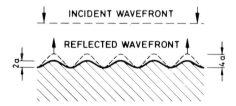

FIG. 14. Reflection of a plane wavefront of light from a sinusoidally corrugated surface. The corrugation of the reflected wavefronts leads to optical diffraction orders.

In this case the intensity I_{nr} of the nth diffraction order reflected from a perfectly reflecting surface is

$$I_{nr}/I_0 = J_n{}^2(2ak_L) \tag{30}$$

where J_n is the nth order Bessel function, a is the amplitude of the surface corrugation, and $k_L = 2\pi/\lambda_L$ is the wave vector of the light.

For small amplitudes a the Bessel function can be expanded, and the relative intensity of the first diffraction order becomes

$$I_{1r}/I_0 = (ak_L)^2 \qquad (31)$$

From the measured intensity of the first diffraction order the amplitude of the surface corrugation due to a Rayleigh wave can easily be calculated.

In a similar way, the wavefronts of light *transmitted* through a surface supporting Rayleigh-wave propagation become corrugated, which gives rise to similar diffraction effects (see, e..g., Korkstad and Svaasand, 1967; Salzmann and Weismann, 1969).

For the optical detection of high-frequency Rayleigh waves it is important to find the conditions under which surface phonons offer the largest cross section for light scattering. Therefore, the sensitivities of the three main experimental arrangements for optical detection of Rayleigh waves by diffraction of light will be compared in the following paragraphs (Salzmann and Weismann, 1969). The experimental results were obtained with crystalline quartz, but they also apply to other transparent solids.

Pulsed Rayleigh waves propagating along the X axis on a Y surface of a prismatic quartz block were excited at 316 MHz ($\lambda = 10\mu$) by means of an appropriate transducer grating (see Section III,B1; see also Arzt *et al.*, 1966). The beam of a He–Ne laser (diameter 2 mm, polarized perpendicular to the plane of incidence) was indicent on this surface and the first diffraction order was monitored with a photomultiplier. The multiplier signal was displayed on an oscilloscope and processed with a boxcar integrator. For the three cases considered the directions of the incident and diffracted beams are sketched in Figs. 15 and 16.

The corresponding magnitudes of the received signals are shown in Figs. 15 and 16 as a function of the angle of incidence φ. When the laser beam is incident through the quartz block (Figs. 15 and 16a) a maximum appears at the critical angle for total internal reflection φ_c. This dependence on φ can be explained by assuming that the incident light is scattered only due to the surface corrugation. The index of refraction is taken to be constant everywhere inside the solid.

With this approximation a simplified calculation yields the energy P_{1t} scattered into the first diffraction order divided by the incident energy P_0:

$$P_{1t}/P_0 = (1 + R)^2 J_1^2[ak_L(n \cos \varphi - \cos \vartheta_0)] \cos \vartheta_1/(n \cos \varphi) \qquad (32)$$

where R is the reflectivity of the surface, depending on φ according to the Fresnel formulae (see, e.g., Born and Wolf, 1959), J_1 is the first-order Bessel function, a is the amplitude of the Rayleigh wave, k_L is the wave vector of the light, and n is the refractive index. The factor $\cos \vartheta_1/(n \cos \varphi)$ expresses the limitation of the diffracted beam for large angles φ. For $\varphi > \varphi_c$, the term $\cos \vartheta_0$ vanishes.

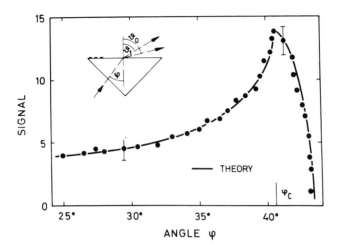

FIG. 15. Optical detection of Rayleigh waves by observing the energy scattered into the transmitted first diffraction order as a function of the angle of incidence φ. The magnitude of the received signal is given in relative units. (Salzmann and Weismann 1969.)

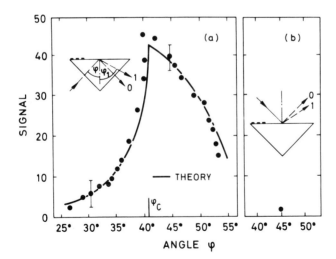

FIG. 16. Optical detection of Rayleigh waves by observing the energy scattered into the reflected first diffraction order as a function of the angle of incidence φ. (a) Reflection from within the quartz; (b) reflection from outside the quartz. The received signal is plotted in the same relative units as in Fig. 15. (Salzmann and Weismann 1969.)

The relative energy scattered into the first *reflected* diffraction order (Fig. 16a,b) is

$$P_{1r}/P_0 = R^2 J_1{}^2[2ak_L n \cos \varphi] \cos \varphi_1/\cos \varphi \tag{33}$$

From Figs. 15 and 16(a) satisfactory agreement of the theoretical angular dependence (solid curves) with the experimental results (dots) is obvious. If φ is chosen close to φ_c, the total energy scattered into the reflected diffraction order (Fig. 16a) is 13.4 times larger than at $\varphi = 0°$. The transmitted diffraction order (Fig. 15) increases by a factor of 5.7 as φ goes from $0°$ to φ_c.

In both cases considered so far the laser beam has to penetrate the layer in which the Rayleigh wave travels, so that a rigorous treatment of the diffraction problem would also involve diffraction resulting from the periodic change of the refractive index in this layer.

If the surface corrugation is neglected and only the periodic change of the refractive index is assumed to contribute to the diffraction of light (Fig. 17), the intensity of the first diffraction order is, for normal incidence

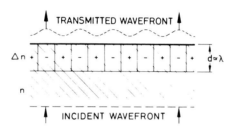

FIG. 17. Transmission of a plane wavefront of light through a surface layer with a periodically changing refractive index n. Diffracting orders occur due to the corrugation of the transmitted wavefronts. (The amplitude of the corrugation is Δnd, and d is the penetration depth of the Rayleigh wave).

of the beam of light,

$$I_1/I_0 = J_1{}^2(\Delta ndk_L) \tag{34}$$

Here Δn is the amplitude of the change of the refractive index and d is the penetration depth of the Rayleigh wave, so that Δnd expresses the corrugation of the optical wavefront.

It is not possible to treat diffraction due to the surface corrugation and to the periodically changing refractive index separately and then simply to add the intensities of the resulting diffraction orders. Therefore, it is interesting to check the accuracy of the predictions of Eq. (33), in which diffraction due to the periodically changing refractive index was neglected, by comparing the experimental results of Fig. 16(a) with the results of an experiment in which the laser beam was directed on the surface from outside of the crystal (Fig. 16b): it is obvious that in this case diffraction results mainly from the surface corrugation. For $\varphi = 45°$, the signal obtained with the arrangement of Fig. 16(b) is compared with the signal received at the same angle of incidence, but with the arrangement of Fig. 16(a). From the known values for R and n the ratio of these signals can be calculated from Eq. (33) to be 1/24. The measured ratio is 1/28 (compare Figs. 16a and b), which shows that the influence of the changing refractive index is indeed negligible within the range of experimental error.

Suppression of stray light, which reduces the signal-to-noise ratio, is accomplished best with the arrangements of Figs. 15 and 16(a) with φ slightly greater than φ_c: in the first case the strong zero order does not penetrate the surface, and in the second case the ideally reflecting surface ($R = 1$) results in diffraction orders having intensities which are usually much higher than the intensity of the stray light. Comparing also the relative magnitudes of the received signals, the arrangement of Fig. 16(a) appears to provide the most sensitive arrangement for an optical detection of Rayleigh waves.

The arrangement of Fig. 15, with φ slightly greater than φ_c, has been successfully used to detect Rayleigh waves of frequencies up to 1047 MHz (Salzmann *et al.*, 1968) having a particle displacement of approximately 1 Å.

In many cases optical detection of Rayleigh waves is performed by utilizing the fact that the frequency Ω_1 of the diffracted light is shifted relative to the frequency Ω_0 of the incident light by an amount equal to the frequency ω of the acoustic wave ($\Omega_1 = \Omega_0 \pm \omega$ for the diffraction order scattered into, or opposite to, the direction of Rayleigh wave propagation, respectively). After optical heterodyning of the diffracted beam ($\Omega_0 \pm \omega$) with a reference beam of the unshifted frequency Ω_0, the amplitude of the resulting beam is modulated with the frequency ω of the Rayleigh wave, and the modulation depth is proportional to the (optical) amplitude of the diffracted light. Applying a frequency analyzer, the intensity of the diffraction order can be measured with high sensitivity (Ippen, 1967).

All the preceding methods give information about the amplitude of a Rayleigh wave, but not about its phase. The latter can be obtained by comparing the signal obtained from probing the surface on which a Rayleigh wave propagates, with an appropriate reference signal carrying known phase information. Two methods will be described briefly in the following paragraphs: in the first method the signals which are to be compared are two beams of light; in the second method a beam of light is reflected from the surface on which the Rayleigh wave propagates, and after detecting the reflected beam with a photomultiplier the resulting electric signal is heterodyned with an electric reference signal.

Figure 18 shows an interferometer where the reference mirror has been mounted on a piezoelectric transducer (compare Ash *et al.*, 1969). The incident beam of light is split into two parts, one of which is reflected from the surface of the crystal on which the Rayleigh wave propagates, and the other of which is reflected from the periodically moving mirror. (It is important that the beam that probes the surface wave have a diameter small compared to the surface wavelength, which restricts this method to frequencies below approximately 1 GHz.)

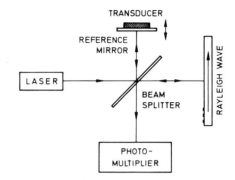

FIG. 18. Interferometer for optical detection of surface waves. The reference mirror is vibrating with the frequency of the surface wave in order to obtain phase sensitivity. (After Ash *et al.*, 1969.)

The phase of each of the two reflected beams is determined by the positions of the crystal surface and the reference mirror, respectively, at a given time. If the reference mirror is fixed and a surface wave is present on the crystal, the normal excursions of the crystal surface introduce a periodic phase shift of the probing beam, which results in an amplitude modulation of the output beam. Only if both the reference mirror and the crystal surface are vibrating with the same frequency and amplitude is the periodic modulation of the output beam cancelled. In this case a constant phase shift remains between the two reflected beams, which determines the (constant) amplitude of the output beam. This relative phase shift is determined by the phase of the Rayleigh wave at the point where the probing beam is incident on the surface. Therefore, a maximum output amplitude changes to a minimum if that point is moved on the surface by a half Rayleigh wavelength along the direction of Rayleigh-wave propagation.

In the above considerations it has been assumed that the amplitude of the normal excursions both of the crystal surface and the reference mirror are smaller than the wavelength of the incident light. This is certainly true for the usual magnitude of acoustic energy transported in a Rayleigh wave.

The sensitivity of the interferometer is increased when the beam splitter is replaced by a Bragg cell (Fig. 19; Whitman *et al.*, 1968; Ash *et al.*, 1969). This arrangement reduces alignment problems and provides a carrier frequency for narrowband amplification and detection of the received signal. In addition, synchronous detection is possible by modulating the amplitude of the carrier. The remarkable theoretical minimum surface excursion that can be detected with this arrangement is as small as 2.6×10^{-3} Å (Whitman *et al.*, 1968).

Adler *et al.* (1968) describe an optical detection method that renders a traveling surface wave as a standing picture on a tv screen. The stroboscopic effect is achieved by electrically heterodyning the photomultiplier signal

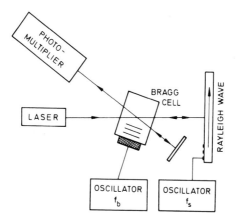

Fɪɢ. 19. Interferometer for optical detection of surface waves, where the beam splitter has been replaced by a Bragg cell. As in Fig. 18, the reference mirror may be vibrated to obtain phase sensitivity. (After Whitman *et al.*, 1968, and Ash *et al.*, 1969.)

with an electric reference signal. Figure 20 shows the schematic of the arrangement: A laser flying spot scanner is used to sweep a beam of light across the surface at normal tv rates. An optical correction system (not shown) ensures that the beam reflected from the undisturbed surface strikes the edge of the obstacle in front of the multiplier, independent of the location of the spot where the laser beam is incident on the surface. If a Rayleigh wave is present, the reflected beam swings periodically across the edge of the obstacle, thus modulating the amplitude of the multiplier signal with the frequency of the Rayleigh wave. This signal is heterodyned with a reference signal obtained from the same generator that drives the Rayleigh-wave transducer. By analogy to optical heterodyning described above, the mixer output is constant for a stationary laser beam and its magnitude depends only on the relative

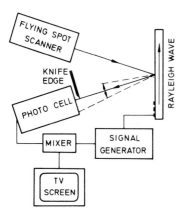

Fɪɢ. 20. Schematic arrangement for making surface waves visible on a television screen. (After Adler *et al.*, 1968.)

phase difference between the two heterodyned signals: it is a positive maximum when the phase difference is an integral multiple of 2π and a negative maximum for $(2n + 1)\pi$. When the laser beam is scanning along the surface in the direction of Rayleigh-wave propagation, positive values of the mixer output signal alternate with negative values. This signal is used to control the intensity of an electron beam scanning across a tv screen. In the resulting picture bright portions alternate with dark portions with a spacing between successive bright (or dark) parts corresponding to the wavelength of the Rayleigh wave.

Displaying an elastic surface wave on a tv screen makes it possible to directly observe its propagation characteristics: (1) its wavelength, from which the phase velocity can be calculated, (2) the direction of energy transport (group velocity), which in crystals is not necessarily parallel to the direction of the phase velocity (see Section II,B), (3) propagation along surface waveguides (see Section V), and (4) diffraction or scattering from obstacles on the surface.

IV. Absorption of Rayleigh Waves

In this section several interaction mechanisms are discussed which contribute to the attenuation of Rayleigh waves:

    (a) Scattering from surface imperfections (scratches, polishing defects);
    (b) Scattering from crystal defects (dislocations and grain boundaries in polycrystalline materials);
    (c) Interaction with thermal phonons (surface and bulk phonons);
    (d) Interaction with electrons in metals;
    (e) Interaction with drifting carriers in semiconductors.

While the mechanisms listed above lead to an attenuation of Rayleigh waves propagating on a free surface, an additional attenuation occurs when the surface is in contact with, e.g., air or a liquid helium film. Therefore;

    (f) The attenuation by ambient media (liquids or gases) will also be discussed.

A. Scattering from Surface Imperfections

As soon as the size of the imperfections approaches the Rayleigh wavelength, scattering from surface imperfections becomes important. For a polished surface with optical finish this effect is practically negligible up to gigahertz frequencies, as will be seen below. This is somewhat astonishing, since the polishing disorder is known to extend from the crystal surface to a depth of up to 5 μ (Lamb and Seguin, 1966), which is the same order of magnitude as the penetration depth of Rayleigh waves at frequencies of about 500 MHz. A rough surface, however, or small droplets of water condensed on the surface will prevent Rayleigh wave propagation almost completely:

for example, a Rayleigh wave at 316 MHz can be made to suffer an attenuation of more than 80 dB/cm by simply breathing on the surface of propagation.

B. Scattering from Grain Boundaries

The effect of grain boundaries in polycrystalline materials on the attenuation of Rayleigh waves can only be neglected if the crystallites are acoustically isotropic, which for most materials is not the case. This attenuation mechanism, e.g., confines the utilization of most piezoelectric ceramics to frequencies below a few hundred megahertz. Even on ideal crystal surfaces, Rayleigh waves can still be strongly attenuated if a thin, but polycrystalline film is evaporated onto the surface (see Section IV,D).

C. Interaction with Thermal Phonons

In dielectric single crystals the main contribution to the attenuation of Rayleigh waves arises from the interaction of ultrasonic Rayleigh phonons with thermal phonons. Since the attenuation caused by the defects mentioned above does not depend on the crystal temperature, interaction with thermal phonons can be studied separately by varying the temperature of the crystal between room temperature and liquid-helium temperature ($4.2°K$).

Thus, the total attenuation α at a fixed frequency can be expressed as the sum of a temperature-independent part α_0 and a contribution α_T changing with temperature,

$$\alpha(\omega) = \alpha_0(\omega) + \alpha_T(\omega, T) \tag{35}$$

The *temperature-dependent* absorption of Rayleigh waves propagating on a crystalline quartz surface between two transducers (Fig. 21) has been

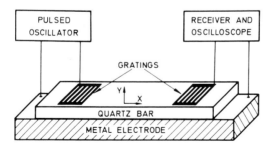

Fig. 21. Experimental arrangement for excitation and detection of Rayleigh waves. (Arzt *et al.*, 1967.)

derived from the measured temperature dependence of the direct transit signal at 316 MHz and at 1047 MHz. The resulting absorption α_T is shown in Fig. 22 (Salzmann *et al.*, 1968). Also included are the room-temperature values for the *total* absorption α which were found by the optical technique described in Section III,E.

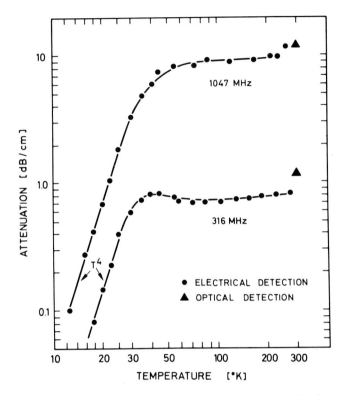

FIG. 22. Attenuation of Rayleigh waves propagated along the X axis on a surface normal to the Y axis of crystalline quartz. The dots represent the temperature-dependent part α_T (electrical detection); the triangles at room temperature show the total attenuation $(\alpha_0 + \alpha_T)$ (optical detection). (Salzmann *et al.*, 1968.)

It is interesting that, according to these data, the absorption of Rayleigh waves shows the same temperature dependence as the absorption of bulk waves: At high temperatures, i.e., above approximately 60°K, the attenuation is independent of temperature and varies with frequency as ω^2. At temperatures below 40°K, on the other hand, the absorption disappears rapidly, and at these low temperatures it is possible to observe echoes which are reflected back and forth between the two transducer gratings (Fig. 23).

A simple argument shows that the temperature-dependent attenuation due to phonon–phonon interaction occurs mainly by interaction with thermal *bulk* phonons, and that the interaction with thermal Rayleigh phonons is negligible (see, e.g., Maradudin and Mills, 1968): Under the usual experimental conditions $(T > 1°K)$ the frequency of ultrasonic Rayleigh waves in the gigahertz range is considerably smaller than the frequency of the thermal phonons (kT/h). Consequently, thermal surface phonons of energy kT have

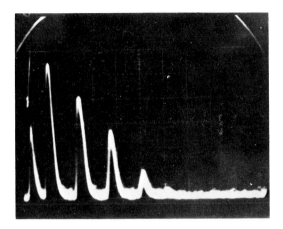

FIG. 23. Echoes of Rayleigh waves reflected back and forth between the transducers of Fig. 21. Frequency, 1047 MHz; time between successive echoes, 7 μsec.

a penetration depth λ_{th} which is only a small fraction of the penetration depth
λ of the ultrasonic Rayleigh waves,

$$\lambda_{\text{th}}/\lambda = \hbar\omega/kT \ll 1 \tag{36}$$

Even at temperatures as low as $10°K$, this ratio is of the order of 10^{-2}, which
means that the ultrasonic Rayleigh wave can interact with thermal Rayleigh
phonons only within a very small volume near the surface, compared to the
volume in which the ultrasonic energy is transported. Thermal bulk phonons,
however, are able to interact with the ultrasonic wave within the entire
volume in which its energy is transported.

The observed absorption at high temperatures can best be explained
by a process first proposed by Akhiezer (1939). This process is based on the
assumption that a periodic strain of frequency ω causes a periodic perturbation of the distribution of thermal phonons. The equilibrium distribution is
reestablished only after a relaxation time τ, which, at high temperatures, is
much shorter than the acoustic period, $\omega\tau \ll 1$. The relaxation to the state of
equilibrium leads to an increase of entropy and thus to an attenuation of
ultrasonic (bulk and Rayleigh) waves. For crystalline quartz the attenuation
coefficient for this relaxation process is given at high temperatures ($\omega\tau \ll 1$)
by

$$\alpha = (1.1\gamma^2 c_{\text{v}} T/\rho v^3)\omega^2\tau \quad [\text{cm}^{-1}] \tag{37}$$

(Bömmel and Dransfeld, 1960), where γ is an average Grüneisen constant
depending on the third-order elastic constants,[4] c_{v} is the specific heat, ρ is

---

[4] The third-order elastic constants also lead to second-harmonic generation. Due
to the high energy density in a Rayleigh wave (compared to bulk waves of the same
intensity), this effect could be measured in crystalline quartz (Løpen, 1968).

the mass density, and ω and v are, respectively, the frequency and velocity of the ultrasonic wave. At sufficiently high temperatures c_v is constant and $\tau \propto T^{-1}$ (the mean phonon lifetime is essentially determined by *umklapp* processes), so that $c_v T \tau$, and therefore α, is independent of the temperature.

It seems quite certain that this process, which is known to lead to the strong absorption of bulk waves at gigahertz frequencies (Bömmel and Dransfeld, 1960), is also responsible for the absorption of high-frequency Rayleigh waves. This follows not only from the nearly identical *temperature dependence* just discussed, but also from the similar *magnitude* of the absorption of bulk and surface waves:

For comparison, Fig. 24 shows the absorption of two transverse bulk waves (propagating along the AC and BC directions) and of Rayleigh waves (RW) in quartz at the same frequency of 1 GHz. In the first two columns of Table I the high-temperature values of the absorption are listed for the three waves together with their propagation velocities. In the third column the

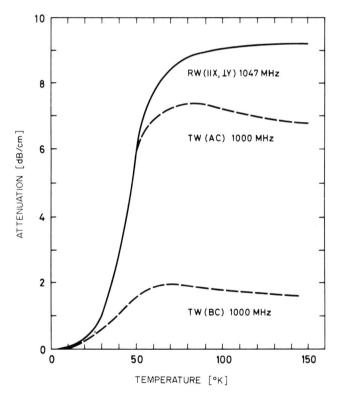

FIG. 24. Temperature-dependent part of the absorption of transverse (TW) and Rayleigh waves (RW) in quartz. Transverse waves along the AC and BC axis, respectively; Rayleigh waves along the X axis on a Y surface. (After data of Bömmel and Dransfeld, 1960; and Salzmann *et al.*, 1968.)

TABLE I

ABSORPTION OF RAYLEIGH WAVES (RW) AND TRANSVERSE WAVES
(TW) IN CRYSTALLINE QUARTZ AT 140°K AND 1 GHz

Wave	v(cm/sec)	α_{T}(dB/cm)	$\alpha_{\mathrm{T}} v^3$
RW (X direction on Y surface)[a]	3.16×10^5	8.2	2.6×10^{17}
TW (AC direction)[b]	3.32×10^5	6.7	2.5×10^{17}
TW (BC direction)[b]	5.04×10^5	1.5	2.0×10^{17}

[a] For the Rayleigh wave α_{T} at 1 GHz was calculated from the measured values at 1047 MHz (see Fig. 22a) assuming an ω^2 dependence.

[b] The values for the transverse waves were taken from Bömmel and Dransfeld (1960).

product αv^3 is shown, which, according to Eq. (37), is proportional to $\gamma^2 \tau$, all other parameters being invariant for different polarizations. It is interesting that αv^3 has almost the same magnitude for the two shear waves and the Rayleigh wave. This is compatible with the assumption that (1) Rayleigh waves and bulk waves are dissipated by the same mechanism, (2) for quartz the relaxation time τ has the same value inside the sample and in the surface regions (in spite of the penetrating polishing disorder), and that (3) it should be possible to predict the absorption of Rayleigh waves if the absorption is known for bulk waves of a few appropriate polarizations. [For cubic crystals, this property has been rigorously derived by Maris (1969). (See also Lamb and Richter 1966.[5])

It is therefore not surprising that $LiNbO_3$ crystals, because of their low loss for bulk waves (Smith *et al.*, 1967), also exhibit only a small absorption for Rayleigh waves. Using this material, it became possible to generate and detect Rayleigh waves up to 3 GHz at room temperature (Slobodnik, 1969).

At temperatures below 40°K, the attenuation shows an ωT^4 dependence (between 316 and 1047 MHz), which is the same as was found by Landau and Rumer (1937) for the attenuation of transverse bulk waves. As the Landau and Rumer mechanism is based on a three-phonon collision, the temperatures must be low enough to distinguish between single phonons. This means that the energy uncertainty ΔE of the thermal phonons—caused by their finite lifetime τ—has to be small compared with the energy $\hbar\omega$ of the ultrasonic phonons,

$$\Delta E \approx \hbar/\tau \ll \hbar\omega \qquad (38)$$

or

$$\omega\tau \gg 1 \qquad (39)$$

In a particle picture, the attenuation process is thus described as a collision of an ultrasonic surface phonon with a thermal bulk phonon which

[5] This reference was brought to the authors' attention by A. Maris.

leads to the excitation of a new thermal phonon of higher energy. The attenuation of ultrasonic waves due to this process is proportional to the number N of the thermal phonons, their energy kT, and the energy $\hbar\omega$ of the ultrasonic phonons. Therefore, since $N \propto T^3$, it follows that $\alpha \propto \omega T^4$.

The absorption of Rayleigh waves due to this three-phonon collision has been calculated by Maradudin and Mills (1968) for a model of a cubic crystal, and these authors have demonstrated that it should vary with frequency and temperatures as ωT^4, in agreement with the experimental results shown in Fig. 22. Preliminary observations by de Klerk (1969), also on quartz but at higher frequencies, seem to indicate a weaker frequency dependence than expected theroetically.

D. Interaction with Electrons in Metals

In solid-state physics Rayleigh waves may become increasingly useful as a tool to investigate the dynamical behavior of thin, evaporated metal films or surface layers. For example, if a thin film of superconducting metal is deposited on a polished piezoelectric crystal between two surface-wave transducers, it is possible to measure the absorption of Rayleigh waves caused by the metal film. The feasibility of this experimental technique has recently been demonstrated by Akao (1969), who measured the absorption of Rayleigh waves at 316 MHz propagating through films of indium and lead evaporated on crystalline quartz. His first results (Fig. 25a,b) show a temperature dependence for the absorption of Rayleigh waves which does not quite agree

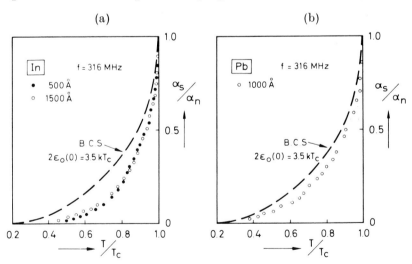

FIG. 25. (a) Attenuation of Rayleigh waves due to a thin film of superconducting indium between the transducers of Fig. 21 (T_c is the critical temperature for superconductivity) (Akao, 1969). (b) Attenuation of Rayleigh waves due to a thin film of superconducting lead. (Akao, 1969.)

with the predictions of the Bardeen–Cooper–Schrieffer theory for the absorption of longitudinal bulk waves. It is also noteworthy that there is no discontinuity of the absorption at the transition temperature, as is found for shear waves, although the strain amplitude of Rayleigh waves has a strong shear component. It would be interesting to investigate how the absorption of Rayleigh waves changes when the superconducting film is exposed to magnetic fields or currents in the film. At present, however, no results are available.

Apart from the temperature-dependent absorption discussed above, the evaporated film also causes a temperature-independent absorption, which in the case of aluminum was found to be relatively small (about 1 dB/cm at 316 MHz for a film 1000 Å thick; see also White and Voltmer, 1965). For indium and lead, however, the absorption is quite considerable (>20 dB/cm for the same frequency and film thickness). This extra absorption probably arises from the scattering of Rayleigh waves at the grain boundaries of the polycrystalline films. If the small crystallites are elastically nearly isotropic ($c_{11} = c_{12} + 2c_{44}$), the scattering is small, as in the case of aluminum, but for larger anisotropies, as in the case of In or Pb, this absorption is very serious. From Table II it can be seen that the "anisotropy factor" $(c_{12} + 2c_{44} - c_{11})/c_{11}$ for aluminum is about five times smaller than for lead.

TABLE II

ELASTIC ANISOTROPY OF CUBIC CRYSTALS AT ROOM TEMPERATURE

	Stiffness constants[a] (10^{12} dyn/cm^2)			Anisotropy factor
	c_{11}	c_{12}	c_{44}	$(c_{12} + 2c_{44} - c_{11})/c_{11}$
Aluminum	1.068	0.607	0.282	0.096
Lead	0.495	0.423	0.149	0.456

[a] The values of the stiffness constants were taken from Kittel (1966).

Another effect that could be studied with Rayleigh waves is surface superconductivity (see Saint James and deGennes, 1963). It is known that below the transition temperature superconductivity can be destroyed by a strong magnetic field. However, even for fields which destroy the superconducting state in the bulk of the sample ($H_{c2} < H < H_{c3}$), superconductivity can still persist in the surface layer. Only if the magnetic field is further increased ($H > H_{c3}$) does the surface layer also become normal. At fields $H_{c2} < H < H_{c3}$, Rayleigh waves are the only acoustic waves having a low loss. The thickness of the superconducting surface sheath increases in the vicinity of the transition temperature, and Rayleigh waves may be a useful probe in order to measure the thickness and properties of the superconducting surface layer.

E. Interaction with Drifting Carriers in Semiconductors

In piezoelectric semiconductors there is a strong interaction between the electric fields associated with an acoustic wave and carriers of electrical charge, e.g., electrons in the conduction band. Hutson *et al.*, (1961) demonstrated for the first time that ultrasonic bulk waves can be considerably amplified by an electric current, provided the drift velocity of the carriers exceeds the velocity of the ultrasonic wave.

White and Voltmer (1966) also succeeded in amplifying Rayleigh waves traveling along the surface of a cadmium sulfide crystal at a frequency of 8 MHz. Free carriers were created at the surface by illuminating the crystal, and a dc drift field was applied between the transmitting and the receiving transducer grating.

As mentioned in Section II,C, the electric fields associated with a Rayleigh wave in piezoelectric crystals also extend to outside of the crystal. This property can be used to amplify Rayleigh waves by interaction with externally drifting carriers. This makes it possible to independently select a piezoelectric material with high electromechanical coupling and low propagation losses for the Rayleigh wave, and a semiconductor with optimal electric performance. Collins *et al.* (1968a) describe an amplifier in which the Rayleigh wave propagates on a surface of lithium niobate and the carriers propagate in a silicon bar (Fig. 26). Since the external piezoelectric fields

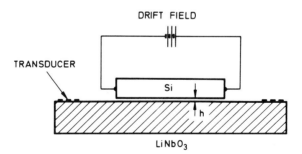

Fig. 26. Amplification of Rayleigh waves by interaction with drifting carriers. (After Collins *et al.*, 1968a.)

decrease exponentially with distance (see Fig. 4 of Section II), it is important to bring the silicon bar as close as possible to the $LiNbO_3$ surface. However, direct mechanical contact must be avoided in order to prevent an attenuation of the Rayleigh wave. The gain obtained with this amplifier is shown in Fig. 27 as a function of the electric drift field.

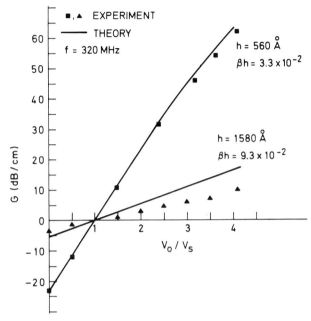

FIG. 27. Gain of the amplifier for Rayleigh waves (shown in Fig. 26) as a function of the electric drift field. (Collins *et al.*, 1968a.)

F. ATTENUATION BY AMBIENT MEDIA

After studying the attenuation of a Rayleigh wave propagating along a free crystal surface, it is interesting to evaluate the additional attenuation which occurs when the surface is surrounded by a fluid or gas (see Arzt *et al.*, 1967; see also Carr *et al.*, 1969). It is especially important to find out if experiments with Rayleigh waves of high frequencies have to be done in vacuum or if the crystal can be exposed to air or liquid helium without causing serious attenuation of the Rayleigh wave.

In the following discussion it will be assumed that the density of the ambient medium is much lower than the density of the solid, which results in a strong acoustic mismatch between the solid and the ambient fluid.

The particle motion in a Rayleigh wave is partially normal and partially parallel to the surface (see Section II). The *normal* component causes density changes in the ambient medium, and therefore, an emission of compressional waves, which leads to an attenuation of the Rayleigh wave. This attenuation mechanism will be discussed first. The *parallel* component of the particle motion at the surface leads to frictional losses which are determined by the viscosity of the surrounding medium. The attenuation of Rayleigh waves due to this frictional loss will be treated later.

1. *Attenuation by Emission of Compressional Waves*

The periodic density changes which are caused in the ambient medium by the normal component of the particle motion associated with a Rayleigh wave lead to the radiation of compressional waves into the fluid (Fig. 28a) at

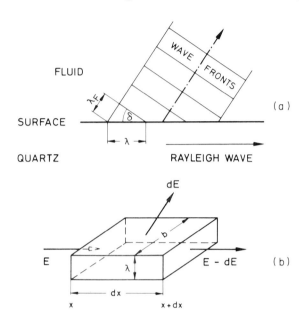

Fig. 28. (a) Attenuation of Rayleigh waves due to radiation of compressional bulk waves into an adjacent fluid or gas. Here $\sin \delta = \lambda_F/\lambda$, where λ_F is the longitudinal wavelength in the fluid and λ is the Rayleigh wavelength. (b) Energy flow through an element at the surface of the solid, describing the attenuation of the Rayleigh wave by radiating logitudinal waves into an adjacent fluid.

an angle δ if the longitudinal velocity v_F of sound in the fluid is smaller than the velocity of the Rayleigh wave v,

$$v_F/v = \sin \delta < 1 \tag{40}$$

This inequality is satisfied, e.g., for air or liquid helium surrounding the quartz sample. In these cases v_F is so much smaller than v that the longitudinal waves are emitted almost normal to the surface. The attenuation coefficient α_L arising from the emission of longitudinal waves can be estimated as follows: If a is the normal component of the particle displacement amplitude of the Rayleigh wave, λ its wavelength, v its velocity, and ρ the solid density, one can express the energy E which is transported per second in a surface layer of thickness λ and of width b (Fig. 28b) as

$$E = \lambda b 2\pi^2 \rho v^3 (a/\lambda)^2 = 2\pi^2 b \rho v^3 a^2 / \lambda \tag{41}$$

(see, e.g., Bergmann, 1954). At the surface the amplitude a of the Rayleigh wave equals the amplitude of the longitudinal wave in the adjacent fluid. Therefore the acoustic energy dE emitted into the fluid per second from the surface area $b\,dx$ is

$$dE = 2\pi^2(b\,dx)\rho_F V_F{}^3(a/\lambda_F)^2$$
$$= 2\pi^2(b\,dx)\rho_F v_F v^2(a/\lambda)^2 \tag{42}$$

where ρ_F, v_F, and λ_F are the density, sound velocity, and acoustic wavelength in the fluid, respectively. This energy loss results in the following absorption coefficient of the Rayleigh wave,

$$\alpha_L = (1/E)\,dE/dx = \rho_F v_F/\rho v\lambda \quad \text{cm}^{-1} \tag{43}$$

or multiplied by a factor 4.3 if α_L is to be expressed in dB/cm).

This estimate has bee made under the assumption that the fluid or gas can be considered as a continuum having an atomic mean free path l much smaller than the Rayleigh wavelength. It can be shown by gas-kinetic arguments that in the reverse case of a very long atomic mean free path $(l \gg \lambda)$ Eq. (43) remains valid with the mean thermal velocity of the gas atoms replacing v_F, both being of a very similar magnitude at room temperature.

2. Frictional Loss

If a' denotes the particle displacement amplitude parallel to the surface, the corresponding velocity amplitude becomes $w_0 = \omega a'$ where ω is the angular frequency of the Rayleigh wave. The viscous force F exerted by the fluid on an area $b\,dx$ is

$$F = \eta(b\,dx)w/l \tag{44}$$

where w is the particle velocity at the surface ($w = w_0 \sin \omega t$), and η is the viscosity of the fluid. Now, l is the penetration depth of a shear wave into the fluid,

$$l = 2\pi(2\eta/\rho_F \omega)^{1/2} \tag{45}$$

(see, e.g., Hertzfeld and Litovitz, 1965), and therefore w/l is approximately the velocity gradient in the fluid. The energy dE' dissipated per second by the viscous force acting on the surface area $b\,dx$ (see Fig. 28) is

$$dE' = \tfrac{1}{2}(\eta/l)w_0{}^2 b\,dx \tag{46}$$

Using $a' = w_0/\omega$ in Eq. (41), the energy flow in the Rayleigh wave can be expressed as

$$E' = \tfrac{1}{2}b\rho v w_0{}^2 \lambda \tag{47}$$

From Eq. (45)–(47) the absorption coefficient of Rayleigh waves due to frictional losses is calculated to be

$$\alpha_S = \frac{dE'}{E'\,dx} = \frac{(\rho_F \eta \omega^3/2)^{1/2}}{4\pi^2 \rho v^2} \quad \text{cm}^{-1} \tag{48}$$

In liquid helium of temperatures above the λ-point and at frequencies in the higher gigahertz range the viscous penetration depth l (Eq. 45) is considerably smaller than the thickness of a helium film absorbed at the saturated vapor pressure [approximately 400 Å (Wilks, 1967)]. Consequently, for the frictional losses of Rayleigh waves propagating on a crystalline surface at these high frequencies it makes no difference whether the surface is in contact with bulk liquid helium or only with an adsorbed helium film.

3. Numerical Examples

In Table III calculated values for the absorption of Rayleigh waves

TABLE III

CALCULATED VALUES FOR THE ATTENUATION OF RAYLEIGH
WAVES IN QUARTZ BY AMBIENT MEDIA[a]

		Attenuation (dB/cm) at		
Constant	Medium	1 GHz	3 GHz	30 GHz
$\alpha_L{}^b$	Atmospheric air	0.7	2.1	21
$\alpha_L{}^b$	Liquid helium	60	180	1800
$\alpha_S{}^c$	Liquid helium	0.34	1.8	56

[a] Measured value for a surface wave propagating along the Z axis of Y-cut LiNbO$_3$: 0.5 dB/cm at 905 MHz (Carr et al., 1969).
[b] Attenuation due to radiation of longitudinal waves.
[c] Attenuation due to frictional loss.

on quartz due to ambient atmospheric air and liquid helium are given for comparison. It is seen that in both cases $\alpha_S \ll \alpha_L$ i.e., the attenuation due to radiation of shear waves can be neglected compared with the longitudinal contribution to the attenuation. Furthermore, it is seen that liquid helium surrounding the sample causes the most serious attenuation. At 30 GHz, even a helium film leads to large *frictional* losses.

In order to measure the attenuation by liquid helium (Arzt et al., 1967), the quartz is immersed vertically so far into the fluid that only the lower transducer grating is covered by it. By further lowering the crystal a definite distance, the additional attenuation due to the fluid can be measured.

In superfluid helium the attenuation due to radiation of longitudinal waves was measured to be 18 dB/cm at 316 MHz, which is in good agreement with the estimated value. (In the case of air at 300°K a similar loss is to be expected at a pressure of 100 atm.) Although the shear viscosity of liquid helium is a function of temperature, no change of the attenuation could be observed for temperatures below and above the λ-point, respectively. This behavior is to be expected, since $\alpha_S \ll \alpha_L$ (Arzt et al., 1967).

The presence of atmospheric air (or similar gases) at 300°K in contact with the solid surface only leads to a relatively small attenuation coefficient.

Carr *et al.* (1969a,b) have verified experimentally the magnitude, frequency, and pressure dependence of the absorption of Rayleigh waves on lithium niobate by air in the frequency range 1–3 GHz (see Table III).

V. Applications

As already mentioned in the introduction, elastic surface waves at megahertz frequencies were first used as a means for ultrasonic flaw detection, since they are reflected from minute surface cracks, which only need to be wide enough to prevent interaction by van der Waals forces. For this application Rayleigh waves are usually generated by wedge transducers (see Section III,A).

To the present Rayleigh waves have been generated by means of interdigital transducers (see Section III,B2) up to a frequency of 3 GHz (see, e.g., Carr *et al.*, 1969), corresponding to a wavelength of approximately 1 μ. At still higher frequencies, i.e., when the penetration depth becomes smaller than 1 μ, it might become interesting to use Rayleigh waves as a research tool in surface physics. As a first example to their application in solid-state physics, Rayleigh waves have recently been used to study the acoustic absorption in thin superconducting films (Akao, 1969). These experiments have already been described in Section IV,D.

From a technical viewpoint, application of surface elastic waves appears to be most promising in electronic devices. For this field of applications it is an additional advantage over the use of bulk waves that generation and detection of surface elastic waves do not require new manufacturing techniques. Instead, the highly developed microelectronic technology—such as thin-film deposition and photoetching processes—can be applied directly.

A. Transducer Fabrication

The most widely used transduction method for high-frequency Rayleigh waves is direct piezoelectric coupling to grating transducers on piezoelectric substrates (see Section III,B). These transducers are usually fabricated using one of the following photoetching techniques:

1. The transducer areas on the crystal surface are covered with a thin metal film (e.g., by evaporation in vacuum). This film is coated with a photoresist which, after exposure through an appropriate photomask and development, leaves only the transducer fingers covered with an etch-resistive film. The uncovered parts of the metal film are etched away and afterwards the photoresist can be removed from the remaining transducer grating.

2. The clean crystal surface is covered with a photoresist which is exposed through a photomask. After development, the remaining parts of the photoresist form a negative of the transducer grating, i.e., a "window" leaving the crystal surface free at those places where the transducer fingers are to be deposited. This is done by evaporating a metal film over the entire

transducer area (the free parts of the surface and the remaining photo-resist). Then the photoresist is dissolved and the metal film adheres to the crystal surface at those places which were free during the evaporation process.

With these techniques minimum linewidths of approximately 1 μ can be obtained, corresponding to a fundamental interdigital transducer frequency of somewhat less than 1 GHz. In order to manufacture transducers for higher frequencies, resistive films are used which are developed after exposure by the beam of a scanning electron microscope. Using this technique, Broers *et al.* (1969) have fabricated interdigital transducers consisting of lines 0.3 μ wide, separated by 0.4-μ spacings, which corresponds to a fundamental frequency of 1.75 GHz (on LiNbO$_3$).

B. Delay Lines and Filters

Two transducer gratings deposited parallel to each other on a piezo-electric substrate can be used as a delay line operating at the center frequency of the transducers (Fig. 21). Since surface waves are the slowest elastic waves, large delay times can be achieved with comparatively small physical dimensions.

A comomn feature of all utrasonic delay lines is the existence of echoes (see Fig. 22b), which in many cases are undesired. For matched Rayleigh-wave transducers the signal of the first echo, i.e., the signal corresponding to a triple transit of the acoustic pulse between the two transducers, is still relatively strong, as will be shown below.

When a Rayleigh wave is incident on an electrically matched transducer from one side, one quarter of the acoustic energy is reflected, one quarter is transmitted to the other side of the transducer, and the remaining half is converted into electrical energy. In order to derive this property, Collins *et al.* (1968b) used a scattering-matrix analysis of the transducer:

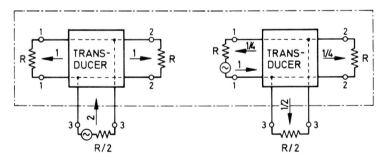

Fig. 29. Energy flow in a Rayleigh-wave delay line with the transducers matched to the impedance of the electrical generator and receiver, respectively: the electric energy incident at port 3 of the transmitting transducer distributes equally between two Rayleigh waves propagating away from the transducer. At the receiving transducer one quarter of the incident acoustic energy is reflected, one quarter passes to the other side, and the remaining half is converted into electric energy. (After Collins *et al.*, 1968a.)

Each transducer is regarded as a three port junction consisting of one electrical port 3 and two acoustic ports 1 and 2 (Fig. 29). The scattering matrix **S** links the amplitude a_i of a signal *incident* at port "i" (e.g., an electric voltage amplitude or a mechanical stress amplitude) with the amplitudes b_i of the signals *emitted* from the respective ports i. Since the transducer is a reciprocal element (i.e., electric energy incident at port 3 divides up equally between the acoustic ports 1 and 2, and, conversely, two acoustic signals of equal energy and of the same phase incident at ports 1 and 2 add up to an electric signal of twice the acoustic energy at port 3), the scattering matrix has to be of the form

$$\begin{pmatrix} b_1 \\ b_2 \\ b_3 \end{pmatrix} = \begin{pmatrix} S_{11} & S_{12} & S_{13} \\ S_{12} & S_{22} & S_{23} \\ S_{13} & S_{23} & S_{33} \end{pmatrix} \begin{pmatrix} a_1 \\ a_2 \\ a_3 \end{pmatrix} \tag{49}$$

Symmetry with respect to the identical acoustic ports 1 and 2 requires

$$\mathbf{S} = \begin{pmatrix} S_1 & S_2 & \pm S_4 \\ S_2 & S_1 & S_4 \\ \pm S_4 & S_4 & S_3 \end{pmatrix} \tag{50}$$

Since the transducer itself does not consume or create energy, the scattering matrix must be unitary,

$$\mathbf{S\hat{S}^*} = \mathbf{1} \tag{51}$$

For a matched transducer (i.e., no energy is reflected at port 3, therefore $S_3 = 0$), this condition yields

$$2|S_4|^2 = 1 \tag{52}$$

$$|S_1|^2 + |S_2|^2 + |S_4|^2 = 1 \tag{53}$$

$$S_4(S_1^* \pm S_2^*) = 0 \tag{54}$$

It follows from (52) that

$$|S_4|^2 = \tfrac{1}{2} \tag{55}$$

Using this in (53),

$$|S_1|^2 + |S_2|^2 = \tfrac{1}{2} \tag{56}$$

or, considering (54),

$$|S_1|^2 = |S_2|^2 = \tfrac{1}{4} \tag{57}$$

If the only signal incident on the transducer is an acoustic signal at port 2, $a_1 = a_2 = 0$. Therefore, Eqs. (49) and (50) reduce to

$$b_1 = S_2 a_2, \qquad b_2 = S_1 a_2, \qquad b_3 = S_4 a_2 \tag{58}$$

Squaring these expressions (in order to get the respective expression for the energies) and using the above results for **S** [Eqs. (57) and (55)], the energies which are emitted at ports 1, 2, and 3 become

$$b_1^2 = \tfrac{1}{4} a_2^2, \qquad b_2^2 = \tfrac{1}{4} a_2^2, \qquad b_3^2 = \tfrac{1}{2} a_2^2 \tag{59}$$

Thus, one quarter (-6 dB) of the acoustic energy incident at port 2 is reflected ($b_2{}^2$), one quarter is transmitted to port 1 ($b_1{}^2$), and the remaining half (-3 dB) is converted into electric energy available at port 3 ($b_3{}^2$).

In conjunction with the trivial result that the electric energy incident at port 3 divides up equally between the acoustic ports 1 and 2, it is obvious that for an electrically matched delay line the signal of the first echo is only 12 dB below the main transit signal: the main transit signal suffers a conversion loss of 3 dB at each of the transducers (i.e., a total loss of 6 dB), while the first echo suffers two additional reflections at the transducers and thereby an additional loss of 6 dB for each reflection, or a total additional loss of 12 dB.

The insertion loss of a delay line (i.e., the sum of the conversion loss in the transducers and the propagation loss in the elastic medium) can be reduced by 6 dB (arising from the bidirectivity of the transducers) by means of transducers which radiate acoustic energy only in one direction and, conversely, detect Rayleigh waves only when they are incident from one distinguished direction. Such unidirectional transducers would also avoid spurious echoes. In analogy with directional electrical antennas, directional Rayleigh wave transducers consist of two grating transducers separated by a relative distance of a quarter wavelength. They are driven from the same rf source, but with a phase difference of $\pi/2$. From measurements on lithium niobate (at frequencies of about 100 MHz) Auld *et al.* (1968) conclude that delay lines having a minimum insertion loss of only 4 dB at a relative bandwidth of 20% should be feasible.

If spurious echoes are undesired, their amplitudes can be reduced in several ways: (1) by lowering the acoustic reflectivity of the transducers [S_1 and S_2 in Eq. (57)], which is done by increasing the impedance of the electrical termination above the radiation resistance of the transducers (see Collins *et al.*, 1968b; this also introduces a mismatch and increases the insertion loss), and (2) by use of a lossy material for Rayleigh-wave propagation. In both cases an unwanted reduction of the main transit signal also occurs; however, the level of the triple transit echo is reduced by twice this amount. A third method to reduce the spurious echo level is to amplify the wave in forward direction and to attenuate the wave traveling backward by interaction with drifting electric carriers (see Section IV,E).

Dispersive delay lines (for which the delay time depends on the frequency) used for pulse compression are easily realized by means of a transducer having a variable spacing over its length (Fig. 30; e.g., Tancrell *et al.*, 1969; or

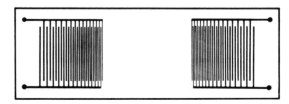

FIG. 30. Dispersive delay line. The delay time decreases for increasing frequency.

Teodori, 1969). The receiving transducer can be either the same kind of grating with its spacings in opposite sequence (Fig. 30), or one transducer can be replaced by a broadband grating consisting of only a few lines (see Fig. 33).

For many applications in digital signal processing, delay lines are required which transform a short input pulse into a delayed pulse of the same shape. The response signal of a usual nondispersive delay line to a short input pulse is a wave train with a carrier frequency equal to the fundamental frequency of the transducers and a triangular envelope of the amplitude (Fig. 31).

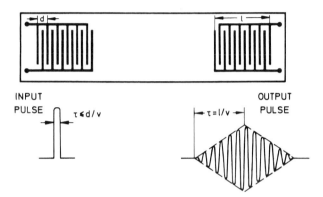

FIG. 31. Impulse response of a Rayleigh-wave delay line.

The time of the sloping parts of the wave train is determined by the length of the transducer gratings divided by the Rayleigh-wave velocity. The demand for a single output pulse means that a wide frequency band has to be transmitted. This can be achieved, e.g., with two single electrode-pair transducers (Section III,B6)—however, at the cost of reduced efficiency.

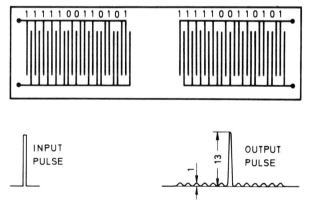

FIG. 32. Impulse response of a delay line using Barker-code transducers. (See White, 1967.)

Higher efficiencies are obtained with transducer gratings having inter-
connections between their lines according to a Barker code (Whitehouse,
1968; White, 1967). An example for a delay line utilizing Barker-code
transducers is given in Fig. 32 (White, 1967). Since 13 electrode pairs are
active per-transducer, the efficiency is appreciably increased, whereas the
output signal is still nearly a single pulse.

Instead of using a device consisting of two identical grating transducers
on a piezoelectric substrate as a delay line, it can also be regarded as a band-
pass filter: for gratings with uniform spacings the frequency response has a
maximum at the fundamental transducer frequency with a -3 dB bandwidth
determined by the inverse number of grating elements (see Section III,B6).
In order to obtain a flat frequency pass-band, the transmitting and receiving
gratings may, e.g., have slightly different spacings, which leads to stagger
tuning. A more complex frequency response can be realized by means of
transducer gratings having variable spacings as well as variable finger
lengths (Fig. 33). These "weighted transducers" are also used in delay lines
for pulse compression (see, e.g., Hartemann and Dieulesaint, 1969, or Tancrell
et al., 1969).

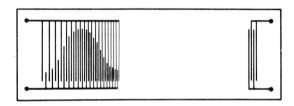

FIG. 33. Dispersive delay line using a "weighted" transducer.

Application of Rayleigh waves of band-pass filtering devices may be of
particular interest in the frequency range between 100 MHz and 1 GHz,
where the performance of electrical LC filters is not ideal and, on the other
hand, waveguide filters are still very bulky components because of the
relatively long electric wavelengths involved.

C. GUIDED SURFACE WAVES

Many electrical waveguide components could be replaced by corresponding "microsound" components having physical dimensions smaller by a
factor of 10^5—the ratio of the velocity of light to the acoustic surface-wave
velocity—if guided surface waves were used. (See, e.g., Stern, 1968. For a
detailed discussion of guided surface elastic waves see Tiersten, 1969; see
also Ash *et al.*, 1969).

The simplest way of guiding elastic surface waves is to cut two parallel
slots into the surface so that the surface wave propagates along the remaining
flat ridge between the slots (Ash and Morgan, 1967).

Usually, however, guiding is performed by utilizing the fact that the velocity of an elastic surface wave can be easily influenced, e.g., by coating the surface with a thin metallic film. If the velocity along the waveguide is lower than in the adjacent parts of the surface, total internal reflection confines the energy transport to the waveguide structure. This relation between the velocities can be realized in two ways (Tiersten, 1969): (1) by lowering the velocity in the waveguide area (Fig. 34a), or (2) by increasing the velocity in the surface area outside the waveguide (Fig. 34b).

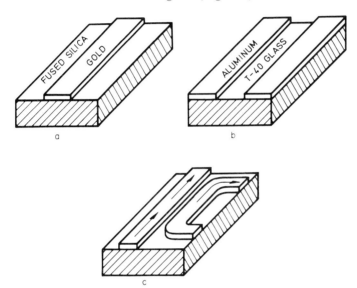

FIG. 34. Guiding of surface elastic waves (after Tiersten, 1969). (a) Lowering the velocity along the waveguide by means of a "loading" film. (b) Increasing the velocity on the surface adjacent to the waveguide by means of a "stiffening" film. (c). Schmatic of a directional coupler for guided surface waves.

The guiding structure shown in Fig. 34a consists of a thin gold strip on a fused silica substrate (Tiersten, 1969). The gold film leads to a mass loading of the surface, so that the resulting lower velocity of the surface wave can be expressed by the relevant elastic constant c_{eff} as $v = (c_{\text{eff}}/\rho_{\text{eff}})^{1/2}$.

On the other hand, an aluminum film on a T-40 glass substrate (Tiersten, 1969; Fig. 34b) *stiffens* the surface (the effective elastic constant c_{eff} is increased), so that the velocity is higher in the coated area of the surface. An advantage of this type of waveguide is that the surface wave in the waveguide is nondispersive, whereas dispersion occurs when the surface is covered with a thin film of different material (see, e.g., Section III,C).

With the aid of waveguides for surface elastic waves, complex electrical circuits can be transformed into microsound circuits: amplifying sections can be linked with filters, directional couplers, delay lines, etc. An example for a directional couplers, delay lines, etc. An example for a cirectional coupler is shown in Fig. 34c: since the surface wave is confined to the waveguide due to total internal reflection, it is known that mechanical stress fields also extend to *outside* of the waveguide. However, these fields decay rapidly with increasing distance from the waveguide area, so that coupling to a second waveguide is posible only when the latter is placed close to the first one. It is also obvious that the radius of a curve in a waveguide is determined by the critical angle for total internal reflection between the waveguide and the adjacent parts of the surface.

ACKNOWLEDGMENTS

The authors gratefully acknowledge many helpful discussions with Dr. R. M. White (University of California, Berkeley) during his stay in München. They also would like to thank Dr. F. Akao (Osaka University, Japan) and Dr. H. Thomann (Siemens Forschungslabor, München) for contributing figures prior to publication. During the preparation of the manuscript one of the authors (K.D.) benefitted greatly from a sabbatical leave in the USA made possible by the hospitality of the Physics Department of Cornell University.

REFERENCES

Adler, R., Korpel, A., and Desmares, P. (1968). *IEEE Trans. Sonics Ultrasonics* **SU-15**, 157.
Akao, F. (1969). *Phys. Letters* **30A**, 409.
Akheizer, A. (1939). *J. Phys. (USSR)* **1**, 277.
Arzt, R. M., and Dransfeld, K. (1965). *Appl. Phys. Letters* **7**, 156.
Arzt, R. M., Salzmann, E., and Dransfeld, K. (1967). *Appl. Phys. Letters* **10**, 165.
Ash, E. A., and Morgan, D. (1967). *Electronics Letters* **3**, 464.
Ash, E. A., De La Rue, R. M., and Humphryes, R. F. (1969). *IEEE Trans. MTT.* (to be published).
Auld, B. A., Collins, J. H., and Shaw, H. J. (1968). IEEE Ultrasonic Symposium, New York, Paper A-2.
Auth, D. C., and Mayer, W. G. (1967). *J. Appl. Phys.* **38**, 5138.
Bergmann, L. (1954). "Der Ultraschall und seine Anwendung in Wissenschaft und Technik." Hirzel, Stuttgart.
Bleustein, J. L. (1968). *Appl. Phys. Letters* **13**, 412.
Born, M., and Wolf, E. (1959). "Principles of Optics," p. 403. Pergamon Press, Oxford.
Broers, A. N., Lean E. G., and Hatzakis, M. (1969). *Appl. Phys. Letters* **15**, 98.
Bömmel, H. E., and Dransfeld, K. (1960). *Phys. Rev.* **117**, 1245.
Cady, W. G. (1964). "Piezoelectricity." Dover, New York.
Campbell, J. J., and Jones, W. R. (1968). *IEEE Trans. Sonics Ultrasonics* **SU-15**, 209.
Carr, P. H., Slobodnik, A. J., and Sethares, J. C. (1969). *IEEE Group on Microwave Theory and Techniques Symposium Digest*, **354**.
Collins, J. H., Lakin, K. M., Quate, C. F., and Shaw, H. J. (1968a) *Appl. Phys. Letter.* **13**, 314.
Collins, J. H., Shaw, H. J., and Smith, W. R. (1968b). M. L. Report No. 1692, Microwave Laboratory, Standford University (unpublished).
Collins, J. H., Gerard, H. M., and Shaw, H. J. (1968c). *Appl. Phys. Letters* **13**, 312.
Collins, J. H., Gerard, H. M., Lakin, K. M., and Shaw, H. J. (1968d). *Proc. IEEE Letters* **56**, 1635.

Cook, G. E., and Valkenberg, H. E. (1954). *Am. Soc. Testing Mater. Bull.* **198**, 81.

Coquin, G. A., and Tiersten, H. F. (1967). *J. Acount Soc. Amer.* **41**, 921.

de Klerk, J. (1969). Private communication.

de Klerk, J., and Kelly, E. F. (1965). *Rev. Scient. Instr.* **36**, 506.

Engan, H., Ingebrigtsen, K. A., and Tonning, A. (1967). *Appl. Phys. Letters* **10**, 311.

Ewing, W. M., Jardetzky, W. S., and Press, F. (1957). "Elastic Waves in Layered Media." McGraw-Hill, New York.

Farnell, G. W. (1970). *Physical Acoustics* **6**, 109.

Firestone, F. A., and Frederick, J. A. (1946). *J. Acoust. Soc. Amer.* **18**, 200.

Foster, N. F., (1965). *Proc. IEEE* **53**, 1400.

Hartemann, P. and Dieulesaint, E. (1969). *Electronics Letters* **5**, 219.

Hertzfeld, K. F., and Litovitz, T. A. (1965). "Absorption and Dispersion of Ultrasonic Waves." Academic Press, New York and London.

Humphryes, R. F., and Ash, E. A. (1969). *Electronics Letters* **5**, 175.

Hutson, A. R., McFee, J. H., and White, D. L. (1961). *Phys. Rev. Letters*, **7**, 237.

Ippen, E. P. (1967). *Proc. IEEE* **55**, 248.

Joshi, S. G., and White, R. M. (1969). *J. Acoust. Soc. Amer.* **46**, 17.

Kittel, C. (1966). "Introduction to Solid State Physics," p. 122. Wiley, New York.

Korpel, A., Laub, L. J., and Sievering, H. C. (1967). *Appl. Phys. Letters* **10**, 295.

Krokstad, J., and Svaasand, L. O. (1967). *Appl. Phys. Letters* **11**, 155.

Lamb, J., and Richter, J. (1966). *Proc. Roy. Soc.* A **293**, 479.

Lamb, J., and Seguin, H. (1966). *J. Acoust. Soc. Amer.* **39**, 752.

Landau, L. D., and Lifshitz, E. M. (1963). "Fluid Mechanics," Course of Theoretical Physics, Vol. 6. Pergamon Press, Oxford, London, Paris, Frankfurt.

Landau, L. D., and Lifshitz, E. M. (1959). "Theory of Elasticity," Course of Theoretical Physics, Vol. 7. Pergamon Press, Oxford, London, Paris, Frankfurt.

Landau, L., and Rumer, G. (1937). *Physik. Z. Sowjetunion* **11**, 18.

Lee, R. E., and White, R. M. (1968). *Appl. Phys. Letters* **12**, 12.

Lim, T. C., and Farnell, G. W. (1968). *J. Appl. Phys.* **39**, 4319.

Loftus, D. S. (1968). *Appl. Phys. Letters* **13**, 323.

Løpen, P. O. (1968). *J. Appl. Phys.* **39**, 5400.

Maradudin, A. A., and Mills, D. L. (1968). *Phys. Rev.* **173**, 881.

Maris, A. (1969). Private communication.

Mason, W. P. (1950). "Piezoelectric Crystals and their Application to Ultrasonics." Van Nostrand, Princeton, New Jersey.

Mason, W. P. (1958). "Physical Acoustics and the Properties of Solids." Van Nostrand, Princeton, New Jersey.

Mayer, W. G., Lamers, G. B., and Auth, D. C. (1967). *J. Acoust. Soc. Amer.* **42**, 1255.

Minton, C. (1954). *Nondestructive Testing* **12**, 13.

Raman, C. V., and Nath (1935). *Proc. Indian Acad. Sci.* **2A**, 406.

Rayleigh, Lord (1885). *Proc. London Math. Soc.* **17**, 4.

Rischbieter, F. (1965/66). *Acoustica* **16**, 75.

Rollins, F. R., Jr., Lim, T. C., and Farnell, G. W. (1968). *Appl. Phys. Letters* **12**, 236.

Saint James, D., and de Gennes, P. G. (1963). *Phys. Letters* **7**, 306.

Salzmann, E., and Weismann, D. (1969). *J. Appl. Phys.* **40**, 3408.

Salzmann, E., Plieninger, T., and Dransfeld, K. (1968). *Appl. Phys. Letters* **13**, 14.

Schnitzler, P. (1967). *Appl. Phys. Letters* **11**, 273.

Smith, A. B., Kestigian, M., Kedzie, R. W., and Grace, M. I. (1967). *J. Appl. Phys.* **38**, 4928.

Slobodnik, A. J., Jr. (1969). *Appl. Phys. Letters* **14**, 94.

Sokolinskii, A. G. (1958). Author's Certificate No. 19297.

Stern, E. (1968). Technical Note 1968–36. Lincoln Laboratory, Massachusetts Institute of Technology.

Stoneley, R. (1955). *Proc. Roy. Soc.* **232A**, 447.

Tancrell, R. H., Schulz, M. B., Barret, H. H., Davis, L., Jr., and Holland, M. G. (1969). *Proc IEEE* **57**, 1211.

Teodori, E. (1969). *Electronics Letters* **5**, 334.

Thomann, H. (1969). *Electronics Letters* **5**, 652.

Tiersten, H. F. (1969). *J. Appl. Phys.* **40**, 770.

Tseng, C.-C. (1968). *IEEE Trans. Electron Devices* **ED-15**, 586.

Victorov, I. A. (1967). "Rayleigh and Lamb Waves." Plenum Press, New York.

von Gutfeld, R. J., and Nethercot, A. H., Jr. (1964). *Phys. Rev. Letters* **12**, 641.

Wauk, M. T., and Winslow, D. K. (1968). *Appl. Phys. Letters* **13**, 286.

Weis, O. (1969). *Z. Angew. Phys.* **26**, 325.

White, R. M. (1963). *J. Appl. Phys.* **34**, 2123.

White, R. M. (1967). *IEEE Trans. Electron Devices* **ED-14**, 181.

White, R. M., and Voltmer, G. W. (1965). *Appl. Phys. Letters* **7**, 314.

White, R. M., and Voltmer, F. W. (1966). *Appl. Phys. Letters* **8**, 40.

Whitehouse, H. J. (1968). 6th International Congress on Acoustics, Tokyo, Japan, Paper K-4-4.

Whitman, R. L., Laub, L. J., and Bates, W. J. (1968). *IEEE Trans. Sonics Ultrasonics* **SU-15**, 186.

Wilks, J. (1967). "The Properties of Liquid and Solid Helium." Oxford Univ. Press (Clarendon), London and New York.

−5−

Interaction of Light with Ultrasound : Phenomena and Applications

R. W. DAMON, W. T. MALONEY, and D. H. McMAHON

Sperry Rand Research Center
Sudbury, Massachusetts

I. General Introduction

The recent development of techniques for the generation of high-frequency coherent elastic waves, combined with the ready availability of sources of coherent light, has given rise to a fertile new field of research and exploitation based on the interaction of these two coherent waves. The study of the diffraction of light beams by elastic waves includes such diverse phenomena as Raman–Nath diffraction. optical Bragg diffraction, and the diffraction of light from both surface elastic waves and bulk magnetoelastic waves. While the purpose of this chapter is to outline the progress made in applying the principle of light diffraction by elastic waves to the development of practical devices, some space will be devoted to recent techniques. A discussion of these new techniques is merited on the ground that they may soon lead to further applications.

The diffraction of light beams by elastic waves is useful as a tool predominantly because of an inherent quantum amplification process whereby the presence of one low-energy phonon is detected by the presence of a corresponding high-energy photon of diffracted light. This quantum gain is simply the ratio of the frequency of the diffracted photon to the frequency of the elastic-wave phonon. This quantum gain process permits one to use a light probe to monitor the presence of an elastic wave without, in the process, significantly altering the properties of the probed elastic wave. This is a standard requirement of a research tool and it has a great significance in terms of device applications. For example, one may note that, since 100% modulation of a light beam requires the removal of an insignificant number of phonons, the process may be repeated many times without depleting the power of the elastic wave. The fact that the spatial intensity distribution of the elastic wave is virtually independent of the spatial distribution of diffracted light results in a great mathematical simplification.

The diffraction of light by elastic waves is useful in device applications predominantly because of the large disparity between the velocity of light and the velocity of sound. The consequence of this disparity is that the instantaneous spatial structure of the elastic wave can be probed everywhere as if the elastic wave were stationary. This added feature introduces to the real-time capability of elastic-wave devices the concept of coherent simultaneous parallel processing. As will be shown below, the possibility of parallel processing leads to many important applications.

The use of a laser source of illumination is particularly convenient from several standpoints. The monochromaticity and collimation of a laser source yield the best possible optical illumination system, which may be limited only by diffraction effects. In addition, the coherence of lasers facilitates greatly the use of heterodyne techniques for improving the signal-to-noise characteristics and for producing other desired effects.

For the convenience of the reader, this chapter has been divided in the following manner. Section II develops the basic concepts necessary for understanding the description of the device applications. It is intended as an elementary description of the process of light diffraction by coherent elastic waves. Section III provides a more complete mathematical description of the diffraction process based on two distinct and complementary approaches, the differential equation method and the integral equation method. In Section IV, we discuss in detail some of the many applications that have already been suggested for the diffraction of light by elastic waves, including the measurement of material parameters, light modulation, and light beam deflection. Section V continues the presentation of applications with a discussion of optical information-processing techniques using diffraction in both the Bragg and Raman–Nath limits.

II. Basic Concepts

A. The Elastooptical Interaction

It is not difficult to understand in some cases why a beam of light should be diffracted by a column of acoustic waves. The simplified explanation proceeds as follows. The acoustic wave produces a spatial variation in the density of the crystalline medium. This spatial density variation produces in turn a concomitant perturbation in the index of refraction. The spatial variations in index of refraction produce at any given instant a phase grating which then diffracts the incident light beam into one or more directions.

The process of diffraction from an acoustic column is, of course, more complicated than this picture because both the index of refraction and the crystalline strains produced by the elastic wave are tensor quantities having off-diagonal elements. The explanation of how a transverse or shear acoustic wave can cause the diffraction of a light beam without a corresponding change in density is hidden in the complexity of the elastooptical interaction.

The elastooptical interaction is characterized by the formula (AIP Handbook, 1963)

$$\Delta\left(\frac{1}{n^2}\right)_{ij} = \sum_{k,l} p_{ijkl}\, e_{kl} \tag{1}$$

where the e_{kl} are cartesian strain components, the $(1/n^2)_{ij}$ are coefficients of the optical index ellipsoid, and the p_{ijkl} are the (strain) photoelastic or elastooptical constants. Crystal symmetry determines which of the tensor components p_{ijkl} are nonzero. In general, the nonzero components of the photoelastic tensor are the same as the nonzero components of the elastic stiffness constants, although the matrix of photoelastic coefficients is not symmetrical about the main diagonal.

The index ellipsoid in the absence of strains can be written in an arbitrary cartesian frame

$$\sum_{ij}\left(\frac{1}{n^2}\right)_{ij} x_i x_j = 1, \qquad i,j = 1,2,3 \tag{2}$$

The index ellipsoid is a convenient mathematical representation which describes the behavior of light when propagating through a birefringent medium and allows one to determine the two indices of refraction (the two eigenpolarizations) which correspond to an arbitrary direction of propagation. For a given propagation direction, the directions of polarization of the extraordinary and ordinary rays are coincident with the major and minor axes of the index ellipse which passes through the center of the index ellipsoid in a plane perpendicular to the propagation vector $\mathbf{k}$. The lengths of the semimajor and semiminor axes of the index ellipse are equal to the index of refraction for the corresponding polarization.

The presence of strains caused by an elastic wave modifies the index ellipsoid so that it becomes

$$\sum_{ij} \left[\left(\frac{1}{n^2}\right)_{ij} + \Delta\left(\frac{1}{n^2}\right)_{ij} \right] x_i x_j = 1 \tag{3}$$

Theories which treat the interaction of coherent elastic and light waves generally assume that the effect of the elastic wave propagating through a medium is the production of a propagating sinusoidal variation in the dielectric constant. This approach is found to be convenient because the dielectric constant transforms in a different, more convenient way than the index of refraction when the coordinate axes are rotated relative to the principal crystal axes (n^2, however, transforms like ε). One therefore assumes that the perturbed value of the dielectric constant is expressed by the formula

$$\varepsilon_{ij} = \varepsilon_{ij}^0 + \delta\varepsilon_{ij} \cos(\mathbf{K} \cdot \mathbf{r} - \Omega t) \tag{4}$$

The present problem consists of relating the change in a component of the dielectric-constant tensor to the corresponding change in the coefficients of the optical index ellipsoid. The relationship is established through the equation

$$(n^2)_{ij} = \varepsilon_{ij} \tag{5}$$

Thus, differentiating the inverse quantities yields

$$\delta(1/n^2)_{ij} = \delta(\varepsilon^{-1})_{ij} \tag{6}$$

The relationship between the tensor components of ε and ε^{-1} is, in turn, given by the following mathematical manipulations. By definition, the inverse dielectric tensor satisfies the equation

$$\sum_k \varepsilon_{ik}(\varepsilon^{-1})_{kj} = \delta_{ij} \tag{7}$$

Differentiating this equation and multiplying through on the right by ε_{jl} then gives

$$\delta\varepsilon_{il} = \sum_{k,j} -\varepsilon_{ik}^0 \, \delta(\varepsilon^{-1})_{kj}\varepsilon_{jl}^0 \tag{8}$$

Thus, using Eqs. (1) and (5), the relationship between the elastic strain produced by the sound wave and the change in the dielectric constant is given by the equation

$$\delta\varepsilon_{il} = \sum_{k,j,m,n} -\varepsilon_{ik}^0 \varepsilon_{jl}^0 \, p_{kjmn} \, e_{mn} \tag{9}$$

This relationship provides the link between the elastooptical interaction and theories of light diffraction by elastic waves.

B. The Bragg Equation

One may in principle conceive of the diffraction of a light beam by a sound column as a coherent sum of one or more three-particle scattering processes. It is evident that conservation of momentum should occur for each step in the sequence of three-particle interactions. On the other hand, while one may also expect the principle of energy conservation to apply to multiple scattering processes, the conservation of energy in view of the Heisenberg uncertainty principle need not apply between intermediate scattering processes if each process occurs in rapid succession. We will initially ignore this possibility and will consider one unit of the scattering process, i.e., the three-particle interaction, assuming that conservation of energy and conservation of momentum apply.

For simplicity of discussion, we will often apply the phrase "conservation of energy" when we mean "conservation of frequency." The proportionality between the energy and frequency of bosons is given by the formula

$$E = \hbar\omega \tag{10}$$

It is also convenient to paraphrase conservation of momentum in terms of conservation of "propagation constant." The proportionality between these two quantities is given by the de Broglie relationship

$$p = h/\lambda = \hbar k \tag{11}$$

It is, however, common usage to deal with the propagation constant rather than with the wavelength. One may alternatively view conservation of propagation constant as the production of an induced wave, i.e., spatial distribution of phase, by the interaction of an incident electromagnetic wave with an elastic wave. This induced wave can propagate through the crystal only if its spatial distribution of phase has a wavelength equal to the wavelength of a freely propagating wave in the given direction.

To Eqs. (10) and (11), we also add the formula relating wavelength and frequency

$$v = \lambda\nu = \omega/k = E/p \tag{12}$$

Since Eqs. (10)–(12) apply equally well to electromagnetic and acoustic waves, we will henceforth distinguish between the two types of waves by applying uppercase letters to phonon parameters and lowercase letters to photon parameters. Thus, for example, K, Ω, V, etc., and k, ω, v, etc., are quantities referring to phonon and photon characteristics, respectively.

The three-particle scattering process is diagramed in Fig. 1. To distinguish between incident and scattered photons, we add a prime to the parameters of the latter. Inspection of Fig. 1 shows that the diagram can represent two distinguishable processes corresponding to the absorption or the emission of a phonon. (The direction of $\mathbf{K}$ is reversed for the latter process.) These two distinguishable processes are described by the following

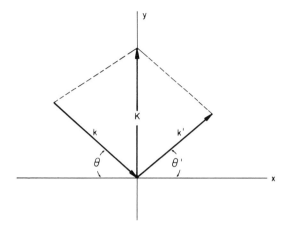

Fig. 1. Momentum-conservation diagram for photon–phonon scattering. One phonon is annihilated.

conservation-of-energy equations:

$$\omega + \Omega = \omega' \tag{13a}$$

$$\omega = \Omega + \omega' \tag{13b}$$

What then is the probability of the phonon absorption process relative to phonon creation? Quantum mechanics shows that the probabilities for the absorption and emission of a phonon are proportional to $N^{1/2}$ and $(N+1)^{1/2}$, respectively, where N is the number of phonons per mode (box normalization). Thus, the ratio of the intensities of the scattered photons produced by absorption and emission of a phonon is given by $N/(N+1)$. Furthermore, the number of thermal phonons per mode at a temperature T is given by statistical mechanics by the formula

$$N = \exp[\hbar\Omega/kT - 1]^{-1} \tag{14}$$

If one substitutes into this formula the values appropriate for a microwave phonon propagating through a crystal at room temperature, the result is

$$N = kT/\hbar\Omega \gg 1 \tag{15}$$

This result is then also true for an injected elastic wave because the effective temperature of the excited modes is much higher than the temperature of the lattice itself. Thus, under normal circumstances, the probabilities for emission and absorption of a phonon are essentially equal and the intensity of the sum-frequency diffraction process, Eq. (13a), is equal to the intensity of the difference-frequency process, Eq. (13b). We have been careful to delineate between these two processes because they may both occur simultaneously if oppositely directed acoustic waves are simultaneously present in the crystal.

Since the frequency of the injected elastic wave is no higher than the microwave frequency range, the frequency of the phonons is very small compared with the frequency of light. The conservation-of-energy equations, Eqs. (13), then simplify to the approximate relationship

$$\omega \cong \omega' \qquad (16)$$

If we use this approximation and that of optical isotropy, the immediate result is

$$k = k' \qquad (17)$$

We point out, however, that it is only the latter equality which is used here, and indeed, will show below that the difference between ω and ω' is essential to the operation of coherent optical modulators and processors.

Using the notation of Fig. 1, one obtains the following equations for conservation of momentum components:

$$k \cos \theta = k' \cos \theta' \qquad (18a)$$

$$-k \sin \theta + K = k' \cos \theta' \qquad (18b)$$

corresponding to the absorption of a phonon. For the case of phonon generation, Eq. (18b) is replaced by the algebraically identical equation

$$-k \sin \theta = -K + k' \sin \theta'$$

Therefore, the set of simultaneous equations (17) and (18) is equally applicable to both the absorption and the emission of a phonon. The general solution of these equations for arbitrary propagation directions in a birefringent medium is by no means an easy task, since, in such a case, k, k', and K are all angular-dependent. We will consequently restrict our discussion to the readily soluble case of isotropic materials. For isotropic materials, $v = v'$ and $k = k'$; hence, Eq. (18a) gives $\theta = \theta'$. Combining these results with Eq. (18b) yields the simple result

$$\sin \theta = \tfrac{1}{2}K/k = \tfrac{1}{2}\lambda/\Lambda \qquad (19)$$

This relationship between the angle θ and the ratio of the wavelengths has been named the "Bragg equation" because of the close correspondence between this equation and a similar equation describing the scattering of x-rays by a crystal lattice. In the present case, the crystal lattice spacing is replaced by an equivalent grating whose spacing is equal to the acoustic wavelength. It should be noted that the Bragg condition limits the wavelength of sound which may be scattered to a value greater than $\Lambda = \tfrac{1}{2}\lambda$, a case which corresponds to backscattering of light ($\mathbf{k} = -\mathbf{k'}$).

It may be further noted that the diffraction effect in isotropic materials is independent of the refractive index of the material, provided that the sides of the sample are parallel to the direction of propagation of the elastic wave. The Bragg condition, Eq. (19), must be satisfied within the elastic material, and the appropriate optical wave vector is that within the material, $k_i = nk_e$.

Snell's law gives $\sin \theta_e = n \sin \theta_i = nK/2k_i = K/2k_e$, so that the Bragg condition for the beam outside the sample is determined by the optical wavelength in air.

C. The Diffraction of Light and Sound

We have thus far treated the diffraction of a light beam by an acoustic beam as a sum of separate three-particle interactions. We would now like to consider effects which arise when finite-aperture, coherent beams of light and sound interact. It is well known that a plane wave is inherently of infinite aperture and that, if the aperture of the wave is restricted to a finite dimension, the wave field beyond the aperture must be represented as a sum of plane-wave components traveling in slightly different directions. The amplitude of each of these plane waves is then determined by the boundary conditions in the plane of the aperture. If the plane-wave amplitudes are suitably arranged, the total wave amplitude may be made zero in this plane except across the aperture. Since an apertured beam consists of an angular distribution of plane waves and therefore a distribution of propagation constants $\mathbf{k}$, it is clear that the beam will spread as an expanding cone at distances sufficiently far from the aperture. On the other hand, if the aperture is large compared with the wavelength of the wave, there will be a region near the aperture where no diffraction spreading will be apparent and the width of the beam is essentially constant. The effects of diffraction on the propagation of an apertured beam are illustrated in Fig. 2.

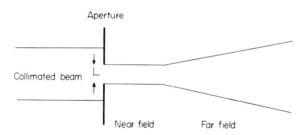

FIG. 2. Diffraction effects produced by aperturing a plane wave.

In discussing the effects of diffraction, it is useful to borrow terminology common to laser physics. That region or limit close to the aperture, where geometrical optics holds sway, where the wavefront is essentially flat (assuming of course that a collimated beam illuminates the aperture), and where the beam does not spread, is termed the near-field of the aperture. The region or limit far from the slit, where the beam expands via diffraction with a spherical wavefront, is correspondingly referred to as the far-field of the aperture.

The angle of spreading for a uniformly illuminated rectangular aperture of width w can be characterized by the angle φ from the normal at which the first dark line of the far-field diffraction pattern occurs. This angle is given by simple diffraction theory as

$$\varphi = \lambda/w, \qquad \Phi = \Lambda/L \tag{20}$$

For a circular aperture, the corresponding half-angle, delineated by the first Airy dark zone, is given by $\varphi = 1.22\lambda/w$.

Consider now the implications of using finite-width beams of sound and light. The principal result of using such finite-width beams is that the three-particle Bragg condition is relaxed, i.e., the Bragg condition becomes an angular distribution of three-particle scattering processes which permit scattering to occur over an enlarged angle, roughly equal to $\varphi' = \varphi + \Phi$, of the diffracted beam of light. The more thorough mathematical treatments of the next section will include directly the effects of beam diffraction on the scattering of light by sound.

An angular distribution in the propagation vectors of the acoustic beam also makes possible multiple or successive three-particle scattering processes. Figure 3 shows the conservation-of-momentum diagram for a two-stage scattering event. Conservation of energy enters into Fig. 3 by virtue of the

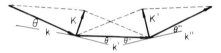

FIG. 3. Momentum-conservation diagram for successive scattering of a photon by two phonons.

fact that $k = k' = k''$ and $\theta = \theta' = \theta'' = \theta'''$. It is clear from this diagram that conservation of energy and momentum can occur in a two-step process only if there are two phonons traveling in the correct different directions. It is also clear that the angular distribution of phonons required in the two-step process is not as great if the Bragg angle θ is reduced. It is consequently apparent that two or more successive scattering processes will be possible at low microwave frequencies (θ small) and that energy conservation and momentum conservation will not both occur if multiple scattering takes place at sufficiently high microwave frequencies.

It is customary to distinguish between the regimes for which multiple scattering is very probable and the regime for which multiple scattering is relatively improbable. Multiple scattering will readily occur if the angular distribution of propagation vectors of the elastic wave is large compared to the Bragg angle, and will not readily occur in the opposite limit. If the sound beam has aperture L, the limit between the two regimes can be expressed by the equation

$$\Phi = \theta \tag{21}$$

or, using Eqs. (19) and (20), $\Lambda^2 = L\lambda/2$. Historically (Klein and Cook, 1967), the boundary between the two regimes has been defined in terms of the quantity

$$Q = 2\,\pi\lambda L/\Lambda^2 = K^2 L/nk \tag{22}$$

by the equation $Q = 1$. We shall henceforth make use of this latter definition. One defines the limit $Q > 1$ as the Bragg region of optical diffraction. In this limit, the paramount consideration for efficient diffraction is the satisfaction of the Bragg condition. In this limit, only one order of diffracted light is produced. Conversely, one defines the limit $Q < 1$ as the Raman–Nath region of optical diffraction in honor of the first authors to successfully explain the variety of phenomena which occur in this limit (Raman and Nath, 1935a, 1935b, 1936). In the Raman–Nath limit, the effects of multiple scattering are readily observed.

We have thus far been careful to avoid the conclusion that multiple scattering is impossible in the Bragg limit, because in this limit, multiple scattering is not impossible, if, in fact, relatively improbable. In this limit, a two-stage scattering becomes a forbidden process because energy conservation cannot occur during the intermediate step (conservation of momentum always must occur). However, the Heisenberg uncertainty principle implies that conservation of energy need not occur in a transitory intermediate step of a multifold interaction. It is, moreover, apparent, because of the very fast propagation velocity of light, that, if successive scatterings occur at all, they must occur in rapid succession as the light beam passes through a normal-width acoustic beam. Consequently, a twofold or multifold scattering process becomes in the Bragg limit merely a relatively forbidden, higher-order process. The forbiddenness of higher-order scattering increases as the elastic frequency is raised above the low-frequency limit of the Bragg regime, and the forbiddenness increases as the order of the process increases.

D. A Simple Calculation of the Intensity of the Diffracted Light Beam

Before passing on to a more thorough theoretical treatment of the diffraction of light by elastic waves, we would like to present a simplified picture of the phenomena which permits one to make quantitative calculations of the intensity of the diffracted light. Such a simplified picture will add a great deal of insight and make the understanding of the following treatments easier. This method of calculation is valid for low acoustic intensities in the Raman–Nath region, but the result applies equally well to the Bragg region.

Assume as in Fig. 4 that the sound beam travels vertically and that the light beam is incident upon the crystal in a horizontal direction. The essence of the approach is to hypothesize that the total effect of the elastic wave is to produce a sinusoidal phase corrugation in the wavefront of the light beam in the air adjacent to the (light) output side of the crystal (assumption of low acoustic intensity). For simplicity, we also assume that the medium is optically and elastically isotropic and that the elastic and optical beams

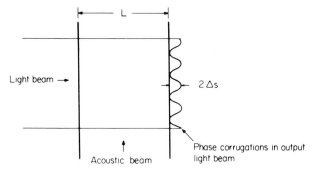

FIG. 4. Geometry for diffraction of light beam by a low-frequency elastic wave.

propagate in such a direction and are so polarized that only one component of the appropriate tensor quantities will come into play. Reflection losses at surfaces and acoustic attenuation will similarly be ignored.

The replacement of a volume distribution of oscillations in the index of refraction by an "equivalent" planar distribution cannot be applied to the Bragg region for two reasons. First, a planar grating can in fact diffract light whatever the angle between the incident light beam and the planar grating, whereas diffraction of light by a volume elastic beam in the Bragg region requires a definite relationship of directions as given by the Bragg condition. Second, the equivalent phase grating in the Bragg case cannot be sinusoidal because in the Bragg limit only one order of diffraction, for example, the +1 order, is produced, whereas a sinusoidal grating symmetrically produces both the +1 and −1 orders with equal probability (as well as other orders).

If we assume that the amplitude of phase corrugation is α, then the amplitude of distance corrugation Δs is given by $\alpha = 2\pi\, \Delta s/\lambda$. Noting further that the optical thickness of the crystal is $s = nL$, one obtains the fractional change in index of refraction

$$\Delta n/n = \Delta s/s = \alpha\lambda/2\pi n L \qquad (23)$$

To proceed further, we must express α in terms of the intensity of the light beam diffracted into the +1 order. This is accomplished indirectly by calculating the intensity of the undiffracted light beam and assuming that the missing power appears equally in the +1 and −1 orders of diffraction. If E is the electric field of the incident beam and E_0 is the electric field of the undiffracted beam (zeroth order) after passing through the crystal, the electric field of the undiffracted beam is given by the following phase integral:

$$E_0 = \frac{\Lambda}{E}\int_0^\Lambda \exp\left(i\alpha\sin\frac{2\pi y}{\Lambda}\right) dy = \frac{E}{2\pi}\int_0^{2\pi}\exp(i\alpha\sin\nu)\, d\nu \qquad (24)$$

If $\alpha \ll 1$, the result is

$$E_0 = E[1 - (\alpha^2/4)] \qquad (25)$$

Thus, the ratio of intensities of the undiffracted beam to the incident beam is given by

$$E_0 E_0^* / EE^* = 1 - (\alpha^2/2) \tag{26}$$

Of the power removed from the diffracted beam, one half appears in the $+1$ order of diffraction. [This can be easily proved by calculating the appropriate integral in place of Eq. (24). The appropriate integrand is multiplied by a phase factor representing the propagation of power in a direction given by the Bragg condition.] Consequently,

$$I_{+1}/I_0 = \alpha^2/4 \tag{27}$$

Using Eq. (23), the result is

$$(\Delta n/n)^2 = (I_{+1}/I_0)(\lambda/\pi n L)^2 \tag{28}$$

We now relate the amplitude of the index-of-refraction change to the intensity of the elastic wave. Let I be the acoustic intensity, U the energy density of the elastic wave, V the acoustic velocity, C the elastic stiffness constant, e the strain amplitude caused by the elastic wave, and ρ the density of the medium. These quantities are related in the following way:

$$I = UV = \tfrac{1}{2}CVe^2 = \tfrac{1}{2}\rho V^3 e^2 \tag{29}$$

Eliminating e between this equation and Eq. (1), one obtains

$$\frac{\Delta n}{n} = -\frac{n^2}{2}\,pe = -\frac{n^2}{2}\,p\left(\frac{2I}{CV}\right)^{1/2} \tag{30}$$

Eliminating $(\Delta n/n)$ from Eqs. (28) and (30) gives the final result

$$\frac{I_{+1}}{I_0} = \frac{\pi^2}{2}\left(\frac{I}{\rho V^3}\right)\left(\frac{n^3 pL}{\lambda}\right)^2 \tag{31}$$

If one now substitutes the following typical values for the quantities of this equation, $I = 20 \text{ mW}/10^{-2} \text{ cm}^2 = 2 \times 10^{10} \text{ erg sec}^{-1}\text{ cm}^{-2}$, $L = 0.2 \text{ cm}$, $V = 7 \times 10^5 \text{ cm}$, $C = 10^{12} \text{ dynes cm}^{-2}$, $p = 0.05$, $\lambda = 6 \times 10^5 \text{ cm}$, $n = 1.5$, the fraction of light diffracted into first order is $I_{+1} = I_0 = 5 \times 10^{-3}$.

One therefore concludes that an easily detectable amount of light power is diffracted into one order by an elastic wave.

E. ANISOTROPIC MEDIA

The Bragg condition is sufficiently modified in the case of anisotropic materials that several interesting phenomena occur which have no correspondence to isotropic materials. The following short discussion of anisotropic materials rests heavily on the work of Dixon (1967a) (see also Lean *et al.*, 1967). The starting point for a derivation of the modified Bragg equation is the conservation-of-energy equation, Eq. (16), and the conservation-of-momentum equation, Eqs. (18). Here, we assume explicitly that the indices of refraction are unequal for the incident and diffracted laser beams, a phenomenon which occurs in all optically birefringent materials. With this in mind, the conservation of energy can be rephrased in terms of the propagation constant of free space k_0 as the following pair of formulas:

$$k = nk_0, \qquad k' = n'k_0 \tag{32}$$

which are equivalent to the equation $k'/k = n'/n$.

To solve for the angles of the incident and diffracted light beams relative to the direction of the elastic wave, we square Eq. (18a), use the trigonometric identity $\cos^2 \theta = 1 - \sin^2 \theta$, substitute from Eq. (18b), and make use of Eqs. (32). The result is

$$\sin \theta = \frac{K}{2nk_0}\left[1 + \left(\frac{k_0}{K}\right)^2 (n^2 - n'^2)\right] \tag{33}$$

$$\sin \theta' = \frac{K}{2n'k_0}\left[1 - \left(\frac{k_0}{K}\right)^2 (n^2 - n'^2)\right] \tag{34}$$

The interesting features of these equations can be most readily observed by plotting θ and θ' as a function of K/k_0. This graph is plotted in Fig. 5

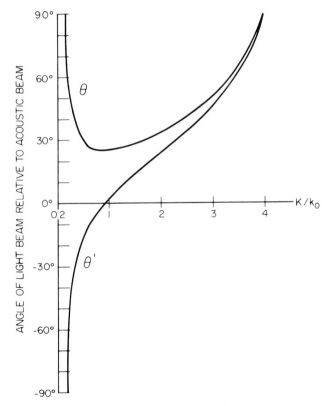

FIG. 5. Curves showing the angle of incident light θ, and the angle of the diffracted light θ', relative to the equiphase fronts of the elastic wave in an anisotropic material. The graph assumes angle-independent indices of refraction of $n = 2.1$ and $n' = 1.9$.

assuming angle-independent indices of refraction of $n = 2.1$ and $n' = 1.9$. One can see from the graph that the curve of θ versus K/k_0 approaches the line $K/k_0 = (n - n') = 0.2$ uniformly as the value of K decreases. Thus, for a given material having unequal indices of refraction, there is a minimum elastic-wave frequency which can be used to diffract light. This minimum-frequency case, shown in Fig. 6, occurs for collinear waves. Using the schematic diagram shown in Fig. 6, one can readily show that $K_{\min} = k - k' = (n - n')k_0$. For this limit, the angle θ and θ' are equal to $\pm 90°$, respectively, since the angles are defined to be zero when the incident and diffracted light beams propagate perpendicular to the elastic wave.

It is important also to note that the curve of θ versus K/k_0 is double-valued, i.e., there are two values of K (elastic-wave frequency) which satisfy the modified Bragg condition for a given angle of incidence θ. This situation is pictured diagrammatically in Fig. 7. The existence of two possible diffraction geometries dramatizes the difference between the anisotropic and isotropic cases.

One may also note that the minimum value of θ represents a stationary condition for which the modified Bragg equations may be satisfied over a range of elastic-wave frequencies even though the angle between the incident light beam and the acoustic beam is held nearly constant. This situation is displayed schematically in Fig. 8. The angle at which the diffracted light

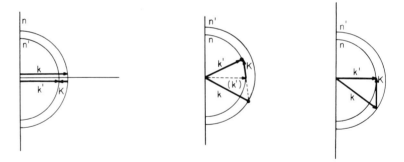

Fig. 6. Collinear forward photon–phonon scattering in an anisotropic material. When, as shown, the incident and diffracted photons have orthogonal polarizations, there exists a minimum value for the propagation constant of the elastic wave.

Fig. 7. Noncollinear photon–phonon scattering in an anisotropic material. For an incident light beam and an elastic wave of specified directions, there exist two distinct elastic-wave frequencies which satisfy the Bragg condition.

Fig. 8. A stationary condition of the modified Bragg equation in an anisotropic crystal. The Bragg condition can be satisfied for a given direction of incident light and elastic waves over a range of acoustic frequencies.

propagates is, however, a rapidly varying function of K/k_0, as Fig. 5 clearly shows. This physical situation, as will be described below, represents a possible means for light-beam deflection.

III. Theory

In this section, we are primarily concerned with supplying a more detailed basis for the theory of light diffraction by coherent elastic waves. There are, historically speaking, two fundamentally distinct approaches which have been adopted for the solution of this problem: (1) the differential equation approach and (2) the integral equation method. The differential equation method is treated in Born and Wolf (1965), Cohen and Gordon (1965), and Quate *et al.* (1965). This treatment stems originally from analyses of Brillouin (1922), David (1937), and Raman and Nath (1935a,b, 1936). The integral equation method was initially developed by Bhatia and Noble (1953). As we shall see below, the two methods, in terms of applicability, are complementary in nature, each being particularly applicable to the solution of a different type of problem. Where both methods overlap, they give equivalent results.

The following analyses will be limited to a description of the basic nature of the interaction between coherent light and elastic waves and will avoid unnecessary complications. Thus, for example, the reflection of light at material boundaries will be ignored and crystal isotropy will be assumed. The reader is referred to Hope (1968) and Lean (1967) for a treatment which includes these latter effects using the integral equation approach.

A. The Differential Equation Method

The differential equation method is based on Maxwell's equations for a nonmagnetic, nonconducting medium whose dielectric constant ε is assumed to be a function of position and time. The use of Maxwell's equations gives the differential equation for the electric field of light as

$$\nabla^2 \mathbf{E} - \frac{1}{c^2} \frac{\partial^2}{dt^2} (\boldsymbol{\varepsilon} \cdot \mathbf{E}) = 0 \tag{35}$$

To proceed further, we use the particular interaction geometry illustrated in Fig. 9. For this geometry, the elastic wave propagates in the x direction (vertically), the elastic wave has an infinite depth in the y direction, and the width of the elastic wave extends from $-L/2$ to $L/2$ in the z direction. The geometry assumes further that a monochromatic (infinite-aperture) plane wave of light is incident from the left downward onto the sound column at an angle θ from the z axis. This geometry eliminates the y-axis dependence of the problem. The time and spatial dependence of the dielectric constant due to the presence of an elastic wave can then be written in the form

$$\varepsilon = \varepsilon_0 + \delta\varepsilon \sin(Kx - \Omega t), \qquad -L/2 < z < L/2 \tag{36}$$

Although ε is a tensor quantity, we will assume, for simplicity, that it can adequately be represented by a scalar. The conversion of the following formulas back into vector notation is not difficult. With the present geometry, the electric field of the incident light beam can then be written in the

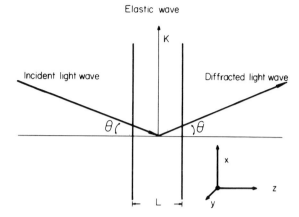

FIG. 9. Interaction geometry for optical diffraction by an elastic wave.

form

$$E = U_0 \exp i(kx \sin \theta - kz \cos \theta - \omega t) \qquad (37)$$

One then assumes that the solution for the diffracted light is given by the sum

$$E = \sum_l U_l(z) \exp -i[(\omega + l\Omega)t - (k \sin \theta - lK)x + kz \cos \theta] \qquad (38)$$

This sum represents a series of plane waves whose amplitudes $U_l(z)$ vary within the crystal along the z coordinate. Each plane wave, except U_0, originates from the absorption or emission of one or more phonons by the incident light beam in the interaction volume. Clearly, the quantum nature of the diffraction process with conservation of momentum and energy allows only a discrete number of possible directions and frequencies for the diffracted light-wave components. The trial solution has therefore been written with due regard for the conservation rules.

This solution is substituted into Eq. (35). If the amplitude of each of the diffracted plane waves increases slowly with distance z (so that the amplitude change is small in one wavelength of light), the resulting terms in $d^2 U_l/dz^2$ can be neglected. Furthermore, one can neglect terms which are relatively small by the factor $\Omega/\omega \ll 1$ and the factor $V/c \ll 1$. Using the substitutions $k = \omega/c$ and $K = \Omega/V$, the resulting equations for the amplitude factors $U_l(z)$ are

$$U_l'(z) + i\beta U_l(z) + \xi[U_{l+1}(z) - U_{l-1}(z)] = 0 \qquad (39)$$

where we have set

$$\beta = -\frac{Kl}{\cos \theta}\left(\sin \theta + \frac{lK}{2k}\right), \qquad \xi = \frac{1}{4}\left(\frac{\delta\varepsilon}{\varepsilon_0}\right)\frac{k}{\cos \theta} \qquad (40)$$

The general solution for the set of coupled equations (39) is indeed very difficult. We will consider the problem in two particular limits. First, restricting the discussion to positive values of l, assume that the elastic-wave amplitude is sufficiently small that $U_{l+1} \ll U_l$ and that initially only $U_0 \neq 0$. The equation for U_l can then be written as

$$\partial U_l/\partial z + i\beta U_l = -\xi U_{l-1} \tag{41}$$

and, as one can readily see by inspection, the solution of this differential equation can be written in the form

$$U_l = -(\exp -i\beta_l z)\int_{-\infty}^{z} \xi U_{l-1} \exp(i\beta_l z')\, dz' \tag{42}$$

Consider now the case $l = 1$, corresponding to the first order of diffraction, which is produced physically by the absorption or emission of a single phonon by each light-beam photon. If the elastic-wave amplitude is uniform and nonzero only in the range $-L/2 < z < L/2$, then ξ is constant and nonzero only within the same limits. Furthermore, since $U_1 \ll U_0$, it is assumed that the diffraction process removes a negligible fraction of the incident light-beam power. Thus, U_0 is essentially constant in value and the amplitude of the first-order diffracted light is

$$U_1 = -(\exp -i\beta_1 z)\xi U_0 \int_{-L/2}^{L/2} \exp(+i\beta_1 z')\, dz'$$

$$= \xi U_0 L (\exp -i\beta_1 z) \frac{\sin(\beta_1 L/2)}{\beta_1 L/2} \tag{43}$$

where

$$\beta_1 = -\frac{K}{\cos\theta}\left(\sin\theta + \frac{K}{2k}\right) \tag{44}$$

Since the factor $\sin(\beta_1 L/2)/(\beta_1 L/2)$ is large only when $\beta_1 \cong 0$, the amount of diffracted light will be insignificant unless the momentum-conserving condition $\sin\theta = -\frac{1}{2}K/k$ is satisfied. This condition

$$\sin\Theta = \frac{1}{2}K/k = -\sin\theta \tag{45}$$

is used to define the Bragg diffraction angle Θ.

Equation (43) gives the fractional amount of light intensity which is diffracted out of the incident beam direction by the elastic wave as

$$\frac{I_1}{I_0} = \frac{U_1 U_1^*}{U_0 U_0^*} = \xi^2 L^2 \frac{\sin^2(\beta_1 L/2)}{(\beta_1 L/2)^2} \tag{46}$$

The maximum amount of power diffracted into one order occurs when $\beta_1 = 0$. For this condition,

$$\frac{I_1(\max)}{I_0} = (\xi L)^2 = \left[\frac{1}{4}\left(\frac{\delta\varepsilon}{\varepsilon_0}\right)\frac{kL}{\cos\theta}\right]^2 \tag{47}$$

Second-order diffraction occurs when the light beam is successively scattered by two elastic phonons. In the limit of small acoustic intensities, the amplitude of the second-order diffracted light can be found by eliminating U_1 from the pair of differential equations (41) in dU_1/dz and dU_2/dz. This gives

$$\frac{d^2 U_2}{dz^2} + 2i\beta_1 \frac{dU_2}{dz} - \beta_1\beta_2 U_2 = \xi^2 U_0 \qquad (48)$$

We comment here only on the solution obtained by neglecting the first term of Eq. (48) in comparison with the second. This situation will occur at small acoustic intensities if β_1 is sufficiently large. The factor β_1 will, however, assume progressively larger values as the difference between the Bragg angle Θ and the angle θ of the incident light beam increases, and this more readily occurs for large Θ, i.e., when the elastic-wave frequency increases. Dropping the term $d^2 U_2/dz^2$ and assuming that the amplitude of the elastic wave is uniform and nonzero between $-L/2 < z < L/2$ gives the result

$$U_2 = \frac{\xi^2 L U_0}{2i\beta_1}\left(\exp -\frac{i\beta_2 z}{2}\right)\frac{\sin(\beta_2 L/4)}{(\beta_2 L/4)} \qquad (49)$$

With the assumption $\beta_1 \neq 0$, relatively large amounts of second-order diffraction will occur only if the overall momentum-conserving condition $\beta_2 = (2K/\cos\theta)(2\sin\Theta - \sin\theta) \simeq 0$ is satisfied. Thus, the angle for second-order diffraction is approximately twice the angle for first-order diffraction. Furthermore, it is apparent that, if $\beta_1 \neq 0$, the value of β_1 must increase with phonon frequency, and consequently, the "nonresonating" denominator factor β_1 decreases the likelihood of second-order scattering at higher elastic frequencies. Thus, if the geometry is arranged to satisfy conservation of energy and momentum for the total second-order process (and this is the only way to obtain second-order diffraction at higher acoustic frequencies), one can easily verify that energy cannot be conserved in the intermediate state. As a consequence of this, second-order diffraction becomes a forbidden process in the Bragg frequency limit.

We now return to Eq. (39) and consider the limit of low elastic frequency. In this limit, one need not assume that the amplitude factors U_l decrease rapidly with l. If Eq. (39) is rewritten in terms of the new variable, $\chi = 2\xi z$, one obtains the result

$$2U_l{}'(\chi) + U_{l+1}(\chi) - U_{l-1}(\chi) = (-i\beta_l/\xi)U_l(\chi) \qquad (50)$$

Following Raman and Nath (1935a), the term on the right-hand side of this equation is set equal to zero. The resulting set of equations are the recursion relations satisfied by Bessel functions of integral order. If, as above, one assumes that only $U_0 \neq 0$ at $z = -L/2$ and that ξ is uniform and nonzero for $-L/2 \leq z \leq L/2$, the intensity of the lth-order diffracted wave is given by

$$\frac{I_l}{I_0} = J_l{}^2\left(\frac{1}{2}\frac{\delta\varepsilon}{\varepsilon_0}\frac{kL}{\cos\theta}\right) \qquad (51)$$

It can be seen that the approximation made by Raman and Nath, i.e.,

$$\frac{\beta_l}{\xi} = 4K_l \left(\frac{\delta\varepsilon}{\varepsilon_0}\right)^{-1} \left(\sin\theta + \frac{lK}{2k}\right) \ll 1$$

must occur in the limit as $K \to 0$ for a given value of $\delta\varepsilon/\varepsilon_0$. However, as the elastic-wave amplitude decreases, the factor $(\delta\varepsilon/\varepsilon_0)^{-1}$ increases in magnitude. Consequently, the Bessel-function solution to Eq. (39) can be expected to be most applicable to light diffraction from large-amplitude, low-frequency elastic waves. Extermann and Wannier (1936) have shown that the Bessel-function solution in practical cases overestimates the intensities of the higher-order diffracted light waves.

It should be noted that Eqs. (46) and (51) give identical results for I_1/I_0 if $\beta_1 = 0$ in the limit of low-elastic-wave intensity.

To conclude the discussion of the differential equation method, we emphasize that the finite aperture of the sound column is taken into account in only one dimension, the z direction of the interaction medium, but that this method of calculation requires an infinite light-beam aperture. In practice, the necessity for infinite apertures is removed by assuming the electric field of the diffracted light is observed at a point sufficiently close to the acoustic column that the diffraction effects caused by a finite aperture are negligible. This restriction asserts that the method of calculation can be used for the case of finite apertures (i.e., practical cases) if the observation point lies at certain positions of the near-field of the diffracted light wave.

B. The Integral Equation Method

The integral equation method assumes that the simultaneous presence of both the incident sound and light waves produces an induced electric-field polarization in the overlap region of the waves. This "nonlinear" induced polarization oscillates at the sum and difference frequencies of the incident waves and represents the source of the radiated diffracted signal.

The approach is based on the formula for the electromagnetic radiation field produced by a distribution of oscillating electric dipoles. If $\mathscr{P}(\mathbf{r}, t)$ represents the phase and amplitude of the induced dipole moment density as a function of position and time, the electric field of the diffracted light wave is given by the formula

$$\mathbf{E}(\mathbf{r}', t) = \int_V \frac{[[\mathscr{P}(\mathbf{r}, t) \times \mathbf{k}'] \times \mathbf{k}']}{n^2} \frac{\exp -ik'|\mathbf{r}' - \mathbf{r}|}{|\mathbf{r}' - \mathbf{r}|} dV \qquad (52)$$

where, as above, k' is the propagation constant of the diffracted wave. Here, for simplicity of presentation, the formula has been written with all variables taking values which exist within the medium.

The induced dipole moment density at the sum or difference frequency of the two incident waves can be calculated using the well-known definition of the electric susceptibility

$$\mathscr{P}_i = \chi_{ij} E_j = \frac{\mathscr{E}_{ij} - \delta_{ij}}{4\pi} E_j \tag{53}$$

The tensor notation used here emphasizes the relationship between the vector components of induced polarization and the polarization of the incident light and elastic beams. To determine the amplitude and phase of the induced dipole moment density corresponding to the diffracted light radiation source, one adds an explicit time and spatial dependence to the dielectric constant and the incident light wave. Thus, by assuming plane-wave motions and by expressing these quantities as

$$\mathscr{E}_{ij} = \varepsilon_{ij} + \delta\varepsilon_{ij} \cos(\Omega t - \mathbf{K} \cdot \mathbf{r}), \qquad E_j = E_j{}^0 \cos(\omega t - \mathbf{k} \cdot \mathbf{r}) \tag{54}$$

one obtains

$$\mathscr{P}_i = \sum_j \frac{\varepsilon_{ij} - \delta_{ij}}{4\pi} E_j{}^0 \cos(\omega t - \mathbf{k} \cdot \mathbf{r})$$

$$+ \sum_j \frac{\delta\varepsilon_{ij} E_j{}^0}{4\pi} \cos(\omega t - \mathbf{k} \cdot \mathbf{r}) \cos(\Omega t - \mathbf{K} \cdot \mathbf{r}) \tag{55}$$

Consequently, if one isolates the amplitude of the induced dipole moment at the sum and difference frequencies, the result is

$$\mathscr{P}_i(\omega \pm \Omega) = \sum_j \frac{\delta\varepsilon_{ij} E_j{}^0}{8\pi} \cos[(\omega \pm \Omega)t - (\mathbf{k} \pm \mathbf{K}) \cdot \mathbf{r}] \tag{56}$$

Using the amplitude and phase dependence of the induced polarization given by Eq. (56), suppressing the explicit time dependence, and substituting into Eq. (52) gives the result

$$\mathbf{E}(\mathbf{r}') = \int \frac{[(\mathscr{P}_0 \times \mathbf{k}') \times \mathbf{k}']}{n^2} \exp{-i(\mathbf{k} + \mathbf{K}) \cdot \mathbf{r}} \frac{\exp{-ik'|\mathbf{r}' - \mathbf{r}|}}{|\mathbf{r}' - \mathbf{r}|} dr^3 \tag{57}$$

Since the volume of integration extends over many wavelengths of light in all dimensions, and $\mathscr{P}_0$ is assumed to be a slowly varying function of position, it is readily seen that the exponential phase factor of Eq. (57) will lead to nearly complete cancellation over the range of integration except under very special circumstances. The only condition which will not give cancellation occurs when the induced polarization has a spatial and temporal distribution which matches the polarization of a freely propagating light wave. This condition is equivalent to the replacement

$$\exp{-i(\mathbf{k} + \mathbf{K}) \cdot \mathbf{r}} = \exp{-i\mathbf{k}' \cdot \mathbf{r}} \tag{58}$$

and represents nothing more or less than conservation of (pseudo) momentum.

However, to make the treatment more general, we will make a somewhat less restrictive assumption,

$$\exp{-i(\mathbf{k} + \mathbf{K} - \mathbf{k}') \cdot \mathbf{r}} \equiv \exp{-i\,\Delta\mathbf{k} \cdot \mathbf{r}} \tag{59}$$

With this replacement, Eq. (57) gives the result

$$\mathbf{E}(\mathbf{r}') = \frac{1}{n^2}\int [\mathbf{k} \times [\mathbf{k} \times \boldsymbol{\mathscr{P}}_0(\mathbf{r})]]\exp(-i\mathbf{k}'\cdot\mathbf{r})\exp(-i\,\Delta\mathbf{k}\cdot\mathbf{r})\frac{\exp-i\mathbf{k}'|\mathbf{r}'-\mathbf{r}|}{|\mathbf{r}'-\mathbf{r}|}dV$$

(60)

We now choose a specific coordinate system in which to evaluate the integral, namely, a coordinate system in which the z axis lies along the direction of the propagation vector of the diffracted light. With this choice of coordinate system, one can make the replacements

$$\mathbf{k}'\cdot\mathbf{r} = k'z$$

$$|\mathbf{r}'-\mathbf{r}| = (z'-z) + \frac{(x'-x)^2}{2(z'-z)} + \frac{(y'-y)^2}{2(z'-z)} + \cdots$$

$$\Delta\mathbf{k}\cdot\mathbf{r} = \Delta k_z z + \Delta k_x x$$

with [using Eq. 59)]

$$\Delta k_z = k(\cos 2\theta - 1) + K\sin\theta = 2k\sin\theta\left(\frac{K}{2k} - \sin\theta\right)$$

$$\Delta k_x = K\cos\theta - k\sin 2\theta = 2k\cos\theta\left(\frac{K}{2k} - \sin\theta\right)$$

(61)

We assume also for simplicity that $\boldsymbol{\mathscr{P}}_0 \perp \mathbf{k}$, so that $[\mathbf{k}'\times[\mathbf{k}'\times\boldsymbol{\mathscr{P}}_0(\mathbf{r})]] = -k^2\boldsymbol{\mathscr{P}}_0(\mathbf{r})$. Setting $\alpha = k'/2(z'-z)$ then yields

$$\mathbf{E}(\mathbf{r}') = -\frac{k^2}{n^2}\frac{\exp-ik'z'}{z'}\int\frac{\boldsymbol{\mathscr{P}}_0(\mathbf{r})}{1-(z/z')}$$
$$\times\exp-i\Delta k_z z\,\exp-i\Delta k_x x\,\exp-i\alpha(x'-x)^2\,\exp-i\alpha(y'-y)^2\,dV$$

(62)

The general utility of the integral equation technique lies in the ability to evaluate the electric field of the diffracted light in the far-field limit relative to the interaction region of the incident light and elastic waves. Using the far-field or Fraunhofer approximation simplifies considerably the evaluation of the integral. The Fraunhofer approximation consists of setting the terms αx^2, αy^2, z/z', which are small in the far-field, equal to zero. It is nevertheless apparent that the number of photons of diffracted light must be the same in the near-field as in the far-field. Consequently, evaluating the electric field of the diffracted light in the far-field enables one to calculate the fraction of incident light power which is diffracted into one order. Although the diffracted-light-power measurement would be the same at any distance from the interaction region, the electric field distribution of the diffracted light wave is much more complicated and difficult to calculate in the near-field. The Fraunhofer approximation yields the simplified integral

$$\mathbf{E} = -\frac{k^2}{n^2}\frac{\exp-ik'\rho'}{\rho'}\int\boldsymbol{\mathscr{P}}_0(\mathbf{r})$$
$$\times\exp-i\Delta k_z z\,\exp-i\Delta k_x x\,\exp-2i\alpha xx'\,\exp-2i\alpha yy'\,dV$$

(63)

where

$$\rho' = z' + \frac{x'^2 + y'^2}{2z'} \quad \text{and} \quad \alpha \simeq \frac{k'}{2z'}. \tag{64}$$

To illustrate the use of this technique, the integral is evaluated for a situation analogous to the case shown in Fig. 9 (which used the differential equation technique). Thus, we assume that the incident light beam is of uniform intensity and square in cross section with width w. The elastic wave is also of uniform intensity and consists of a layer of thickness L. In evaluating the integral, one must give due regard to the fact that the coordinate system of the integral is aligned along the direction of the diffracted light wave. If one assumes that $L \ll w$, and that the magnitude of the Bragg angle is small, one can with little error "reshape" the interaction region to represent more simplified boundary conditions in this altered frame of reference. The integral then becomes, in effect,

$$\int_{-w/2}^{w/2} \exp -2i\alpha yy' \, dy \int_{-w/2}^{w/2} \exp -i\Delta k_x x$$

$$\times \exp -2i\alpha xx' \, dx \int_{-L/2\cos\theta}^{L/2\cos\theta} \exp -i\Delta k_z z \, dz \tag{65}$$

The integration over z is evaluated first with the result

$$\int_{-L/2\cos\theta}^{L/2\cos\theta} \exp(-i\,\Delta k_z z)\,dz = L\frac{\sin(\Delta k_z\,L/2\cos\theta)}{(\Delta k_z\,L/2\cos\theta)} \tag{66}$$

Obviously, the z integral will be very small except for small values of Δk_z. However, in view of Eq. (61), the condition $\Delta k_z = 0$ is satisfied only when the Bragg condition $\sin\theta = K/2k$ is satisfied. Thus, near $\Delta k_z = 0$, one can make the approximations

$$\Delta k_z = K\left(\frac{K}{2k} - \sin\theta\right), \qquad \Delta k_x = 0 \tag{67}$$

Using these approximations and performing the remaining two integrations gives the electric field of the diffracted light in the far-field limit,

$$\mathbf{E}(\mathbf{r}') = -\frac{k^2\mathcal{P}_0}{n^2}\frac{\exp -ik'\rho'}{\rho'} Lw^2 \frac{\sin(\beta L/2)}{\beta L/2}\frac{\sin(\alpha x'w)}{\alpha x'w}\frac{\sin(\alpha y'w)}{\alpha y'w} \tag{68}$$

where

$$\beta = \frac{K}{\cos\theta}\left(\frac{K}{2k'} - \sin\theta\right) \quad \text{and} \quad \alpha = \frac{k'}{2z'}.$$

As one would expect, the diffracted field resembles the Fourier transform (far-field picture) of a light beam which has passed through a rectangular aperture.

The maximum power in the diffracted light beam is calculated for $\beta = 0$ using the well-known formula

$$P = (nc/8\pi) \iint EE^* \, dx' \, dy' \tag{69}$$

If one substitutes Eq. (68) into (69) and replaces $\mathscr{P}_0$ by the value given in Eq. (56), one obtains for the power of the diffracted light beam

$$P_{\text{diff}} = \frac{nc}{8\pi} \left(\frac{\delta\varepsilon_{ij} E_j{}^0 kwL}{4n} \right)^2 \tag{70}$$

The power of the incident laser beam can also be calculated using Eq. (69), with the result

$$P_{\text{incid}} = (nc/8\pi)(E_j{}^0 w)^2 \tag{71}$$

The ratio of diffracted light power to incident light power is then given by

$$\frac{I_1}{I_0} = \frac{P_{\text{diff}}}{P_{\text{incid}}} = \frac{k^2 L^2}{16} \left(\frac{\delta\varepsilon_{ij}}{n} \right)^2 \tag{72}$$

If one compares this result with Eq. (47), one finds substantial agreement between the two techniques.

IV. Some Applications of the Elastooptical Effect

A. Introduction

After the original suggestion of Brillouin (1922) that elastic waves would diffract a light beam, ten years elapsed before the experimental observation of this effect, independently by Debye and Sears (1932) and Lucas and Biquard (1932). Subsequent investigation and application were rapid, however. Reviews of early experiments to measure material properties and to investigate the diffraction pattern under various conditions are given by Bergmann (1938), Born and Wolf (1965), and Hargrove and Achyuthan (1965). These experiments generally verified the intensity distribution of the diffracted light, the transition from the Raman–Nath to the Bragg region, and the frequency shift of the diffracted light by multiples of the elastic-wave frequency. Although adequate to verify most of the theoretical predictions, the accuracy of these results was poor, owing to the relatively low frequency of the elastic waves and the broad spectrum and low intensity of the light sources.

Despite these limitations, measurements of material parameters were obtained and several useful devices were constructed. Schaefer and Bergmann (1934) and Hiedemann (1935) developed techniques to measure the elastic constants of transparent solids, and Mueller (1938) showed how these techniques could be used to measure elastooptical constants. In the Schaefer–Bergmann method, the sample is excited in a high overtone so that elastic waves propagate in many directions, forming a space grating of elastic strain. When the sample is illuminated by a beam of light, the image of the diffracted light exhibits the elastic symmetry of the crystal and forms a pattern which can be related to the anisotropic elastic constants. A significant advantage of the method is that all of the elastic constants can be obtained from a single sample. The technique has been described in detail by Bergmann (1938) and more recently by Spencer *et al.* (1967). Becker *et al.* (1936) constructed light modulators by passing the central beam through a slit and blocking the diffracted light, or conversely, by blocking the central beam and collecting the diffracted light. The design and performance of these modulators have been reviewed by Bergmann (1938). Probably the most ambitious effort was the development of the Scophony projection television system [Okolicsanyi (1937), Lee (1938), and Robinson (1939). For a review, see Bergmann (1938), Adler (1967), or Zworykin and Morton (1954)]. In this system, a 10-MHz elastic wave, intensity-modulated by the video signal, was generated in a liquid cell. The cell was long enough to contain an entire television line and was uniformly illuminated by a light beam propagating parallel to the elastic wavefronts. The undiffracted light was removed by focusing onto an opaque bar; the diffracted light which passed the bar provided an image of the video signal. This image, which traveled parallel to the cell axis at the speed of sound, was made stationary by reflection from a rotating mirror. By this means, an entire television line could be projected onto the screen at one time. The performance is described succinctly by Zworykin and Morton (1954): "The amount of light that can be obtained is sufficient, when a high-intensity moving-picture type arc serves as a light source, so that the image formed on a 5 by 6 foot screen can be viewed without discomfort."

The recent advances in this field have resulted principally from the availability of lasers, which produce intense, coherent light; from the development of efficient broadband transducers which generate elastic waves up to microwave frequencies; and from the discovery of materials which have excellent elastic and optical properties. In this section, we shall emphasize applications in the Bragg region, because these most fully exploit the latest experimental capabilities.

The use of Bragg diffraction is based on the results obtained earlier in Eqs. (42) and (63), which express the diffracted light amplitude in terms of the scattering parameters and the experimental conditions. The salient features can be summarized by using the special case of a wide optical beam passing through a uniform-intensity elastic beam of width L. For this configuration, the diffracted light amplitude is [Eq. (43), repeated here for convenience]

$$U_1 = \xi U_0 L (\exp -i\beta_1 z) \frac{\sin(\beta_1 L/2)}{\beta_1 L/2} \tag{73}$$

where

$$\beta_1 = -\frac{K}{\cos\theta}\left(\sin\theta + \frac{K}{2k}\right), \qquad \xi = \frac{1}{4}\left(\frac{\delta\varepsilon}{\varepsilon_0}\right)\frac{k}{\cos\theta} \tag{74}$$

It should be recalled at this point that U_1 is an amplitude factor and does not include the explicit time dependence. The electric field of the incident light U_0 varies as $e^{-i\omega t}$ and the strain parameter ξ varies as $e^{-i\Omega t}$, so, from Eqs. (38) and (56), the electric field of the diffracted light varies as $e^{-i(\omega \pm \Omega)t}$ for the ± 1 orders of diffraction.

The important parameters of the Bragg diffraction are illustrated in Fig. 10. Note that the Bragg angle Θ' shown here is the external angle, re-

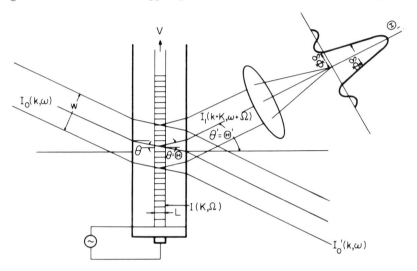

FIG. 10. Bragg diffraction of light by a sound column of width L occurs with highest efficiency when the angles of incidence and diffraction are symmetrical and equal to the Bragg angle. Zeros of the diffraction pattern are at angles $\pm\delta\Phi$ from the Bragg angle. The frequency of the diffracted light is shifted by the acoustic frequency $\Omega/2\pi$.

lated to the internal Bragg angle by $\Theta' \approx n\Theta$. In what follows we will generally use internal angles; the transformation to the externally observed angle is evident. The central maximum of the diffraction pattern occurs at the Bragg angle

$$\theta = \Theta = \sin^{-1}(K/2k) \tag{75}$$

and the relative peak intensity of the diffracted beam, in the small-signal approximation, is [Eq. (47)]

$$I_1(\text{max})/I_0 = \xi^2 L^2 \tag{76}$$

The zeros of the diffracted light intensity occur at $\beta_1 L/2 = \pm\pi$, corresponding, for small angles, to the deviation from the Bragg angle

$$(\theta - \Theta)_0 = \pm(\Lambda/L)\cos\theta = \pm\delta\Phi\cos\theta \approx \delta\Phi \tag{77}$$

The angle $\delta\Phi = \Lambda/L$ is just the Rayleigh criterion for the diffraction spreading

angle of the acoustic beam.[1] The intensity of the diffracted light is reduced
to half the peak value at the angles given by $\beta_1 L/2 = \pm 0.45$, corresponding
to a total width at half-maximum of $0.9\pi \times 2 \cos \theta/LK$. We will usually
use the simplified expression for the width

$$2(\theta - \Theta)_{1/2} = (\Lambda/L) \cos \theta = \delta\Phi \cos \theta \approx \delta\Phi \qquad (78)$$

which describes the range of angles for which the intensity is within approximately 4 dB of the central maximum.

These results lead immediately to several important applications:

1. The angle of maximum diffracted intensity Θ is proportional (for
small angles) to the ratio of elastic-wave frequency to velocity. This proportionality can be used to measure the velocity and elastic constants and,
by changing the acoustic frequency, to obtain controllable deflection of the
light beam.

2. The intensity of the diffracted light is proportional to the acoustic
intensity, which provides a means to modulate the light intensity or to probe
the acoustic intensity within the rod.

3. Conversely, measurement of the diffraction efficiency can be used to
determine elastooptical coefficients of materials.

4. The frequency shift of the diffracted light can be used to produce
frequency-modulated light or, by mixing with the undiffracted light, to obtain heterodyne detection.

5. The fact that the dependence of the diffracted light intensity on the
angle between the optical and acoustic wave vectors is proportional to the
Fourier transform of the elastic amplitude distribution can be used to measure
the spatial variation of intensity over the cross section of the elastic beam.

A number of experimental techniques and devices based on these results
are described in this section. Some devices operating in the Raman–Nath
region will also be described for completeness. The application to signal-
processing systems will be reviewed in the following section.

B. Measurement of Properties of Materials

1. *Velocity and Elastic Constants*

The diffraction of light provides a convenient method to measure the
wavelength of sound, and from this to obtain the elastic-wave velocity and
the elastic constants. In the Bragg region, the velocity is given by

$$V = \lambda\Omega/(4\pi \sin \Theta) \approx \lambda f/2\Theta \qquad (79)$$

---

[1] This notation differs from that of Section II because it is necessary in this section
to distinguish among: angles relative to fixed coordinates, which we designate for
optical and acoustic rays by θ and Φ, respectively; deviations of a collimated beam from
a central value (as from the Bragg angle for a particular frequency), designated by
$\Delta\theta$, $\Delta\Phi$; and diffraction spreading angles, designated by $\delta\theta$, $\delta\Phi$.

where 2Θ is the angle between the incident light beam and the diffracted beam for Bragg diffraction. Since the optical wavelength and elastic-wave frequency can be determined to high accuracy, the measurement reduces to an accurate determination of the diffraction angle 2Θ. From Eq. (78), for a wide beam of incident light, the diffracted light intensity as a function of angle of incidence has half-power points separated by $(\Lambda/L)\cos\theta$, which limits the minimum resolvable change in direction $\Delta\theta_{min}$. The velocity can be measured to a fractional accuracy of $\Delta V/V = \Delta\theta_{min}/\Theta < (\Lambda/L)/(\lambda/2\Lambda)$ $= 2\Lambda^2/\lambda L$. For elastic waves of 10^9 Hz and sample dimensions of a few millimeters accuracies of about 0.1% can be achieved (Krischer, 1968). The light-beam diameter can be as small as 1 mm before optical diffraction affects the accuracy, so only small samples are required. Spurious elastic modes can be eliminated by their differing transit times, diffraction angles, and dependence of diffraction efficiency on light polarization.

The same technique can be used to measure the velocity of elastic surface waves. Light diffraction by surface waves was first reported by Ippen (1967). Diffraction occurs for both the reflected light, owing to the periodic mechanical displacement of the surface (Mayer *et al.*, 1967), and for the transmitted light, principally because of the elastooptical effect (Ippen, 1967). The Raman–Nath conditions apply to the diffraction by surface waves, since the effective thickness of the acoustic wave is less than Λ, but the expression for the velocity is still given by Eq. (79). Auth and Mayer (1967) and Krokstad and Svaasand (1967) have determined the velocity of surface waves by measuring the diffraction angle.

2. *Attenuation*

Elastooptical diffraction provides a versatile tool for the measurement of acoustical attenuation, especially for operation in the Bragg region. The experimental techniques for the Raman–Nath case have been reviewed recently by Hargrove and Achyuthan (1965). A typical arrangement for measurements at microwave frequencies is shown in Fig. 11.

The most direct way to measure acoustical attenuation is by observing the exponential decrease of the diffracted light intensity as the laser beam is moved along the crystal. It is necessary to eliminate the effect of reflected acoustic waves, either by using a well-matched acoustical line or by using time-resolution of pulsed acoustic signals. This technique has the advantage, compared to conventional methods, that abnormal changes in elastic strain intensity caused by localized flaws in the sample are immediately apparent. Wilkinson and Caddes (1966) used this technique to measure the attenuation in Z-cut quartz at frequencies up to 3.38 GHz. Korpel *et al.* (1967), Salzmann *et al.* (1968), and Slobodnik (1969a) measured the attenuation of Rayleigh surface waves by elastooptic diffraction probing. Cohen and Gordon (1965) have shown that it is not necessary to use a narrow light beam to measure the acoustical attenuation by diffraction techniques. Including the elastic-wave intensity decay $e^{-2\alpha x}$ in the derivation of Eq. (73), they find that a translation

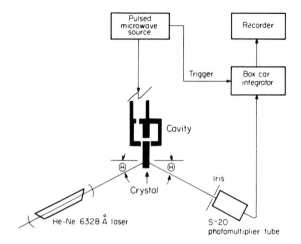

Fig. 11. Basic experimental apparatus used to measure acoustical attenuation by means of acoustooptical Bragg diffraction.

of the light beam in the x direction by an amount Δx results in a change in the diffracted light intensity by a factor $e^{-2\alpha\Delta x}$, independent of the width of the light beam. A limitation to the method of optical-beam probing is that it uses only a single pass of the elastic wave through the sample, so that only a slight change of intensity occurs in small, low-loss samples.

By using samples with parallel, polished end surfaces, multiple pulse echoes can be observed, thereby increasing the path length and the sensitivity for small attenuation. The position of the optical beam remains fixed. The sample fabrication and analysis of data in this case are substantially the same as for ultrasonic pulse-echo measurements, with the advantage that slight nonparallelism of the end faces can be accommodated by rotating the crystal so that the angle of incidence of the light beam is at the Bragg angle for the desired echo. Spencer *et al.* (1966) and McMahon (1967a) have used this technique to measure attenuation in crystals of $LiNbO_3$.

A third method to measure the attenuation uses the sample as an acoustical cavity resonator. For a sample length of several hundred acoustic wavelengths, conditions of constructive and destructive interference can be observed by sweeping the microwave frequency over a small range. The Bragg-diffracted light from the acoustic standing waves goes through corresponding maxima and minima. Typical experimental results are shown in Fig. 12. The attenuation is given by the simple relation

$$I_1(\min)/I_1(\max) = \tanh^2 \alpha D \tag{80}$$

where α is the attenuation per unit length and D is the length of the crystal. McMahon (1967a) has given a detailed description of the technique and representative results for these three methods of measuring attenuation by Bragg diffraction.

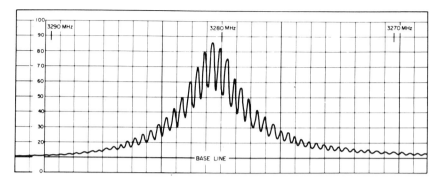

FIG. 12. Acoustical interference as observed by means of Bragg diffraction. A cw acoustic beam was injected into a crystal which had opposite ends fabricated flat and mutually parallel. Destructive and constructive interference are produced alternately as the frequency of the microwave source is swept slowly through the resonance of the microwave cavity.

3. *Harmonic Generation and Third-Order Elastic Constants*

Acoustical attenuation is closely related to the anelastic properties of materials. Except at very low temperatures, the attenuation results from the interaction of ultrasonic and thermal sound waves. The nonlinear elastic properties are of primary importance in determining the magnitude of the interactions. Detailed discussions and further references are given in Mason (1965), Thurston (1965), and Klemens (1965). The elastooptical diffraction provides a useful tool for measuring the third-order elastic constants which describe the anelastic behavior. An excellent review of the theory and experiments, principally on fluids, in the Raman–Nath region is given by Hargrove and Achyuthan (1965). Parker *et al.* (1964) have used this technique, based on a theory of Melngailis *et al.* (1963), to measure one of the third-order elastic constants of NaCl. In the Raman–Nath region, the analysis of experimental data is complicated by the fact that high-order diffraction from the fundamental is observed at the same angle as first-order diffraction from elastic-wave harmonics. The ambiguity between the two sources of diffracted light is eliminated in the Bragg region because multiple scattering is a forbidden process.

Acoustical second-harmonic generation provides a useful method for measuring the nonlinear elastic constants. The analysis is intimately related to the theory of elastic-wave propagation in nonlinear media. The general equations of motion for the propagation of finite-intensity waves have been derived by Thurston (1964). Brugger (1965) has studied the symmetry properties of the second- and third-order elastic constants. As an example of the results, Brugger (1965) finds for a longitudinal wave propagating in a pure mode direction that

$$\frac{\partial^2 u}{\partial t^2} = \frac{C_{xx}}{\rho}\left(1 + F\frac{\partial u}{\partial x}\right)\frac{\partial^2 u}{\partial x^2} \tag{81}$$

where u is the displacement, ρ is the density, C_{xx} is the appropriate second-order elastic constant for propagation in the x direction, and F is a constant which depends on the direction of propagation and on the second- and third-order elastic constants. In the case of propagation along the c-axis of a trigonal material such as $LiNbO_3$, for example, $F = 3 + C_{333}/C_{33}$. Recognizing that the particle motion couples the oscillations at the frequencies Ω and 2Ω, one sees that the last term in the equation of motion represents the nonlinear coupling which generates the second harmonic wave. Thus, a suitable solution of Eq. (81) is of the form

$$u = A_1(x)e^{i(Kx - \omega t)} + A_2(x)e^{i2(Kx - \omega t)} + \text{complex conjugate} \qquad (82)$$

With the approximation of slowly varying amplitude parameters ($\partial A_1/\partial x \ll KA_1$, etc.), substitution of Eq. (82) in Eq. (81) yields a pair of nonlinear coupled equations. Including a phenomenological damping term, these are

$$dA_1/dx - \alpha_1 A_1 = -4FK^2 A_1 \qquad (83a)$$

$$dA_2/dx - \alpha_2 A_2 = FK^2 A_1^2 \qquad (83b)$$

General solutions of Eqs. (83) cannot be obtained, but from numerical calculation, one finds that the amplitude of the fundamental wave decreases with distance, both because of attenuation and transfer of energy to the harmonic wave. The second-harmonic amplitude at first increases with distance, but at large distances, it decreases because of attenuation and decreased excitation by the weakened fundamental. Typical results, for several different initial conditions, are shown in Fig. 13.

Approximate solutions of Eq. (83) can be readily obtained by making certain simplifying assumptions (attenuation varies as Ω^2, relatively small-amplitude fundamental). The initial buildup (for $\alpha x \ll 1$) of the second-harmonic amplitude A_2 is

$$A_2 = FK^2 A_1^2 x \, e^{-2\alpha x} \qquad (84)$$

where A_1, K, and α refer to the amplitude, wave vector, and attenuation constant of the fundamental. At the position of maximum second harmonic, the relation between the amplitudes is

$$A_2(\text{max}) = \tfrac{1}{4}FK^2 A_1^2/\alpha \qquad (85)$$

The measurement of the nonlinearity coefficient F is now straightforward. It is only necessary to determine the strain amplitudes A_1 and A_2 by observing the Bragg diffraction at the angles corresponding, respectively, to the fundamental and harmonic waves, and to calculate the value of F from Eq. (84) or Eq. (85). The accuracy with which the third-order elastic constants can be determined is limited, both because the strain amplitudes can be measured only to the accuracy with which the elastooptical constants are known and because of the result shown above that F often depends on the second-order elastic constant as well as the third-order constant. Richardson *et al.* (1967) have used this technique to estimate the third-order elastic

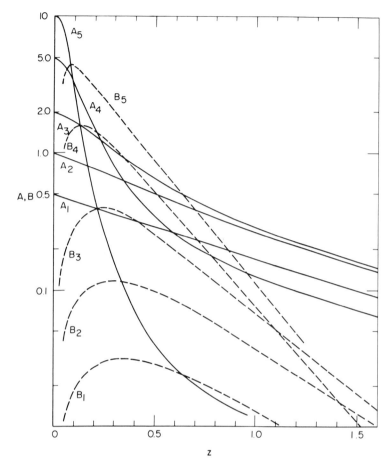

FIG. 13. Spatial dependence of the fundamental (A) and second harmonic (B) acoustic amplitudes. These curves show the solution of the nonlinear coupling equations (83a) and (83b) for five initial conditions of A, labeled 1–5.

constants in quartz and MgO. In these materials, the nonlinearity is significant at an acoustic intensity level of 10 W/cm². McMahon (1968) has described this method in more detail and used the technique to measure the constant C_{333} for LiNbO$_3$.

Dixon (1967d) has used Bragg-diffraction techniques to demonstrate the nonlinear mixing of two collinear elastic waves, generating the sum and difference frequencies. In the former case, the inputs were two transverse elastic waves, at 265 MHz and 60 MHz, which entered a fused quartz rod collinearly from opposite ends. The sum frequency was a longitudinal wave at 325 MHz. The phase-matching conditions for the interaction are satisfied when the input frequency ratio is

$$\Omega_1/\Omega_2 = (V_L + V_T)/(V_L - V_T) \tag{86}$$

as in this experiment. The waves were observed by diffraction of light at the Bragg angles appropriate for each frequency. The diffraction by the input waves could be reduced to low levels by placing the polarization of the transverse elastic waves in the plane of diffraction. Conclusive evidence was obtained for the nonlinear mixing interaction, including observation of the sum signal outside the region of interaction of the two input waves and demonstration that the intensity of the sum-frequency wave varied as $(\sin^2 \kappa x/2)/(\kappa x/2)^2$, where $\kappa = K_3 - (K_1 - K_2)$ is the phase mismatch, when the frequency of one of the input waves changed.

Nonlinear effects in the propagation of elastic surface waves have been investigated by Lean and Tseng (1969) and Slobodnik (1969b) by using elastooptic diffraction to observe the harmonics of the fundamental acoustic frequency.

4. Measurement of Elastooptical Coefficients.

The elastooptical coefficients, which relate changes in the relative dielectric impermeability tensor to elastic strain, were defined in Section II. The efficiency of elastooptical diffraction depends directly on these coefficients and it is important to have means to measure both absolute and relative values of them in various materials. Accurate measurements of the absolute values are not readily obtained by using Bragg diffraction because of the difficulty in accurately determining the elastic strain associated with the acoustic wave. The traditional technique is to measure the optical retardation resulting from a known static stress and then to convert this to the elastooptical (strain) components by using the elastic moduli. Alternatively, for liquids, the elastooptical constant may be calculated from the refractive index using the Lorentz–Lorenz relation, $p = \frac{1}{3}(n^2 - 1)(n^2 + 2)/n^4$. It is approximately $\frac{1}{3}$ for most liquids.

Measurement of the coefficients of the elastooptical tensor in crystals requires careful analysis to establish which coefficient is being measured. The detailed treatment of the elastooptical effect is inappropriate here, but simple examples will serve to illustrate the principle. In an optically isotropic solid (this includes cubic crystals), the index ellipsoid defined in Eq. (2) reduces to a sphere

$$(1/n^2)_1(x_1{}^2 + x_2{}^2 + x_3{}^2) = 1 \tag{87}$$

in the Voigt notation. When a longitudinal elastic wave propagates along a principal axis, this wave produces a change in the components of the index ellipsoid and index ellipse. Using Eqs. (3) and (1) for a strain e_1, the index ellipsoid becomes

$$\left[\left(\frac{1}{n^2}\right)_1 + p_{11}e_1\right]x_1{}^2 + \left[\left(\frac{1}{n^2}\right)_1 + p_{21}e_1\right]x_2{}^2 + \left[\left(\frac{1}{n^2}\right)_1 + p_{31}e_1\right]x_3{}^2 = 0 \tag{88}$$

so that the sample exhibits an induced birefringence. For light propagating along x_3, $\Delta n_2 = \frac{1}{2}n^3 p_{11} e_1$ and $\Delta n_1 = \frac{1}{2}n^3 p_{21} e_1$. Using Eq. (31), the quantity p_{11}/p_{21} can be determined by measuring the ratio of the diffracted light intensities when the axis of polarization is changed from x_1 to x_2. For light polarized at $45°$ to x_1 and x_2, the undiffracted light can be removed with an analyzer while a component of the diffracted light intensity, proportional to $(p_{11} - p_{12})^2$, has orthogonal polarization and is transmitted.

A shear wave, say e_4, which propagates along x_2 with particle motion along x_3, produces a change $\Delta(1/n^2)_4 = p_{44} e_4$, so that the index ellipsoid becomes

$$(1/n^2)_1(x_1{}^2 + x_2{}^2 + x_3{}^2) + p_{44}\, e_4 x_2 x_3 = 1 \qquad (89)$$

The perturbed ellipsoid, shown in Fig. 14, has principal axes at $45°$ to x_2 and x_3. For light propagating along x_1 and polarized along x_2 or x_3, the

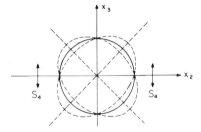

FIG. 14. Index ellipsoid for light propagating along the x_1 axis of a cubic crystal with a shear wave polarized along x_3 and propagating in the x_2 direction.

diffracted light is polarized at $90°$ to the incident light. The intensity of the diffracted light is proportional to p_{44}^2. [In an isotropic solid, $p_{44} = \frac{1}{2}(p_{11} - p_{21})$ and, for equal strains, the maximum diffracted light intensity observed with crossed polarizers is the same for either longitudinal or shear waves.]

The periodic strain of the elastic wave causes the lengths of the principal axes of the index ellipse to oscillate (out of phase) at the elastic-wave frequency, so that the diffracted light is shifted in frequency to the sum and difference frequencies of the two interacting waves, $\omega \pm \Omega$. This frequency shift permits the use of heterodyne techniques to detect the diffracted light. Note that a shear wave polarized parallel to the optical beam direction does not produce any diffraction in a cubic crystal.

Ratios of the elastooptical coefficients in sapphire were determined by Caddes and Wilkinson (1966) by measuring the intensity ratio of the light diffracted by 730-MHz longitudinal waves along the c-axis as the polarization of the incident light was varied. A serious limitation to the use of diffraction techniques for measuring elastooptical constants was the inability to obtain absolute values or accurate comparisons between various materials. Methods have been devised recently to overcome this problem.

Smith and Korpel (1965) described a technique using Bragg diffraction to measure the elastooptical coefficients of solids relative to the (known) values of liquids. The solid sample is immersed in a liquid ultrasonic cell as shown in Fig. 15 and a pulsed acoustic beam is sent through the liquid into the solid and out into the liquid beyond. A laser beam, incident at the proper Bragg angle, is sequentially directed into each of the three regions and the diffracted light is measured. From Eq. (31), the diffracted light I_2 in the liquid in front of the sample is proportional to $(p_l^2 n_l^6 / \rho_l V_l^3)I$, where the subscript l refers to the liquid and I is the acoustic power density in the liquid.

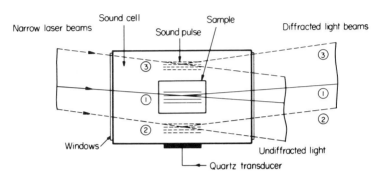

FIG. 15. Apparatus used to measure elastooptical constants of a solid relative to the known value of a liquid. After Smith and Korpel (1965).

If the acoustic power reflection coefficient of the liquid–solid interface is R, the diffracted light from the solid is $I_1 \sim (p_s^2 n_s^6)/(\rho_s V_s^3)RI$, where p_s is a suitable combination of elements of the elastooptical tensor determined by the experimental conditions. By symmetry, the second interface has the same reflection coefficient, so the diffracted light in the liquid beyond the crystal is $I_3 \sim (p_l^2 n_l^6)/(\rho_l V_l^3)R^2 I$. Thus, one obtains

$$\frac{I_1}{(I_2 I_3)^{1/2}} = \frac{p_s^2 n_s^6 / \rho_s V_s^3}{p_l^2 n_l^6 / \rho_l V_l^3} \tag{90}$$

a direct measure of the relative efficiency for Bragg diffraction. The elastooptical coefficient of the solid is determined by using known properties of the reference liquid and of the remaining parameters of the solid. A significant advantage of the technique is that the elastic wave can be directed along each crystal axis of the rectangular sample, without the need to bond transducers. The technique is limited, however, because shear waves cannot be transmitted through the liquid and because liquids have very high attenuation at high frequencies.

Dixon and Cohen (1966) and Carleton and Soref (1966) extended this method to use a solid, usually fused silica, as the reference material. The sample, with flat, parallel end faces, is bonded to one end of the reference rod and a transducer is attached to the other end of the reference rod. Pulsed elastic waves are generated and the laser beam, adjusted to the Bragg angle, is used to probe the reference and the sample. An oscilloscope trace of the Bragg-diffracted light in the reference rod shows three pulses.

The first (I_1) is produced by the outgoing wave, the second (I_2) by sound reflected from the bond, and the third (I_3) by sound which has traversed the sample, reflected from the free end, and reentered the reference. The first two pulses on the trace of Bragg-diffracted light from the sample are (I_4) the transmitted sound beam and (I_5) the wave reflected from the free end. An analysis similar to that for the case of a liquid reference shows that

$$\left(\frac{I_4 I_5}{I_1 I_3}\right)^{1/2} = \frac{p_s^2 n_s^6 / \rho_s V_s^3}{p_r^2 n_r^6 / \rho_r V_r^3} \tag{91}$$

independent of the bond reflection coefficient and loss (provided the bond is uniform over the acoustic beam cross section), of loss at the free end, and of attenuation in either the sample or the reference.

Based on this technique, Dixon (1967b), Maloney and Carleton (1967), and Reintjes and Schulz (1968) have reported the elastooptical coefficients of a number of materials which are of interest for high-frequency devices. Pinnow and Dixon (1968) and Pinnow *et al.* (1969) have recently reported data on α-HIO$_3$ and PbMoO$_4$, respectively. In addition to a large elastooptical diffraction efficiency, low acoustical attenuation is required at the operating frequency. Data on typical materials are given in Table I, including approximate attenuation values at 500 MHz. The utility of the parameters M_1 and M_3 shown in Table I will be discussed in a later section. It is found that the range of variation of p is relatively small and that p is approximately 0.1 in most materials. Since the density also exhibits no extreme variation in different materials, the diffraction efficiency is primarily determined by differences in the refractive index and velocity.

5. *Measurement of Magnetoelastic and Magnetooptical Properties*

Bragg diffraction has been shown to be of considerable use for investigating the properties of elastic waves in magnetic materials. The technique has been applied solely to yttrium iron garnet, up to the present time, because YIG is transparent in the wavelength range 1–6 μ and exhibits very low acoustical and magnetic loss. The experiments have mostly had as a goal the investigation of the magnetoelastic coupling between elastic and spin waves. A recent review by Strauss (1968) describes the theory of this effect and a number of experiments using pulse-echo techniques. The advantages of Bragg diffraction derive, as in the elastic-wave case, from the ability to probe within the sample.

In the course of experiments to measure the elastooptical constants of YIG by the comparison method described in the preceding subsection, Dixon and Matthews (1967) noted that the acoustical Faraday effect could be observed. This effect is a rotation of the plane of polarization of a shear elastic wave propagating along the magnetic field, similar to the analogous optical rotation. The behavior can be investigated by using a YIG rod with a transducer which excites shear waves propagating along the axis and an axial magnetic field. The light is incident from the side of the rod at the Bragg angle. As shown in Eq. (89), for a cubic crystal like YIG, the induced optical

TABLE I

FIGURES OF MERIT FOR ACOUSTOOPTICAL DEVICES

Material	n	ρ (g/cm³)	V (10⁵ cm/sec)	M_1 $(n^7p^2/\rho V)$ [b]	M_2 $(n^6p^2/\rho V^3)$ [b]	M_3 $(n^7p^2/\rho V^2)$ [b]	α_{500} [c] (dB/μsec)
Fused quartz	1.46	2.2	5.95	7.89×10^{-7}	1.51×10^{-18}	1.29×10^{-12}	1.8
GaP	3.31	4.13	6.32	590	44.6	93.5	1.0
TiO₂	2.58	4.6	7.86	62.5	3.93	7.97	0.1
LiNbO₃	2.20	4.7	6.57	66.5	6.99	10.1	0.03
YAG	1.83	4.2	8.60	0.98	0.073	0.114	0.07
LiTaO₃	2.18	7.45	6.19	11.4	1.37	1.84	0.02
As₂S₃	2.61	3.20	2.6	762	433	293	—
β-ZnS	2.35	4.10	5.51	24.3	3.41	4.41	—
α-Al₂O₃	1.76	4.0	11.15	7.32	0.34	0.66	0.05
ADP	1.58	1.80	6.15	16.0	2.78	2.62	—
KDP	1.51	2.34	5.50	8.72	1.91	1.45	—
PbMoO₄ [d]	2.36	6.95	3.75	120	35.8	30	1.0
α-HIO₃ [e]	1.98	4.63	2.44	99	83.4	40	0.62

[a] All values quoted are for longitudinal waves and $\lambda_0 = 0.6328 \mu$m. Except for PbMoO₄ and α-HIO₃, data are selected from Dixon (1967b), where additional data and details on propagation directions and polarization are tabulated.

[b] Figures of merit are given in cgs units.

[c] Approximate attenuation at 500 MHz in dB/μsec.

[d] Values of n, ρ, V, M_2, and α from Pinnow et al. (1969). M_1 and M_3 calculated from M_2 and extrapolated values of n.

[e] Values of V, M_2, and α from Pinnow and Dixon (1968).

birefringence is proportional to the elastooptical constant p_{44} times the component of strain perpendicular to the plane of diffraction. In acoustical Faraday rotation, the plane of polarization of the transverse strain rotates, as a function of position at fixed field or of field at fixed position, from perpendicular to parallel to the plane of diffraction. A series of maxima and minima is observed in the diffracted light intensity. This effect has been studied by Dixon (1967c) and Smith (1967). The period of the alternations depends on the internal magnetic field, which is a function of position in a nonellipsoidal sample. Dixon (1967c) used this dependence to measure the field variation. The acoustic Faraday rotation occurs because the two circularly polarized waves, into which the linearly polarized wave can be decomposed, couple differently to the magnetization and therefore travel with different phase velocities. The group velocities are also different, and under proper experimental conditions the use of short pulses of elastic energy permits time resolution of the two circularly polarized waves. The diffracted light beams from these resolved pulses have equal intensities, because the time-average components of strain are equal (neglecting differential attenuation), but the wave vectors of the two waves are unequal and thus the Bragg angles are different. The group velocity of the circularly polarized wave which is coupled to the magnetization depends on the internal magnetic field and the second magnetoelastic coupling constant b_2. Dixon (1967c) obtained the value of b_2 from his measurements.

As described by Strauss (1968), the magnetoelastically coupled circular polarized wave takes on the character of a spin wave as the magnetic field approaches the resonance value. This change in wave type is the cause of the dispersion and of the reduced group velocity. Clearly, the elastooptical constants cannot describe the Bragg diffraction if the wave has a significant magnetic component. Auld and Wilson (1967a) have calculated the Bragg diffraction by coherent spin waves, using the integral equation technique of Section III. The induced polarization P arises from the magnetooptical effect caused by the transverse magnetization of the spin waves. The intensity of the Bragg-diffracted light is obtained by using measured values of the optical Faraday rotation in YIG. It is found that the diffraction by spin waves is about as efficient as the elastooptic diffraction in quartz, for equal power. The saturation of spin-wave amplitude is well known, however (see, for example, Damon, 1963), and this limits the maximum spin-wave power density in YIG to less than 1 mW/cm² at the frequencies commonly used.

By adjusting the magnetic field so that the light beam illuminated a magnetoelastic wave with significant spin-wave component, Smith (1967) and Dixon (1967c) have observed the change in Bragg angle which results from the change in wavelength. This effect, which changes the Bragg angle by about 3° for a field change of 20 Oe, could be used for a magnetically controlled light deflector were it not for the saturation effect, which limits the intensity of the magnetically diffracted light. The magnetic and elastic components of the wave energy are equal for a change in angle of only about 0.1° (Smith, 1968a).

The theory of Auld and Wilson (1967b) showed that the polarization of light diffracted by spin waves is rotated 90° from the input polarization, and that the intensity of the magnetically diffracted light is independent of the input polarization. This rotation permits the separation of the elastic and magnetic diffraction mechanisms by choosing a suitable input light polarization and using an analyzer in the diffracted beam. Auld and Wilson (1967b) and Smith (1968a,b) have used this technique to measure the relative contributions of the elastic and magnetic components of the magnetoelastic wave. These experiments generally verified the conversion of elastic to spin energy as described by Strauss (1968) and others, and showed clear evidence of spin-wave saturation effects.

Collins and Wilson (1968) have used the diffraction from magnetostatic spin waves at 1.7 GHz to investigate the mechanism by which electromagnetic energy is converted to exchange spin-wave energy. The Bragg angle is small ($<\frac{1}{2}°$) because the magnetostatic waves have long wavelength, but the diffracted beam could be separated from the main beam by means of polarizers because of the 90° polarization rotation. The nature of the spin wave as a function of position in the rod was determined by optical probing at different values of the external field. The properties of the diffracted light provided information on the excitation process, magnetic focusing effects, and spin-wave group velocity.

C. Imaging of Elastic Waves by Bragg Diffraction

The most straightforward way to construct an image of the acoustic strain distribution is to probe the sound field with a light beam, measuring the diffracted intensity as a function of position. As shown in Eq. (73), there is a direct proportionality between the amplitude of the elastic wave and the amplitude of the diffracted light. This is the technique described earlier for measuring the acoustical attenuation and magnetoelastic effects. Maloney et al. (1968) have used this method to measure the radiation pattern of an acoustical transducer, probing in both the longitudinal and transverse directions within the acoustic strain field. Lean (1967) has used optical probing to evaluate inhomogeneities and the quality of bonds between samples.

Another technique for determining the intensity distribution of an acoustic field uses a collimated beam of light; the information is contained in the angular variation of the diffracted light. It was shown in Eq. (42), using the definition of β in Eq. (40), that the first-order, Bragg-diffracted light amplitude is

$$U_1 = -U_0 \exp i \frac{K}{\cos \theta} \left(\sin \theta + \frac{K}{2k} \right) z \int_{-\infty}^{z} \xi(z')$$
$$\times \exp \left[-i \frac{K}{\cos \theta} \left(\sin \theta + \frac{K}{2k} \right) z' \right] dz' \qquad (92)$$

This case was calculated for an acoustic beam of infinite extent in the y

direction, normal to the directions of light and sound propagation. It shows that the angular distribution of light about the Bragg angle, $\Theta = \sin^{-1}(K/2k)$, is proportional to the Fourier transform of the distribution of elastic strain parameter $\xi(z)$ normal to the elastic wave vector. The same result is obtained from Eq. (63), which gives the far-field Fraunhofer diffraction pattern in terms of the induced polarization $\mathscr{P}_0(\mathbf{r})$. As shown in Eq. (68), for a light beam of finite size, the aperture diffraction of the light is superimposed on the Fourier transform of the elastic strain distribution.

Quate *et al.* (1965) have described the effect in a conceptually simple way. In Bragg diffraction, the light is partially reflected from successive waves of the acoustic beam of width L. If N waves contribute to the scattering, then $1/N$ of the intensity is reflected by each wavelength. The Bragg condition is imposed by the requirement that these waves interfere constructively. In other respects, the diffraction pattern of the reflected light is the same as that from a single wave, so that for a uniform intensity distribution the acoustic wave can be replaced by a mirror of the same width, at the Bragg angle to the incident light, as shown in Fig. 16. The mirror, of width L, has a projected area on the light beam of $L \sin \Theta$. The reflected light appears to come from a virtual light source, separated from the actual source by twice the Bragg angle, which passes through a slit of width $D = L \sin \Theta$. It is well known that the aperture diffraction of light passing through a slit of width D produces an intensity distribution (Born and Wolf, 1965, p. 393)

$$I = I_0 \frac{\sin^2[(\pi D/\lambda)(\sin \theta_i - \sin \theta_d)]}{[(\pi D/\lambda)(\sin \theta_i - \sin \theta_d)]^2} \approx I_0 \frac{\sin^2(\pi D/\lambda)(\theta_i - \theta_d)}{[(\pi D/\lambda)(\theta_i - \theta_d)]^2} \tag{93}$$

where θ_i and θ_d are the angles of incidence and diffraction at the slit. This is just the Fourier transform of the uniform amplitude distribution reflected by the rectangular mirror, as illustrated in Fig. 16. For light incident at the Bragg angle $\theta_i = \Theta$, the light reflected from the mirror (the acoustic beam) is at the angle $\theta_d = \Theta$. The total deflection angle is $\theta_i + \theta_d = 2\Theta$. If the mirror is tilted by an angle $\Delta\theta_i = \Delta\theta$, keeping the total deflection angle constant, then $\Delta\theta_d = -\Delta\theta$. The intensity pattern is displaced by an angle $2\,\Delta\theta$, reducing the intensity observed at the Bragg angle by the amount given by Eq. (93) with $\theta_i - \theta_d = 2\,\Delta\theta$. Thus, the intensity distribution of the diffracted light can be measured by rocking the mirror (the elastooptical crystal) about the Bragg angle, maintaining a fixed angle 2Θ between the incident light beam and the ray to the detector.

Few experiments are done with a uniform, rectangular distribution of sound and a uniform, wide light beam. Frequently, the laser beam is of Gaussian cross section, for example. The diffracted beam has a finite spreading angle, which prevents the observation of complete nulls in the far-field pattern. The diffraction spread of the light beam can usually be kept small enough so that the acoustic-beam cross section can be measured to high accuracy.

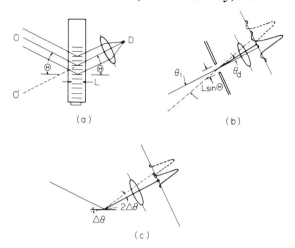

FIG. 16. Fourier-transform properties of Bragg diffraction. (a) Light from O diffracted at the Bragg angle appears at detector D to come from virtual source O'. (b) An equivalent slit produces Fraunhofer diffraction. Light incident at an angle θ_i to the normal produces a diffraction pattern centered at the same angle on the output side. A detector located at position denoted by θ_d observes the corresponding amplitude of diffracted light. (c) Replacing the slit with a mirror shows that the diffraction pattern can be observed by rocking the mirror angle while keeping source and detector positions fixed.

Cohen and Gordon (1965) have demonstrated this technique for elastic waves up to 250 MHz in fused quartz. The $(\sin x/x)^2$ pattern resulting from a single transducer was observed. Experiments were also conducted with two transducers generating parallel sound beams. In this case, evaluation of Eq. (42) shows that the far-field pattern of the diffracted light has the form of a two-slit diffraction pattern, and the $(\sin x/x)^2$ behavior is superimposed on the pattern, $\cos^2 \frac{1}{2}[K(\theta - \Theta)W + \psi]$, where W is the transducer spacing and ψ is the electrical phase shift of the driving signal between the two transducers. Quate *et al.* (1965) have extended these experiments to micro-wave-frequency elastic waves, using single crystals of several materials. Dixon and Gordon (1967) have used optical heterodyne detection of the diffracted light (as will be described below) to observe the acoustic field of a transducer. Heterodyne detection increases the sensitivity and, by homo-dyning the detector output with the input signal to the transducer, the phase as well as the amplitude of the acoustic field can be determined.

Korpel (1966, 1968) has described a technique to produce an optical image of a two-dimensional sound field. The principle of operation is illus-trated in Fig. 17. The acoustic point source S may represent a Huyghens source of the sound field. Rays SA, SB, and SC are representative of the acoustic radiation from this source. An optical source O illuminates the sound field with diverging light extending over a wide range of angles of incidence. Rays OA, OB, and OC intercept SA, SB, and SC at the Bragg

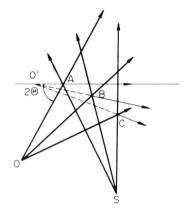

Fig. 17. Optical imaging of an acoustic field by Bragg diffraction. Acoustic rays from source S diffract light rays from source O at points A, B, C, creating an optical virtual image O' of the acoustic source.

angle, respectively, creating diffracted light beams. Extrapolation shows that the diffracted light beams appear to diverge from a virtual source at O', which thus represents an optical image of acoustic source S. A distribution of acoustic sources is imaged by a distribution of virtual optical sources. The diffracted light for each acoustic source point is proportional to the corresponding acoustic amplitude. The phase of the acoustic field is also preserved in the imaging process. A geometrical construction (Korpel, 1966) or two-dimensional wave-interaction theory (Korpel, 1968) shows that the image of the acoustic strain field is decreased in size from the actual acoustic pattern by the wavelength ratio λ/Λ. Similar imaging properties are obtained if the optical illumination is converging to point O, except that O' is then a real optical image of the acoustic source. Korpel (1966), Hance *et al.* (1967), and Landry *et al.* (1969) have used this technique to image the acoustic field in water at frequencies up to 100 MHz. Tsai and Hance (1967) used the same method of imaging at 923 MHz in a rutile crystal.

D. Elastooptical Light Modulators

The elastooptical effect can be used to construct a variety of light modulators. Both amplitude modulation and frequency translation can be achieved. The details of the design depend on whether operation is in the Raman–Nath regime or the Bragg regime, as determined by the parameter Q defined in Eq. (22). Other important parameters include the selection of materials and the experimental conditions, such as those which establish the element of the elastooptical tensor responsible for the modulation. In this section, modulators operating in the Bragg regime $(Q > 1)$ will be emphasized, since only high-frequency modulators can give the very wide bandwidths which are of interest. We begin, however, with a summary of low-frequency modulators. Note that the discussion here concerns devices which uniformly modulate the entire optical-beam cross section, rather than techniques to modulate the optical wavefront with the spatial amplitude variation of the acoustic beam. This latter technology will be described in Section V on optical signal processing.

1. Raman–Nath Modulators

The earliest elastooptical modulators used the Raman–Nath diffraction of light, usually in liquids or isotropic solids. Bergmann (1938) gives a good description of some of the first devices. Consider a traveling sound-wave carrier of angular frequency Ω_0 which is amplitude-modulated at some lower frequency. According to Eq. (51), the diffracted light intensity is a function of the acoustic power. This light is removed from the undiffracted beam and a modulator can be constructed by focusing the undiffracted light on a slit, thereby blocking the diffracted light. All of the light is diffracted when $\frac{1}{2}(\delta\varepsilon/\varepsilon_0)kL/\cos\theta = 2.4$, which is the first zero of the zero-order Bessel function. It is often preferable, in order to obtain 100% modulation at lower power levels, to block the central beam and use the diffracted light. The configuration is shown in Fig. 18.

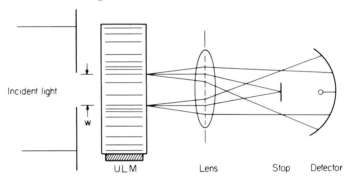

FIG. 18. Ultrasonic light modulator operating in the Raman–Nath region using a stop to block undiffracted light.

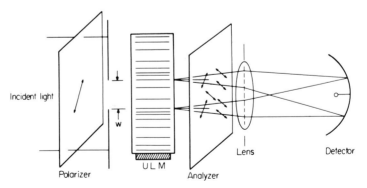

FIG. 19. Ultrasonic light modulator operating in the Raman–Nath region using crossed polarizers to block undiffracted light while transmitting diffracted light of rotated polarization.

A significant advantage is obtained by using an ultrasonic light-modulation scheme that produces a 90° change in the plane of polarization of the diffracted light relative to the incident beam. The configuration is shown schematically in Fig. 19 with the polarizer and analyzer arranged for longitudinal waves in the modulator. In this case, the axes of the perturbed index ellipsoid are at 0° and 90° and light is polarized at 45° to the direction of ultrasonic propagation. The elastic strain wave diffracts light into the perpendicular polarization and this is transmitted by the analyzer. A similar arrangement is used for shear waves except that the polarizer is aligned parallel or perpendicular to the acoustic wavefronts. Carleton and Maloney (1967) have compared the two cases and have shown that, for a given value of elastooptical coefficient, the transverse-wave configuration has a larger figure of merit than the longitudinal-wave device.

The Raman–Nath theory neglected refraction of light by the index gradient caused by the sound wave; hence, the theory applies only to small modulator thicknesses, such that $Q^2 \ll 1$. This restriction has been relaxed by Minkoff *et al.* (1967), who show that the internal refraction establishes an optimum transducer depth, $L_{opt} = \Lambda^2/\lambda$. This value of L yields the largest first-order light intensity. Note that L_{opt} leads to operation at approximately the boundary between the Raman–Nath and Bragg regimes. For 100-MHz longitudinal waves in fused silica, the optimum transducer depth is 0.8 cm.

Adrianova (1962), Gordon and Cohen (1965), and Hance and Parks (1965) showed that the maximum modulation frequency of the ultrasonic light modulator is limited by the time τ required for the sound wave to move across the light beam. It is evident that, if a step-function elastic wave enters the light beam, the transition from no diffraction to full diffraction will occur over a time $\tau = w/V$. The response to a sinusoidally modulated signal is obtained by use of the modulation transfer function; the bandwidth varies inversely with w. Clearly, it is desirable to use a focused light beam of small width. One limitation to the size of w is that the beam must overlap several acoustic wavelengths if the diffraction-grating model is to apply. Unless the diffracted beam can be selected by polarizers, a second limitation to the size of w occurs because the modulation can be observed only if the diffracted order is separated from the main beam. Thus, the diffraction angle must exceed the spreading of the narrow beam, which imposes the limit $\Omega_0/2\pi > 1/\tau$. These requirements usually lead to operation in the Bragg regime and further consideration of the bandwidth will be deferred to the following discussion on Bragg modulators.

Most experiments on Raman–Nath modulators have used isotropic materials such as liquids, fused quartz, or glass. Adrianova (1962, 1963) has measured in some detail the performance of modulators of the types illustrated in Figs. 18 and 19. The modulation bandwidth varied as expected with the transit time of sound across the light beam. The linearity of response to the electrical input signal was also determined, for both traveling-wave and standing-wave modulators. Modulation bandwidths up to 100 kHz with 5-MHz ultrasonic waves were obtained. Raman–Nath intensity modulators have been used primarily in applications where relatively narrow bandwidth is acceptable. De Maria and Gagosz (1962) placed liquid cells, operating with standing waves of about 100 kHz, within the optical cavity of a ruby laser. The diffracted light is removed from the optical feedback system of the laser. In a standing-wave cell, the medium becomes homogeneous twice in each period, so that the gain is modulated at twice the acoustic frequency. By this means, the random output pulses of the free-running laser were synchronized with the modulator frequency. A ruby laser was Q-switched by De Maria *et al.* (1963) by adjusting one laser mirror to an angle $\theta \sim \lambda/\Lambda$, thereby preventing oscillation except in the presence of an ultrasonic wave. De Maria (1963) performed similar experiments by using the diffraction of light by traveling acoustic waves at a frequency of 4 MHz. The ultrasonic modulator has also been used to mode-lock lasers. Hargrove *et al.* (1964) placed an ultrasonic cell inside the cavity of a He–Ne laser and stabilized the oscillation by modulating the optical gain at the mode-locking frequency by diffracting light from the central order. Caddes *et al.* (1968) used the same technique to mode-lock a CO_2 laser. Foster *et al.* (1965) placed an external modulator between the laser cavity and an additional mirror set at the diffraction angle. The mirror returned the diffracted light to the laser and mode-locking was obtained when the frequency shift was equal to the frequency difference between adjacent laser modes.

A number of workers (Broneus and Jenkins, 1960; Liben and Twigg, 1962; Distler, 1966; Brienza and De Maria, 1966, 1967; Cahen *et al.*, 1968) have used acoustooptical modulators to construct variable delay lines, by changing the distance from the transducer to the position of the light beam. By high-speed scanning of the light beam along the acoustic path, Brienza (1968) has performed time compression, expansion, and reversal of the rf modulation. Heterodyne techniques are usually employed to obtain adequate depth of modulation.

2. Bragg Modulators

From Eq. (51), it is seen that the fraction of light diffracted, at a given acoustic power level, increases monotonically as the cell thickness increases provided that the argument of the Bessel function is less than 2.4. The preceding discussion of modulator bandwidth indicated that broad bandwidth is attainable only by operating with high ultrasonic frequencies. Thus, the requirements of efficiency and broad bandwidth both lead toward the condition (defined in Eq. 22) $Q = 2\pi\lambda L/\Lambda^2 > 1$, the criterion for operation in the Bragg regime. In this case, as described earlier, the incident light is diffracted into only one order, and even this occurs with acceptable efficiency only when the directions of incident and diffracted light are approximately symmetrical with respect to the acoustic wavefronts. The angle between the diffracted and undiffracted light beams is twice the Bragg angle.

a. Bandwidth. For infinitely wide beams of sound and light, Eq. (68) shows that the diffracted light is confined to a single direction. For a given angle of incidence and the corresponding angle of diffraction, the Bragg condition can be satisfied at only one elastic-wave frequency. It is important to be able to operate modulators (and also deflectors, to be discussed below) over a wide modulation bandwidth. The operating frequency can be adjusted by changing the angles of incidence and diffraction to satisfy the Bragg condition. By differentiating Eq. (75), one obtains the diffraction bandwidth

$$\Delta f = [2V(\cos \theta)/\lambda]\, \Delta \theta \tag{94}$$

where $\Delta \theta$ is the change in angle of incidence and diffraction required to satisfy the Bragg conditions when the acoustic frequency is changed by Δf. Such adjustment is not feasible if a wide instantaneous bandwidth is required, as in the case of a modulated acoustic wave. One way to relax this limitation is to use the diffraction spreading of finite beams of sound and light. The removal of energy from the undiffracted beam reduces the intensity of this beam in proportion to the acoustic intensity, so the light contains the modulation information. This can be detected with a square-law detector, but the depth of modulation is usually relatively low for reasonable levels of acoustic intensity at high frequency. Although some applications of this type of modulator will be described later, the following discussion considers only modulation of the diffracted light. The argument follows that of Gordon (1966b).

Consider first the case shown in Fig. 20(a) of a wide optical beam with small spreading angle $\delta \theta \approx \lambda/w$ and a narrow acoustic beam with large

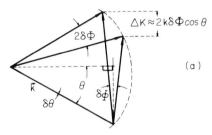

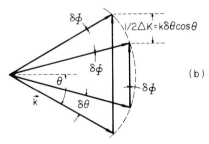

Fig. 20. Vector diagrams for Bragg diffraction by finite beams with acoustic beam spreading angle $\delta \Phi$ and optical beam spreading angle $\delta \theta$. (a) Narrow acoustic beam, $\delta \Phi \gg \delta \theta$. (b) Narrow optical beam, $\delta \theta \gg \delta \Phi$. After Gordon (1966b).

spreading angle $\delta\Phi \approx \Lambda/L \gg \delta\theta$. The diverging beams can be represented by plane waves with propagation directions lying within the respective spreading angles. For a fixed angle of incidence of the light beam, the Bragg condition can be satisfied over a range of frequencies corresponding to the allowed values of $\mathbf{K}$ within $\delta\Phi$. At each acoustic frequency, a different plane wave with a different direction of $\mathbf{K}$ is responsible for the diffraction, and the direction of the diffracted light ray is different for each frequency. At a fixed observation point, according to Eq. (78), the intensity of the diffracted light is reduced 4 dB when $\Delta\theta = \delta\Phi$. Combining Eq. (78) and Eq. (94) leads to the bandwidth between 4-dB points,

$$\Delta f = [2V(\cos\theta)/L]\Lambda/\lambda \tag{95}$$

This configuration of a highly collimated optical beam and a diverging acoustic beam is precisely what is required for light deflectors with a large number of resolvable positions, as will be discussed later. It does not meet the requirements of an intensity modulator, however. The frequency of the light in each separate diffracted beam is shifted by the corresponding acoustic frequency, and, in order to recover the intensity modulation of the diffracted light, these beams must be mixed in a nonlinear detector. This mixing takes place only if the beams are collinear at the detector. Collinear rays of different frequencies exist only within the spreading angle of the light beam, so the modulation bandwidth is determined by the diffraction spread of the light. Thus, although the large divergence of the acoustic beam permits light diffraction over a wide range of acoustic frequencies, it does not increase the modulation bandwidth and wastes acoustic power by diffracting light into unusable directions.

In the converse situation, illustrated in Fig. 20(b), the acoustic beam is well-collimated and the optical beam diverges. Separate rays of the optical beam satisfy the Bragg condition for different acoustic frequencies, but rays diffracted by sound waves of different frequencies travel in different directions. The spreading angle of the diffracted light at each frequency is reduced to approximately the spreading angle of the acoustic beam. This determines the modulation bandwidth and the remaining light is wasted.

It is plausible to conclude from this argument that the optimum condition occurs when the acoustic and optical beams diverge by approximately the same angle. This analysis has been put in quantitative form by Gordon (1966b). The resultant modulation bandwidth is about half the total frequency range for Bragg diffraction with a wide optical beam, obtained in Eq. (95). Taking $\delta\theta = \delta\Phi$, or $\lambda/w = \Lambda/L$, gives the modulation bandwidth

$$(\Delta f)_m = \tfrac{1}{2}\,\Delta f = (V/w)\cos\theta \tag{96}$$

equal to the reciprocal of the acoustic transit time across the optical beam. The maximum fractional bandwidth is usually determined by the condition that the diffracted light does not interfere with the central beam, which requires that $\Delta\theta < \Theta$. Thus, from Eq. (94), approximately $\tfrac{1}{2}\,\Delta f < \Lambda\Theta/\lambda =$

$V/2\Lambda$, or

$$(\Delta f)_m/f_0 \cong \tfrac{1}{2} \Delta f/f_0 < \tfrac{1}{2} \tag{97}$$

In the above discussion of bandwidth limitations, it has been assumed that the incident and diffracted light were of equal wave vector. This is not necessarily the case in birefringent materials. As described in Section II,E, the Bragg condition is modified in these materials if the elastooptical interaction changes the polarization of the diffracted beam from that of the incident beam. In particular, from Fig. 5, it is seen that the angle of incidence may be nearly constant over a broad range of acoustic frequencies, while the angle of diffraction varies according to the conventional Bragg condition. The converse condition, with nearly constant direction of the diffracted light, is also possible. It might appear that anisotropic materials can be used to diminish the requirements of large beam spreading in a broad-bandwidth intensity modulator. It follows from arguments similar to those for the isotropic case that this is not so, however. In the case that the direction of the incident light is constant, the diffracted light is spread over a range of angular deflection given by the Bragg condition and mixing cannot be obtained unless the various components arrive collinearly at the detector. This condition is achieved only for spreading angles of light and sound comparable to the isotropic case. It may be noted, however, that, even for small spreading angles of both the light beam and sound beam, the diffracted light is removed from the direct beam and this modulation could be observed over a large bandwidth. Conversely, if the input beam is focused to a small diameter, corresponding to a spread of the incident light equal to the Bragg angle over the desired bandwidth, the diffracted beam has small spreading angle and can produce wideband intensity modulation. It is only necessary that the spreading of the acoustic beam be equal to the small angular range of the output beam in order to obtain beam overlap and intensity modulation. In this case, however, use of the acoustic power is inefficient because each acoustic frequency component diffracts only light traveling in one direction. Optimum performance is obtained in this case, also, if the acoustic beam spread is equal to the change in Bragg angle over the modulator bandwidth. Thus, birefringent materials offer no significant advantages for intensity modulators.

b. Efficiency. We consider next the question of modulation efficiency. From Eq. (31), the fraction of light diffracted by the sound wave at the Bragg angle is

$$\eta = \frac{I_{+1}}{I_0} = \frac{\pi^2}{2} \left[\frac{P_a/LH}{\rho V^3} \right] \left(\frac{n^3 p L}{\lambda_0 \cos \theta} \right)^2 \tag{98}$$

where $P_a = LHI$ is the acoustic power in a beam of intensity I with width L and height H, and λ_0 is the wavelength of light in air. Equation (31) was derived in a linear approximation. For operation in the Bragg region at larger elastic strain amplitudes, the diffracted light intensity varies as

$\sin^2 \eta^{1/2}$ (Gordon and Cohen, 1965). The quantity

$$M_2 = n^6 p^2 / \rho V^3 \tag{99}$$

has been used by Smith and Korpel (1965) and others as a figure of merit for materials for operation at a single frequency under Bragg conditions. The values of p and V must be properly selected according to the direction and polarization of the light and sound.

A more useful figure of merit should include the modulator bandwidth, since this is often an important design parameter. The diffraction efficiency η increases with the acoustic beam width, according to Eq. (98), but the spreading angle and bandwidth vary inversely with the beam width, as shown by Eq. (94). Thus, some design compromises are possible. The bandwidth of a Bragg intensity modulator is given by Eq. (96). This value is half the available bandwidth between half-power points of the diffracted optical power for isotropic materials, given by Eq. (95). Gordon (1966a) has shown that a quantity independent of the modulator width L is

$$2\eta f_0 \, \Delta f = \left(\frac{n^7 p^2}{\rho V}\right) \frac{2\pi^2}{\lambda_0{}^3 \cos \theta} \frac{P_a}{H} \tag{100}$$

The factor

$$M_1 = n^7 p^2 / \rho V \tag{101}$$

thus provides a figure of merit for materials to be used in modulators (and deflectors). Varying the width of the acoustic beam L at a constant power, center frequency, and beam height allows tradeoff between the diffraction efficiency η and the dynamic bandwidth Δf. It should be emphasized, however, that the bandwidth used here is valid for intensity modulators only if the optical beam spread is less than or at most equal to the acoustic beam spread. Otherwise, M_2 is the appropriate figure of merit.

In both Eq. (98) and Eq. (100), it was assumed that the acoustic beam height H is larger than the optical beam diameter and is fixed. If the acoustic beam height can be as small as the diameter of the light beam, then $H = w = 2V/\Delta f$, from Eq. (96). Using the relation $\Lambda/L = \lambda/w$ for equality of optical and acoustic spreading angles, Dixon (1967b) showed that the quantity

$$f_0 \eta = \left(\frac{n^7 p^2}{\rho V^2}\right) \frac{\pi^2}{\lambda_0{}^3 \cos \theta} P_a \tag{102}$$

is independent of the acoustic and optical beam dimensions, and a third figure of merit is

$$M_3 = n^7 p^2 / \rho V^2 \tag{103}$$

Values of M_1, M_2, and M_3 for a number of materials have been given in Table I. As will be seen in the following section, these parameters are also applicable to acoustooptical light deflectors.

It is evident from this discussion that no single figure of merit is suitable for all applications. Although M_1, M_2, and M_3 are expressed in terms of material parameters, they are not truly figures of merit for materials. The specific combination of material parameters depends on the particular device configuration and specifications of interest, and there are, in fact, a number of alternative formulations which have not been considered here. For example, it was assumed in the derivation of Eq. (100) that $\eta \sim L$ and $\Delta f \sim L^{-1}$. The latter assumption is valid if the bandwidth is determined by the spreading of the acoustic beam. For a focused light beam, however, the beam diameter varies with distance. If L is too large, the beam diameter increases and this limits the modulation bandwidth. Hance and Parks (1965) have shown that the length of the plane-wave region around the focal spot is proportional to $w^2/n\lambda_0$. With this limitation to L, a suitable figure of merit which is independent of L is $\eta(\Delta f)^2$, which is proportional to $n^5p^2/\rho V$. As a further example, it will be shown later [Eq. (114)] that the diffraction bandwidth for anisotropic materials varies as $(n/L)^{1/2}$ and this dependence further modifies the various figures of merit. These alternative formulations do not usually change the ranking of materials, but it is clear that the tabulated figures of merit for various materials must be used with caution and with full consideration of the intended application.

In constructing a modulator for a television display, Korpel *et al.* (1966) have used an adaptation of the Scophony system described earlier to achieve high diffraction efficiency without sacrificing the effective modulation bandwidth. With an acoustic beam 18 mm wide and 3 mm high, in water, more than 95% of the light was diffracted with less than 1 W of power. The video signal was modulated on a carrier at 45 MHz. Using a laser beam with a Gaussian intensity distribution of 2.8 mm diameter, the diffracted light was reduced 6.7 dB at 3 MHz bandwidth. The spreading angles of light and sound were about equal, so increasing the bandwidth by reducing the optical beam diameter would require reduction of L, with concomitant decrease in diffraction efficiency. The spreading angle of the sound could instead be increased by lowering the acoustic carrier frequency, but this approach would require transducers with greater fractional bandwidth, and was rejected. Instead, the light-beam diameter was increased to illuminate several television picture elements in the modulator. The image of these picture elements, which moved at the speed of sound, was kept stationary on the screen by directing the image through a linearly scanning light deflector. This had the effect of reducing the required modulator bandwidth and permitted the use of a thick cell, thereby retaining the high diffraction efficiency.

c. Frequency Translation and Optical Heterodyne Detection. It was noted in Eq. (38) that the diffracted light is shifted in frequency by multiples of the acoustic frequency. The frequency is increased or decreased, depending on whether the acoustic waves are approaching or receding from the light source. The intensity of each optical frequency component depends on the acoustic power at the corresponding acoustic frequency. The narrow spectrum of the laser permits observation of this frequency shift and its use to construct optical frequency translators, or single-sideband, suppressed-carrier modulators. The diffracted light may be mixed with a portion of the undiffracted light to obtain heterodyne detection. In optical heterodyne detection, the output is a linear function of the diffracted light amplitude. Goodwin and Pedinoff (1966) have pointed out that this provides a more sensitive detector than square-law devices which respond to the intensity, provided the shot noise of the local oscillator (undiffracted light) exceeds the detector noise.

Although heterodyne detection can be used in the Raman–Nath region, as demonstrated with a liquid cell by Cummins *et al.* (1963) and De Maria (1963), it is usually of more interest for detecting high-frequency sound waves with light incident at the Bragg angle. The construction of a heterodyne detector for monochromatic ultrasonic waves is relatively straightforward. Cummins and Knable (1963) have used a Mach–Zehnder interferometer, with a collimated optical beam interacting with a collimated acoustic beam.

A novel method for reinserting the undiffracted light beam to lie along the diffracted beam, where it can serve for heterodyne mixing, was described by Dixon and Gordon (1967). The technique used the Kösters prism as a beam splitter and path equalizer to provide optically identical paths for the two beams, giving mechanical convenience compared to the Mach–Zehnder interferometer. This configuration also provides two output signals which can be used in a balanced mixer configuration. Another form of balanced optical heterodyne detector, useful when the signal and local oscillator beams have orthogonal polarization, has been described by Carleton and Maloney (1968).

When the Bragg angle is well-defined, as it is for a single acoustic frequency, the use of optical and acoustic beams with small spreading angles is acceptable. For broad instantaneous bandwidth, however, each frequency component travels in a different direction and heterodyne detection is possible only if part of the undiffracted light is made to be collinear with each component of the diffracted light. It is thus evident that finite beam spread of both optical and acoustic beams is necessary, much as was shown for intensity modulators. A detailed analysis of the performance has been given by Dixon and Gordon (1967). The derivation of the optimum design for a given bandwidth is similar to that described for intensity modulators and, as before, maximum efficiency results when the acoustic beam spreading angle is approximately equal to the optical beam spreading angle. The dynamic bandwidth, as in Eq. (96), is approximately equal to the transit time of the acoustic wave across the light beam.

The principal results of this analysis were verified by Dixon and Gordon (1967). In particular, it was shown that the modulation bandwidth varies inversely with the optical beam diameter and that the detected signal at the modulation frequency goes through a broad maximum when the spreading angles of light and sound are equal. The increased sensitivity of heterodyne detection of the diffracted light compared to square-law detection was also demonstrated. In the small-signal approximation, the depth of modulation for heterodyne detection is approximately $4\eta^{1/2}$, compared to η for square-law detection. Thus, for a diffraction efficiency $\eta = 10^{-4}$, as observed by square-law detection, the optical heterodyne detector yields a depth of modulation of 4%.

The experiments of Dixon and Gordon (1967) used 300-MHz sound waves in fused quartz. Using both fused quartz and rutile crystals, Quate *et al.* (1965) observed frequency translations up to 3000 MHz for the diffracted light. Heterodyne detection was used by Brienza and De Maria (1967) to construct a variable delay line using acoustic waves of frequencies up to 900 MHz. Siegman *et al.* (1964) increased the diffraction efficiency of the Bragg modulator by placing a fused quartz rod inside the resonant cavity of a laser and observing the light which was diffracted so that it passed by the laser mirrors and out of the laser cavity. Similar experiments have been reported by Spencer *et al.* (1967) using lithium niobate crystals placed inside the laser cavity.

E. Optical Beam Deflectors

The elastooptical effect provides some unique possibilities for high-resolution optical beam deflection. The important parameters for most applications are the number of resolvable beam positions and the speed of access to a specified position. Both random-access and continuously scanned systems are of interest. In this section, we shall describe several deflection techniques based on the elastooptical effect. The phenomena will be described in order of increasing elastic-wave frequency, from the low-frequency limit, where the acoustic wavelength is much longer than the optical beam diameter, up to the high-frequency limit or Bragg regime. Emphasis will be placed on the latter case, and especially on techniques based on the elasto-optical effect in solids.

1. Gradient Deflectors

A sinusoidal scanning device can be made using low-frequency elastic waves and an optical beam diameter much less than the elastic wavelength. The configuration is shown in Fig. 21. Kolb and Loeber (1954) showed that light traveling at right angles to a linearly varying index of refraction is bent in a radius of curvature $r = n/(dn/dx)$, where n is the refractive index and x is the direction of the gradient. In traveling a distance z through the material, the beam is deflected through an angle given by $\sin \theta = z/r = (z/n)\, dn/dx$ (Giarola and Billeter, 1963).

The passage of a sound wave through the material produces a spatially varying refractive index, $\Delta n = -\frac{1}{2}n^3 pe$. For a standing elastic wave of the form $e = e_0 \cos Kx \cos \Omega t$, the gradient of the index is

$$dn/dx = \frac{1}{2}n^3 pe_0\, K \sin Kx \cos \Omega t \tag{104}$$

where p is the appropriate elastooptical coefficient.

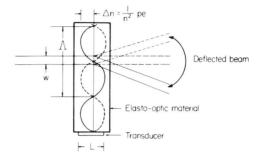

FIG. 21. Configuration for an acoustooptical beam deflector using the time-varying gradient of the refractive index with $\Lambda \gg w$.

In the construction of a light deflector, the light beam is directed at normal incidence through a nodal point of the standing elastic strain field, where the gradient is linear and largest. At this position, the gradient varies periodically at the elastic frequency, sinusoidally scanning the light beam at frequency $\Omega/2\pi$. Including refraction of the exit beam according to Snell's law, the peak deflection angle, for a cell of thickness L, is

$$\theta_0 = \tfrac{1}{2} L n^3 p K e_0 \tag{105}$$

An upper limit to the allowable deflection angle occurs because the beam is bent into regions of decreasing gradient until ultimately the gradient changes sign. This sets a limit $|\theta_{max}| L \leqq \Lambda/4$. It is also necessary to restrict the optical beam diameter so that the entire beam is subjected to a uniform gradient. This condition is $w < \Lambda/4$, and consequently the spreading angle of the light beam is $\delta\theta \approx \lambda/w > 4\lambda/\Lambda$. Combining this result with the limiting deflection angle θ_{max} gives the number of resolvable positions,

$$N_{max} = 2\theta_{max}/\delta\theta < \tfrac{1}{8}\Lambda^2/\lambda L \tag{106}$$

A more detailed analysis by Beiser (1967) has considered additional limitations to the operation of gradient deflectors which arise because of the optical system needed to focus the deflected beam onto the screen. A large optical aperture is required to avoid the limitations of diffraction beam spreading. Large-aperture systems introduce aberrations, however, which reduce the number of resolvable beam positions. If the lens follows the deflector, the aberration results from off-axis propagation through the lens. If the lens precedes the deflector, then the convergent light beam is distorted by the deflection process. The distortions increase with increasing aperture of the optical system, so that there is an optimum f-number, representing a compromise between deflection distortion and beam spreading, which gives the maximum number of resolvable spots. Beiser (1967) has shown that this optimum ranges from about $f/20$ to $f/30$ for scan lengths of 30 to 70 mm and that N_{max} varies under these conditions from about 1000 to 2000 elements per scan.

The principal limitation to useful deflectors of this type resides in the difficulty of attaining the required large gradient of the refractive index. Traveling-wave and standing-wave versions of the gradient deflector have been constructed. Aas and Erf (1964) constructed both a liquid cell and a crystal quartz cell. Operating at resonance, the quartz cell produced a maximum deflection angle slightly above 1° at 145-kHz scan rate before fracture occurred due to the high acoustic strain. The liquid cell produced 6° deflection at 320 kHz. The limiting resolution was not achieved in either case. Lipnick *et al.* (1964, 1965) used various liquids in both traveling-wave and standing-wave modulators, at frequencies up to 1 MHz. About 50 resolution elements were observed, at acoustic power levels of 15 W/cm², in the traveling-wave deflector.

2. Diffraction Deflectors

a. Performance Parameters. Consider now the case where the light beam is wide enough to extend over many wavelengths of the elastic strain. The dominant effect is diffraction of light by the periodic change in the refractive index. From Eq. (75), the angle between the undiffracted beam and the first-order sidebands is $2\theta = 2\Theta = 2\sin^{-1}(\lambda/2\Lambda)$. The diffracted beam is directed to different positions by changing the acoustic frequency. As in Eq. (94), for a frequency change Δf, the corresponding change in deflection angle is

$$2\,\Delta\theta = (\lambda/V\cos\theta)\,\Delta f \qquad (107)$$

For an incident light beam of spreading angle $\delta\theta = \lambda/w$, Eq. (107) gives the number of resolvable positions

$$N = 2\,\Delta\theta/\delta\theta = (w/V\cos\theta)\,\Delta f \qquad (108)$$

This indicates the desirability of using a wide light beam and a large acoustic bandwidth. If the bandwidth over which diffraction occurs is limited by the acoustic beam spread, as in Eq. (95), then

$$N = 2(\Lambda/L)(w/\lambda) \qquad (109)$$

A common figure of merit for a deflector is the number of separate positions to which the beam can be directed in unit time, expressed as a speed–capacity product. The access speed, as for the modulator, is

$$1/\tau = V(\cos\theta)/w \qquad (110)$$

Thus, the speed–capacity product is

$$N\tau^{-1} = \Delta f \qquad (111)$$

The desirability of using high-frequency sound waves for light deflectors is evident, so we henceforth emphasize devices operating in the Bragg regime

Much of the earlier discussion on modulators is applicable to the design of deflectors. The differences arise because it is neither necessary nor desirable in the deflector to have overlap of the light diffracted at different acoustic frequencies. Rather than constructing the system with $\delta\theta = \delta\Phi$, one should use a light beam of minimum spreading angle. In an isotropic material with fixed beam directions, the Bragg condition is satisfied over the bandwidth by the acoustic beam spread, which requires $\delta\Phi \geqq \Delta\theta$.

As an illustration of the performance, consider an acoustical system with bandwidth of 100 MHz. If the acoustic spreading angle is sufficiently large to obtain diffraction over this bandwidth, the speed–capacity product is $N\tau^{-1} = 10^8$. For a typical acoustic velocity of 4×10^5 cm/sec, an optical beam diameter of 1 cm will permit a random access time of 2.5 μsec to 250 positions, and an optical beam diameter of 0.1 cm will give access to 25 positions at a rate of 0.25 μsec. To achieve this performance requires the acoustic beam spread $\delta\Phi \geqq \Delta\theta = 7.5 \times 10^{-3}$ rad. An acoustic center frequency of 200 MHz yields $\Lambda = 2 \times 10^{-3}$, so that an acoustic beam width of 0.2 cm has a spreading angle $\delta\Phi = 10^{-2}$, sufficient to provide the results above. A smaller value of Λ/L would reduce the capacity as in Eq. (109).

Korpel *et al.* (1965) constructed an experimental light-deflection system using Bragg diffraction in water and achieved substantially the expected performance. The bandwidth was 5 MHz and with an effective optical beam diameter $w/\cos\theta = 2.2$ cm, corresponding to $\tau = 14.7 \times 10^{-6}$ sec, they observed about 70 resolvable beam positions. In this case, the maximum deflection angle was limited by the transducer bandwidth and was considerably less than the limit set by acoustic beam spreading.

Pinnow *et. al* (1969) made a two-dimensional deflector using lead molybdate crystals. Two similar stages, oriented orthogonally, were used. The high figure of merit for $PbMoO_4$ was mentioned earlier; the value of M_2 is over five times greater than that of $LiNbO_3$. With 1 W of electrical drive power, more than 50% of the incident light beam at 514.5 nm was deflected by longitudinal sound waves in the crystal. Each stage operated over a bandwidth of 80 MHz, from 90 to 170 MHz. The tradeoffs possible with this device can be obtained from Eq. (111).

The use of acoustic beam spreading to achieve high deflector capacity is limited by the decline in diffraction efficiency, proportional to N^{-1}. One means to avoid this limitation is to steer a collimated elastic beam so that the Bragg condition is satisfied at each frequency. This can be done by constructing an array of acoustical transducers driven by electrical signals with a progressive phase shift between successive transducer elements. The acoustic beam direction then varies with frequency. The phase shift between elements should be a nonlinear function of frequency in order to match perfectly the variation of the Bragg angle, but the Bragg condition is satisfied to first order by using a constant phase shift. It is most convenient to use a phase shift of π rad. This produces two symmetrical beams if a planar transducer is used, but one of these beams can be eliminated by constructing a stepped array with a fixed increase in time delay between successive elements. Korpel *et al.* (1966) have used this technique to obtain a bandwidth improvement exceeding a factor of three compared to the bandwidth available from acoustic beam spreading.

b. *Anisotropic Materials.* A significant improvement in the bandwidth, and thus the capacity, of light deflectors can be achieved by using birefringent materials. As described in Section II, the Bragg condition can then be satisfied over a broad frequency band without resorting to widely diverging or steered acoustic beams. Referring to Fig. 5, we note that the angle of incidence is nearly constant over a broad range of frequencies, while the angle of diffraction varies according to the conventional Bragg condition. The frequency at which this stationary condition occurs is found by differentiating Eq. (33), to be

$$f' = (2nB)^{1/2}V/\lambda_0 \qquad (112)$$

where λ_0 is the free-space wavelength and $B = n' - n$ is the birefringence. As shown in Fig. 8, at the frequency f', the diffracted light is approximately normal to the acoustic beam. Rewriting Eqs. (33) and (34) in terms of frequency, Dixon (1967a) has shown that the approximate relation between the frequency deviation $f - f'$ and the change in angle of incidence $\theta(f) - \theta(f')$ is

$$f - f' \approx 2f'\left\{[\theta(f) - \theta(f')]\cos\theta\,\frac{f'\lambda_0}{VB}\right\}^{1/2} \qquad (113)$$

As before, the range of angular change $\Delta\theta = \theta(f) - \theta(f')$ which yields high efficiency is set by the spreading angle of the sound beam $\Delta\theta = \delta\Phi = \Lambda/L = V/Lf'$. The total bandwidth is $\Delta f = 2(f - f')$, because the variation is an even function, approximately symmetrical about f'. Substitution in Eq. (113) gives

$$\Delta f = 2V\left[\frac{2n\cos\theta}{\lambda_0 L}\right]^{1/2} \qquad (114)$$

The output beam varies over the angle

$$\Delta\theta' = \frac{\lambda_0\,\Delta f}{nV\cos\theta} = 2\left[\frac{2\lambda_0}{nL\cos\theta}\right]^{1/2} \qquad (115)$$

Using $\delta\theta = \lambda_0/nw$, the number of resolution elements is

$$N = \frac{\Delta\theta'}{\delta\theta} = 2w\left[\frac{2n}{\lambda_0 L\cos\theta}\right]^{1/2} \qquad (116)$$

For comparison with the isotropic deflector, the ratio of Eq. (116) to Eq. (108) gives, using Eq. (112),

$$\frac{N_{\text{biref}}}{N_{\text{isot}}} = \frac{\Delta f_{\text{biref}}}{\Delta f_{\text{isot}}} = \frac{f'}{V}\left[\frac{2\lambda_0 L}{n\cos\theta}\right]^{1/2} = 2\left[\frac{BL}{\lambda_0\cos\theta}\right]^{1/2} \qquad (117)$$

so the advantage becomes greater at high frequencies. In most materials, B leads to $f' \approx 1$ GHz, which also necessitates use at high frequencies. The access time τ is set by the same conditions as for isotropic deflectors, so the bandwidth, capacity, and speed–capacity product are all increased by the factor of Eq. (117).

Lean *et al.* (1967) have experimentally demonstrated the improved performance. For the slow shear wave propagating along the x_1 axis in sapphire, the frequency $f' = 1.56$ GHz is the operating point for the diffracted light to be perpendicular to the acoustic beam. In accordance with Eq. (114) and as predicted by Eq. (117), the measured bandwidth was 550 MHz, an increase by a factor of 2.5 compared to an isotropic material. For corresponding conditions using shear waves in $LiNbO_3$ at a frequency of 3.6 GHz, the bandwidth is increased more than 30-fold by using the birefringent diffraction.

For the special case of repetitive scan, such as television, Korpel *et al.* (1966) have shown that the requirements of a light deflector are simplified. The operation is based on the observation by Gerig and Montague (1964) that a sound wave of linearly varying wavelength produces an effective cylinder lens which focuses the light at a position determined by the average acoustic wavelength. The focused spot is swept continuously across the screen as the sound wave passes through the light beam. The number of resolvable positions depends on the total bandwidth and on the diameter of the light beam, as in Eq. (108). For a repetitive scan, the diameter of the light beam can be selected so that the transit time of the sound wave is equal to the retrace time, 12.5 μsec for television images. This long transit time permitted Korpel *et al.* (1966) to achieve a resolution of 200 positions with a frequency sweep of only 16 MHz.

c. Efficiency. Several different measures of the efficiency of the elasto-optic interaction were described in Section IV,C. The figures of merit M_1, M_2, and M_3 were shown to be useful, depending on the conditions of operation. The same considerations apply to light deflectors and the proper figure of merit should be selected to weight appropriately the bandwidth and acoustic beam configuration. Some modification is required when acoustic beam steering is used to obtain large deflection. Gordon (1966a) has shown that the proper figure of merit in this case is $(\Delta f)^2 \eta$, which has the same value as $2f_0 \eta \, \Delta f$ given in Eq. (100) and M_1 is therefore the appropriate figure of merit.

V. Ultrasonics in Information Processing

A. Introduction

In the previous sections, we have seen that the interaction of sound with a light beam results in a modulation of the light beam. If the beam of light is wide and the acoustical line is long, a time-varying signal applied to the transducer at one end of the line results in a spatial modulation of the light beam as it emerges from the line. In short, because the light wave traverses the modulator in a fraction of an acoustic period, an ultrasonic light modulator can write a signal onto an optical wavefront. When this modulator is inserted in a coherent light beam and the combination is followed by appropriate optical filters (lenses, slits, stops, transparencies, etc.), real-time optical processing is possible. In what follows, we shall restrict our treatment of ultrasonic signal processing to these acoustooptical systems. The important, but restricted class of all-acoustical processors exemplified by the perpendicular diffraction delay line, the wedge delay line, and the interdigital-transducer surface-wave delay line will not be discussed.

Proposed applications of acoustooptical processors include real-time optical correlation, radar pulse compression, spectrum analysis, and complex waveform generation. While a digital computer is capable of processing a great deal of information, its speed is inherently limited by the fact that it must perform its operations sequentially. The optical processor, conversely, makes the entire signal available at once as a spatial modulation and is therefore capable of processing information "in parallel." It is this high-information-rate capability which is one of the principal attractions of the acoustooptical processor. Beyond this, some versions are capable of handling a variety of input waveforms, in contrast to dispersive electrical delay lines and the all-acoustic devices in which the waveform design is frozen in. The ability of many acoustooptical processors to change waveforms easily is a feature shared only by the digital machines.

The main application of acoustooptical signal processors to date has been in the field of radar. A radar pulse, to have good range resolution, must have large bandwidth. To have great range, it must have high energy content. When a short pulse is used to achieve large bandwidth, peak power is limited by generating capabilities or transmission-line breakdown. The transmission of a long, moderate-amplitude pulse whose frequency varies linearly in time allows the transmission of large energy with simultaneously large bandwidth. When the received echo (Fig. 22) is passed through a dispersive filter, the

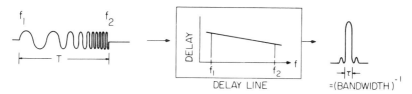

Fig. 22. In passing through a dispersive delay line, a linear fm pulse is compressed by a factor T/τ which is the time–bandwidth product—a figure of merit for processors. The dispersive line is a matched filter for this signal.

long, frequency-swept pulse is compressed to a short, high-peak-power pulse which has excellent range-discrimination capability. The enhancement in peak power is given by a factor variously called the compression ratio or the time–bandwidth product (TW) which is, as the second name suggests, the product of the original pulse length with the transmitted signal bandwidth. The ability of acoustooptical processors to perform the dispersive delay function at large TW products is the first of their advantages.

With such linear fm modulation, there remains an ambiguity between target range and velocity. For example, if the target is moving toward the antenna, the echo spectrum will be Doppler-shifted to higher frequencies and the time of the correlation maximum (interpreted as target range) will be changed. To circumvent this and other limitations, more sophisticated waveforms have been devised. Many forms of acoustooptical processors are able to handle any essentially band-limited signal, and some of these are able to generate the signal themselves. Moreover, the signal waveforms or codes are changeable in certain systems—in some, on a pulse-to-pulse basis. Thus, in a nutshell, the acoustooptical processor offers the advantages of wide bandwidth, large time–bandwidth product, and flexibility of coding.

The optimum[2] signal-to-noise ratio (S/N) is achieved when the input signal is passed through a matched filter, defined for our purpose as a filter whose response is the complex conjugate of the signal spectrum. The output of such a filter is the correlation integral of the input pulse with the impulse response of the filter. A dispersive delay line is a matched filter for the linear fm pulse. It will be shown that modifications of the processors can be operated as wideband spectrum analyzers.

In what follows, we shall take up, in turn: low-frequency $(f \leq 100 \text{ MHz})$ processors, high-frequency processors $(f \geq 100 \text{ MHz})$, and spectrum analyzers. Because so many approaches to low-frequency processors have been devised, it seems worthwhile to try to bring some unity to the field. We shall attempt this unification by leaning heavily on the operator view of optical filters (see Cutrona *et al.*, 1960; Vander Lugt, 1966; Mittra and Ransom, 1967; Papoulis, 1968; and Goodman, 1968). In this approach, we view the optical elements as filters or modulators and deduce the system functions representing them. A discussion of these operators will occupy the first part of the section. Various schemes for calculating the correlation integral will next be treated in terms of these operators. It is possible to make correlators coherent in the sense that the output is linearly proportional to the input signal and the phase of the rf input is preserved. This coherent correlation is achieved by heterodyning two light signals at the photodetector. Because the required signals are nearly always present (i.e., no external local oscillator need be injected) we will restrict our attention to the coherent case. The derivation of the squared envelope from the coherent output is easily accomplished. It should be clear in every case at which point a stop need be inserted to block the reference light so that only the quadratic output is obtained. In all acoustooptical processors, great care must be exercised to guarantee that the optical illumination is uniform across the processing aperture and that the interaction efficiency of light with sound is likewise constant across the aperture. The effect of Fresnel diffraction of the acoustic beam on this interaction is treated by Maloney *et al.* (1968) and Ingenito *et al.* (1967). For the present treatment, we will ignore such considerations and assume that both problems have been adequately solved.

The high-frequency processors will be treated more briefly. A proper

[2] This is strictly true only for random noise. See Turin (1960) and Cook and Bernfeld (1967).

mathematical treatment would begin from the induced polarization of Eq. (52). One would then calculate the first-order diffracted light for a given complex acoustic signal. We will prefer to describe the several approaches more in terms of geometrical optics and refer the reader to the mathematical treatments existing in the literature. With one exception, these processors are incoherent in that the output is proportional to the squared magnitude of the correlation integral. This one coherent correlator, in addition, is capable of handling signal formats other than linear fm.

We will now proceed to consider optical filters.

B. LOW-FREQUENCY PROCESSORS (THE RAMAN–NATH LIMIT)

1. General Discussion

Linear optical processing denotes a technique wherein a signal is modulated onto a collimated, monochromatic light beam so as to impress a spatial modulation $s(x, y)$ in a transverse plane. If the signal is a photographic transparency of uniform optical thickness, a pure amplitude modulation results. An acoustooptical modulator will, on the other hand, produce a phase modulation.[3] In what follows, we will restrict our discussion to one-dimensional signals $s(x, y) = s(x)$ and will be concerned with processing in only one transverse dimension. The modulated light is caused to pass through various optical filters which are realized as photographic transparencies, stops, apertures, corrugated mirrors, etc., all of which are designed to perform linear operations on the complex light amplitude. The emergent filtered light is then collected onto the cathode of a photodetector. A block diagram of an acoustooptical processing system sufficiently general for our purposes is shown in Fig. 23 together with a schematic representation of its functions.

Collimated, monochromatic light is incident from the left on the signal modulator which will be for the present an ultrasonic light modulator (ULM) operating in the Raman–Nath region of parameters ($Q \leq 1$). The Bragg limit of operation will be considered later. Under these conditions, the light is incident parallel to the acoustic fronts and the phase modulation results in output light diffracted symmetrically about the incident light direction. Moreover, we shall assume sufficiently weak acoustical drive so that the light output is adequately described by the lowest three diffraction orders $(0, \pm 1)$. An incident light wave[4] $e^{i(kz - \nu t)}$ is modulated by an applied acoustic signal $s(x - Vt)$ to produce at $z = 0$ a phase-corrugated light front

$$e(x, t) = \exp -i[\nu t - \alpha s(x - Vt)] \approx e^{-i\nu t}[1 + i\alpha s(x - Vt)] \qquad (118)$$

---

[3] We shall assume throughout this section that the Raman–Nath modulator introduces a simple phase modulation or corrugation of the light front. The possibility of gradient beam deflection is excluded (Bhatia and Noble, 1953; Hargrove, 1962; Minkoff *et al*, 1967).

[4] In this section, the symbol ν will be used for radian light frequency in place of the ω of previous sections. The symbol ω will be reserved for spatial frequencies in accordance with custom.

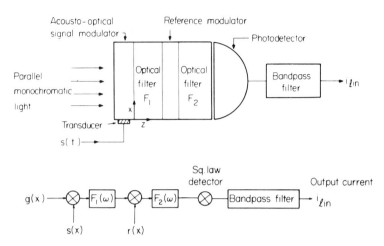

FIG. 23. Block diagram of a general acoustooptical processing system operating in the Raman–Nath limit. Filters F_1 and F_2 represent prereference and postreference filtering, respectively.

where α is defined to be a modulation index. The argument of s indicates a strain field propagating in the $+x$ direction at acoustic velocity V.

The modulated light front of Eq. (118) is then operated upon by filter F_1 which includes the effect of Fresnel diffraction in the region between the modulators as well as other processing functions. The next block represents a reference modulator which may consist of a transparency, coded mirror, or second ULM driven with a proper electrical reference signal. The reference modulator will usually be designed so as to perform the matched filtering function. As in the case of F_1, the second filter F_2 compensates for intervening space as well as more explicit filter functions. A photodetector converts the modulated light into an electrical output. All optical detectors are quadratic in the applied light amplitude. In operation, the detector calculates the light intensity at each point of the photocathode and then adds the intensities from all points. A linear output can be obtained only if two input light signals are incident on the same region of the cathode with light fronts parallel to within a quarter of an optical wavelength. Under these conditions, heterodyning is possible and the " cross term " will be linear in the amplitudes of the incident waves. The bandpass filter is usually employed to select this linear component of the current and to exclude dc and second-harmonic terms. With this diagram, it is possible to understand the basic principles and specific features of a number of coherent Raman–Nath processors.

Before we can proceed to specific systems, however, and in order to provide mathematical expressions for F_1 and F_2, it is necessary first to discuss the functional forms taken by various possible filters. We shall review some of the basic filtering operations rather briefly, since excellent treatments are available elsewhere, notably in Papoulis' fine book (1968). We will consider the following filters in turn: an interval of empty space, a thin convex lens, a slit, a half-plane 180° phase delay, a half-plane stop, the Zernicke phase plate. To simplify the treatment, all lenses will be assumed perfect and infinite in extent, i.e., lens apertures will be neglected. Similarly, the ULM aperture will be taken as infinite in x. These restrictions can be removed by convolving the infinite aperture signal with an appropriate aperture or pupil function $g(x)$.

2. Function Forms of Optical Filters

a. An Interval of Intervening Space of Width z. According to the Kirchhoff approximation to the Fresnel–Kirchhoff formula (Papoulis, 1968), the field $e(x, z, t) = \mathscr{E}(x, z)\,e^{-ivt}$ due to an aperture distribution $e(x_0, 0, t) = \mathscr{E}(x_0, 0)\,e^{-ivt}$ is derived from an integral

$$\mathscr{E}(x, z) \approx C \int_{-w/2}^{w/2} \mathscr{E}(x_0, 0)\,e^{ik\rho}\,dx_0 \tag{119}$$

where $\rho = [(x - x_0)^2 + z^2]^{1/2}$ is the distance from aperture point $(x_0, 0)$ to field point (x, z); $k = v/c$; $C = (k/8\pi\rho)^{1/2}\,e^{-i\pi/4}(1 + z/\rho)$ is a function of ρ and z, varying slowly enough to be taken outside the integral; and w is the width of the aperture. Equation ('119) is an approximation valid when $\rho, z \gg w$, i.e., throughout the Fresnel and Fraunhofer regions of the diffraction field.

At very large z, ρ may be expanded in terms of $r \equiv (x^2 + z^2)^{1/2}$, the distance from the aperture "midpoint" to the field point, retaining only terms linear in x,

$$\rho \approx r + \frac{xx_0}{r}, \qquad C \approx \left(\frac{k}{2\pi z}\right)^{1/2} e^{-i\pi/4} \tag{120}$$

With this approximation,

$$\mathscr{E}(x, z) \approx C e^{ikr} \int_{-w/2}^{w/2} \mathscr{E}(x_0, 0)\,\exp(-ikxx_0/r)\,dx_0 \tag{121}$$

and we obtain the familiar Fraunhofer or far-field diffraction pattern of $\mathscr{E}(x_0, 0)$. Equation (121) also reveals that, on the surface $r = \text{const}$, the far-field diffraction pattern is the spatial Fourier transform of the aperture. (Note: for this to be true, we assume that $\mathscr{E} = 0$ for $|x_0| > w/2$ so that we may extend the limits of integration to $\pm\infty$.) For small x, the transform is approximated in the plane $z = \text{const}$. If we introduce the transform variable $\omega = kx/r \approx kx/z$ and the convention of using capital letters for Fourier-transform quantities, we may write

$$\mathscr{E}(x, z) \approx C e^{ikr} E(\omega, 0) \tag{122}$$

We introduce the notation

$$\mathscr{E}(x) \leftrightarrow E(\omega) \tag{123}$$

to denote a spatial Fourier-transform pair.

It would be more straightforward to denote the transform variable conjugate to x by the symbol k_x rather than ω, especially in view of the fact that ω has dimensions of $(\text{length})^{-1}$. Nevertheless, it is very useful to interpret the transverse spatial position in the far-field as a spatial frequency and the ω-notation is sanctioned by custom. Note that the spatial frequency $\omega[L^{-1}]$ is not the acoustic frequency $\Omega[T^{-1}]$, although they are closely related. The spatial frequency ω is defined by

$$\omega = k_x = k \sin \theta \approx kx/z \tag{124}$$

In the special case of modulation by ultrasound of frequency Ω, the diffraction angle θ is given by $\theta \approx K/k$. We may therefore include this information in Eq. (124) to obtain the relation between ω and Ω,

$$\omega = K = \Omega/V \tag{125}$$

For z less than the far-field distance but still much larger than w, the Fresnel approximation is employed: $\rho \approx z + [(x - x_0)^2]/2z$, in which terms are kept to the second order in x. We obtain

$$\mathscr{E}(x, z) = Ce^{ikz} \int_{-w/2}^{w/2} \mathscr{E}(x_0, 0) \exp[ik(x - x_0)^2/2z] \, dx_0 \tag{126}$$

This expression is more compactly written as the convolution of $\mathscr{E}$ and ϕ,

$$\mathscr{E}(x, z) = \mathscr{E} * \phi \tag{127}$$

where we write $*$ for the convolution operation and define

$$\phi(x, z) \equiv Ce^{ikz} \exp(ikx^2/2z) \tag{128}$$

This quantity is known as the spread function or impulse response of free space. It represents the field due to a point source $\delta(x_0)$ at $z_0 = 0$. If we define the transform pairs

$$\mathscr{E} \leftrightarrow E, \qquad \phi \leftrightarrow \Phi \tag{129}$$

we may rewrite Eq. (127) as

$$E(\omega, z) = E(\omega, 0)\Phi(\omega, z) \tag{130}$$

using the fact that the transform of a convolution is simply the product of the transforms. It can be shown from Eq. (128) that

$$\Phi(\omega, z) = Be^{ikz} \exp(-iz\omega^2/2k) \exp^{-i\pi/4} = B' \exp(-iz\omega^2/2k) \tag{131}$$

with B and B' standing for unimportant constant factors and phase shifts. Thus, the light, in passing a distance z between two reference planes, suffers a phase shift proportional to the square of its spatial frequency. Physically, plane waves of larger ω are propagating at larger θ and therefore travel a longer distance (and accumulate more phase) than paraxial waves. Fresnel diffraction acts as an all-pass filter with quadratic phase. In many cases, this Fresnel diffraction is undesirable. It is, however, essential to the operation of the wide-aperture correlator described below in Section B,3c.

b. *Perfect Lens.* A perfect thin, positive cylinder lens introduces precisely the correct differential phase delay required to convert a plane incident wavefront into a cylindrically converging front, namely, a phase advance quadratic in x. Thus, for $\mathcal{E}(x, 0_-) = 1$ (a collimated beam) at the input plane of the lens, the emergent field at the output plane of the lens $\mathcal{E}(x, 0_+)$ should be $\mathcal{E}(x, 0_-) \exp(-ikx^2/2f)$, where f is the focal length of the lens. From this argument, we infer that the insertion of a perfect thin lens is equivalent to multiplying the incident light field by the factor $\exp(-ikx^2/2f)$.

As an illustration of the two previous concepts, we now combine the free-space problem with a perfect lens. Figure 24 shows a situation in which the input plane is displaced by an amount z to the left of the lens plane.

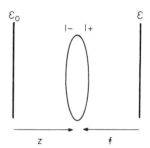

FIG. 24. Sketch for deriving the optical transfer function of an ideal lens.

From this illustration, we will be able to derive the important case of Fourier transformation by a lens operating between focal planes. Using Eq. (127), we find

$$\mathcal{E}(x, 1_-) = \mathcal{E}_0(x, 0) * \phi(x, z) \tag{132}$$

$$\mathcal{E}(x, 1_+) = [\mathcal{E}_0 * \phi(z)] \exp -ikx_1^2/2f \tag{133}$$

$$\mathcal{E}(x, z+f) = \{[\mathcal{E}_0 * \phi(z)] \exp -ikx_1^2/2f\} * \phi(f) \tag{134}$$

$$= e^{-i/\pi 2} \frac{k}{2\pi(zf)^{1/2}} e^{ik(z+f)}$$

$$\times \iint dx_0 \, dx_1 \, \mathcal{E}_0(x_0, 0) \exp ik(x_1 - x_0)^2/2z$$

$$\times \exp -ikx_1^2/2f \exp ik(x - x_1)^2/2f \tag{135}$$

$$= \left(\frac{k}{2\pi f}\right) e^{-i\pi/4} e^{ik(z+f)} \frac{\exp ikx^2(1 - z/f)}{2f} E(kx/f, 0) \tag{136}$$

We see that the field distribution in the back focal plane of a lens is the Fourier transform of that in the input plane, aside from a quadratic phase factor. If the input plane is coincident with the front focal plane, i.e., if

$z = f$, the phase factor vanishes, and $\mathscr{E} \propto E_0(\omega)$. Here, we have again introduced the spatial frequency ω by identifying it with kx/f. The field distribution in the back focal plane is, apart from unimportant constants or constant phase shifts, equal to the spatial Fourier transform of the field distribution in the conjugate focal plane. The positions in this "frequency" plane may be thought of as spatial frequencies ω.

c. *The Slit.* Since it has been established above that the amplitude distribution in the back focal plane of a thin lens is proportional to the spatial Fourier transform of the field distribution in the front focal plane, it is obvious that a slit in the back focal plane will block all but a predetermined range of spatial frequencies $\Delta\omega$ corresponding to the slit width. If the slit is movable, the essential feature of a spectrum analyzer is present.

d. *Modulation Conversion by Means of the Half-Plane Phase Delay and Half-Plane Stop.* Because the ULM is essentially a phase modulator, it frequently occurs that the light front arriving at the photodetector is phase-modulated and therefore, as we will see, invisible. By this we mean that the detector is incapable of producing an output current proportional to the first power of the signal. Several approaches which are available to remedy this situation will be discussed below, namely, the half-plane delay filter, the half-plane stop, the Zernicke phase plate, and Fresnel diffraction. We will take up the first two at once.

We begin by considering a plane monochromatic light front e^{-ivt} carrying a spatial modulation such that the signal at the detection plane may be written

$$e(x, t) = [1 + \mathscr{N}(x, t) \exp[-ivt + i\psi(x, t)] \tag{137}$$

where $\mathscr{N}$ indicates an amplitude modulation and ψ a phase modulation of the light carrier. For weak phase modulation $[\psi(x, t)_{\max} < 0.3]$, $e^{i\psi} \approx 1 + i\psi$ and Eq. (137) may be written

$$e(x, t) \approx [1 + \mathscr{N}(x, t) + i\psi(x, t)]\,e^{-ivt} = [1 + \tilde{\mathscr{N}}(x, t)]\,e^{-ivt} \tag{138}$$

where we have defined $\mathscr{N} + i\psi \equiv \tilde{\mathscr{N}}$. The average output current from the detector is

$$i(t) = \tfrac{1}{2}\operatorname{Re}\int_{-\infty}^{\infty} dx\, ee^{*} \tag{139}$$

The only term in Eq. (139) which is linear in the modulation amplitude is easily seen to be

$$i_{\text{lin}}(t) = \int_{-\infty}^{\infty} dx\, \operatorname{Re}\tilde{\mathscr{N}} \tag{140}$$

which vanishes unless $\tilde{\mathscr{N}}$ has a real part.

Thus, as we said above, only an amplitude modulation is capable of producing a linear output current. Therefore, before a phase-modulated light front can be detected, the modulation must first be converted to an amplitude modulation. To see how this conversion can be accomplished, we look more closely at the form of the modulation signal. We will restrict further consideration in this chapter to signals which are band-limited and whose positive and negative spectra do not overlap. Note that, because $\mathrm{Re}\,e(x, t)$ in Eq. (137) is a monochromatic incident wave modulated by a much lower-frequency elastic wave, it is also a band-limited signal, with spectrum closely concentrated about $\pm\nu$, where ν is the optical carrier frequency. We may similarly express $\tilde{\mathcal{N}}$ as a spatial carrier (really a subcarrier) at ω_0 modulated in both amplitude and phase. This signal is sketched in Fig. 25. The spectrum will be essentially contained in a band $\omega_0(1 - \Delta)$ $< |\omega| < \omega_0(1 + \Delta)$.

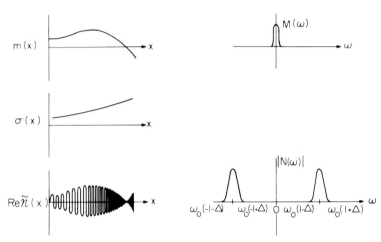

Fig. 25. Time dependence and spectrum of a band-limited, modulated carrier. Here, $m(x)$ and $\sigma(x)$ are the amplitude and phase modulation imposed on a spatial carrier of radian frequency ω_0. $\mathcal{N}(x)$ is the resultant modulated carrier. The corresponding spectra shown at the right demonstrate the band-limited nature of $\mathcal{N}(x)$.

For the case of the acoustooptical modulator (a phase modulator),

$$\tilde{\mathcal{N}}(x, t) = i\psi(x, t) = ias(x - Vt)$$
$$\mathrm{Re}\,\tilde{\mathcal{N}} = 0 \tag{141}$$

and the modulation is invisible. It can be rendered visible. Let us express this invisible modulation as a modulated carrier

$$\tilde{\mathcal{N}}(x, t) = im(x, t)\sin[\omega_0 x + \sigma(x, t)] \tag{142}$$

with $m(x, t)$ and $\sigma(x, t)$ representing, respectively, amplitude and phase modulation of the ω_0 (sub)carrier. The "i" here refers to the phase of the *light* modulation relative to the *light* carrier. It is *this* phase difference, between light carrier and light modulation, represented by "i," which renders the modulation invisible. The (sub)carrier ω_0 may be amplitude or phase modulated and no problem arises. We may rewrite Eq. (142) as

$$\tilde{\mathcal{N}} = \tfrac{1}{2}m[\exp i(\omega_0 x + \sigma) - \exp -i(\omega_0 x + \sigma)]$$
$$= \tfrac{1}{2}m[t_+ \exp i\omega_0 x - t_- \exp -i\omega_0 x] \tag{143}$$

where we have defined $t_+ = e^{i\sigma}$ and $t_- = e^{-i\sigma}$. Defining the following spatial transform pairs

$$\tilde{\mathcal{N}} \leftrightarrow N = N_+ + N_-, \qquad mt_\pm \leftrightarrow P_\pm \tag{144}$$

we find

$$N(\omega, t) = \tfrac{1}{2}[P_+(\omega - \omega_0) - P_-(\omega + \omega_0)] \tag{145}$$

Spectrum P_+ exists only in the range of $\omega > 0$ centered on $\omega = \omega_0$, and P_- exists only in the range of $\omega < 0$ centered on $\omega = -\omega_0$. The spectra P_+ and P_- represent the light diffracted by the modulator into $+1$ and -1 orders, respectively. If a transforming lens is inserted in the optical system after the modulator, these orders will actually appear in the ω-plane at the location given by $x = \omega f/k$. A spatial frequency filter can be inserted directly into the appropriate region of the physical ω-plane. Because P_+ is concentrated near $\omega = \omega_0$ and P_- near $\omega = -\omega_0$ and these plus and minus components do not overlap, we may introduce a 180° phase delay in half of the ω-plane, for example, by inserting a filter defined by

$$U_1(\omega) \begin{cases} = 1, & \omega > 0 \\ = -1, & \omega < 0 \end{cases} \tag{146}$$

Multiplication by $U_1(\omega)$ yields a new spectrum denoted $N'(\omega, t)$

$$N' = U_1 N = \tfrac{1}{2}[P_+ + P_-] \tag{147}$$

which has the inverse

$$\tilde{\mathcal{N}}'(x, t) = \tfrac{1}{2}m(x, t)[t_+(x, t) \exp i\omega_0 x + t_-(x, t) \exp -i\omega_0 x]$$
$$= \tfrac{1}{2}m[\exp i(\omega_0 x + \sigma) + \exp -i(\omega_0 x + \sigma)] \tag{148}$$
$$= m \cos(\omega_0 x + \sigma)$$

From Eq. (140), we at once obtain a nonzero output

$$i_{1in}(t) = \int_{-\infty}^{\infty} dx \, m(x, t) \cos[\omega_0 x + \sigma(x, t)] \tag{149}$$

Thus, the half-plane phase-delay method is capable of converting the modulation to achieve a linear output current. Removing all the negative spatial frequency components of the spectrum will also affect the modulation conversion. Suppose the spectrum N [Eq. (145)] is multiplied by a function U_2,

$$U_2(\omega) \begin{cases} = 1, & \omega > 0 \\ = 0, & \omega < 0 \end{cases} \tag{150}$$

representing a half-plane stop. The new spectrum is

$$\tilde{N}'' = U_2 N = \tfrac{1}{2} P_+ \tag{151}$$

Then

$$\mathscr{N}''(x, t) = \tfrac{1}{2} m(x, t) \exp i[\omega_0 x + \sigma(x, t)] \tag{152}$$

From Eq. (140), we find

$$i_{1\text{in}}(t) = \tfrac{1}{2} \int_{-\infty}^{\infty} dx \, m(x, t) \cos(\omega_0 x + \sigma) \tag{153}$$

which is exactly one-half the result of Eq. (148), as one might expect.

e. *Modulation Conversion by Means of the Zernicke Phase Plate.* Modulation conversion can also be accomplished by the Zernicke phase plate. To understand its operation, we refer to Eq. (138). If the phase of the first term could be shifted 90° relative to the third term, i.e., if a quarter-wave delay were inserted in the undiffracted light beam, a real output

$$i_{1\text{in}}(t) = \int dx \, \psi(x, t) \tag{154}$$

would be obtained. This is accomplished physically by introducing a 90° phase delay at the $\omega = 0$ position of the ω-plane. This principle is derived from the phase-contrast microscope of Zernicke (Born and Wolf, 1965) wherein the undiffracted light from a phase object is delayed 90° relative to the diffracted light to render a phase object visible.

f. *Modulation Conversion by Fresnel Diffraction.* The phenomenon of Fresnel diffraction may be employed to effect modulation conversion in a very direct and natural way (Hiedemann and Breazeale, 1959; Korpel *et al.*, 1967). Consider the modulation of Eq. (142) to exist in a given plane $z = 0$. We will suppress the time dependence but include z explicitly. In a plane $z = z$

$$\mathscr{N}(x, z) = \mathscr{N}(x, 0) * \phi(x, z) \tag{155}$$

or, taking transforms,

$$N(\omega, z) = N(\omega, 0)\Phi(\omega, z)$$

if we invoke the transform equivalence $\mathscr{N} \leftrightarrow N$. We now divide N into its plus and minus frequency components: $N = N_+ + N_-$ and define $\mathscr{N}_\pm \leftrightarrow N_\pm$. We may then write

$$N(\omega, z) = [N_+(\omega, 0) + N_-(\omega, 0)] \exp\left(-i\frac{z}{2k}\omega^2\right) \tag{156}$$

where we ignore constant factors. (We shall ignore all constant phase shifts and unimportant multipliers to simplify the equations.) For band-limited signals where N_+ and N_- do not overlap we may write

$$\omega^2 = \omega_0{}^2 + 2\omega_0(\omega - \omega_0) + (\omega - \omega_0)^2, \qquad \omega > 0 \tag{157}$$

$$\omega^2 = \omega_0{}^2 - 2\omega_0(\omega + \omega_0) + (\omega + \omega_0)^2, \qquad \omega < 0 \tag{158}$$

Then

$$N(\omega, z) = (\exp i \frac{z}{2k} \omega_0^2)\Big\{N_+(\omega, 0) \exp -i \frac{z}{2k} [2\omega\omega_0 + (\omega - \omega_0)^2]$$

$$+ N_-(\omega, 0) \exp i \frac{z}{2k} [2\omega\omega_0 - (\omega + \omega_0)^2]\Big\} \tag{159}$$

We now define a length Z such that, when $z = Z$, $\exp[i(z/2k)\omega_0^2] = e^{i\pi/2}$. That is, $Z = \pi k/\omega_0^2 = \pi k/K_0^2$, where we make use of the fact that the spatial frequency is related to the acoustic wave number by Eq. (125),

$$\omega_0 = k \sin \theta_0 = K_0 = 2\pi/\Lambda_0$$

For any odd number of intervals $z = (2n + 1)Z$, a factor of $e^{i\pi/2}$ enters, which we might suspect will yield the desired conversion. We may set $(\omega - \omega_0)/\omega_0 = \delta$, $\omega > 0$, and $(\omega + \omega_0)/\omega_0 = \delta$, $\omega < 0$, and write

$$N[\omega, (2n + 1)Z] = i^{(2n+1)}\Big\{N_+[\omega_0(1 + \delta), 0] \exp -i(2n + 1) \frac{\pi}{2} (2\delta + \delta^2)$$

$$+ N_-[-\omega_0(1 - \delta)] \exp i(2n + 1) \frac{\pi}{2} (2\delta - \delta^2)\Big\} \tag{160}$$

The terms in δ^2 contribute quadratic phase error (QPE). We may estimate its importance. At the band edges $\delta = \Delta$, we will take QPE to be negligible when $(2n + 1)(\pi/2)\Delta^2 < \pi/8$ (Meltz and Maloney, 1968), which is to say, $\Delta^2 < 1/4(2n + 1)$. With $n = 0$ (first amplitude plane), this condition is satisfied for $\Delta < \frac{1}{2}$—a fractional bandwidth of 50%. We may therefore drop the terms in δ^2 as negligible under normal operating conditions. The inverse transform may then be written from Eq. (159) as

$$\mathscr{N}[x, (2n + 1)Z] = i^{(2n+1)}\Big\{\tilde{\mathscr{N}}_+\Big(x - \frac{2n + 1}{2}\Lambda_0, 0\Big) + \tilde{\mathscr{N}}_-\Big(x + \frac{2n + 1}{2}\Lambda_0, 0\Big)\Big\} \tag{161}$$

This very important result shows that a free-space interval produces an automatic modulation conversion through the action of Fresnel diffraction. A phase (invisible) image at $z = 0$ is converted to an amplitude (visible) image at $z = (2n + 1)Z$ and back to a phase image at $z = 2nZ$. In any amplitude plane,

$$i_{\text{lin}} = (-1)^n \text{Re} \int dx \, i\Big[\tilde{\mathscr{N}}_+\Big(x - \frac{2n + 1}{2}\Lambda_0\Big) + \tilde{\mathscr{N}}_-\Big(x + \frac{2n + 1}{2}\Lambda_0\Big)\Big] \tag{162}$$

This response is the sum of two partial images, a plus-order image and a minus-order image, shifted laterally by one carrier wavelength. For small n, this sum closely approximates $\tilde{\mathscr{N}}_+(x) + \tilde{\mathscr{N}}_-(x)$. For arbitrary z [but small enough to validate neglect of $(\omega - \omega_0)^2$] and for complex modulation in the plane $z = 0$, we may write

$$\tilde{\mathcal{N}}(x, z) = \left(\exp i \frac{\pi}{2} \frac{z}{Z}\right)\left[\tilde{\mathcal{N}}_+\left(x - \frac{\Lambda_0}{2} \frac{z}{Z}, 0\right) + \tilde{\mathcal{N}}_-\left(x + \frac{\Lambda_0}{2} \frac{z}{Z}, 0\right)\right] \quad (163)$$

$$i_{\text{lin}} = \cos \frac{\pi}{2} \frac{z}{Z} \operatorname{Re} \int dx \left[\tilde{\mathcal{N}}_+\left(x - \frac{\Lambda_0}{2} \frac{z}{Z}, 0\right) + \tilde{\mathcal{N}}_-\left(x + \frac{\Lambda_0}{2} \frac{z}{Z}, 0\right)\right]$$

$$- \sin \frac{\pi}{2} \frac{z}{Z} \operatorname{Im} \int dx \left[\tilde{\mathcal{N}}_+\left(x - \frac{\Lambda_0}{2} \frac{z}{Z}, 0\right) + \tilde{\mathcal{N}}_-\left(x + \frac{\Lambda_0}{2} \frac{z}{Z}, 0\right)\right] \quad (164)$$

which more clearly displays the alternation of amplitude and phase planes with increasing z.

We here conclude our discussion of possible filter functions, operators, and modulation conversion techniques.

3. Specific Designs

a. General. Having developed the necessary operators, we will now consider several specific low-frequency optical correlators.

The purpose of any correlator is to calculate the cross-correlation function of a signal $s(x)$ with a reference function $r(x)$,

$$R(-Vt) = \operatorname{Re} \int_{-\infty}^{\infty} dx\, s(x - Vt)r(x) \quad (165)$$

We will now derive an alternative expression for Eq. (165) which will be useful later on. For band-limited signals, we may express s and r as

$$s(x - Vt) = \mathcal{S}_+(x - Vt) \exp i\omega_0(x - Vt) + \mathcal{S}_-(x - Vt) \exp -i\omega_0(x - Vt) \quad (166)$$

$$r(x) = \mathcal{R}_{+1}(x) \exp i\omega_0 x + \mathcal{R}_{-1}(x) \exp -i\omega_0 x + \mathcal{R}_{+2}(x) \exp i2\omega_0 x$$
$$+ \mathcal{R}_{-2}(x) \exp -i2\omega_0 x + \cdots \quad (167)$$

and with the definitions

$$\mathcal{S}_\pm(x) \leftrightarrow S_\pm(\omega), \qquad \mathcal{R}_{\pm l}(x) \leftrightarrow R_{\pm l}(\omega) \quad (168)$$

we have

$$R(-Vt)$$

$$= \int_{-\infty}^{\infty} dx\, [\mathcal{S}_+(x - Vt) (\exp -i\omega_0 Vt)\mathcal{R}_{-1}(x)$$

$$+ \mathcal{S}_-(x - Vt) (\exp i\omega_0 Vt)\mathcal{R}_{+1}(x)]$$

$$+ [\mathcal{S}_-\mathcal{R}_{+2} \exp i(\omega_0 x + Vt) + \mathcal{S}_+\mathcal{R}_{-2} \exp -i(\omega_0 x + Vt)]$$

$$+ [\mathcal{S}_+\mathcal{R}_{+1} \exp i(2\omega_0 x - Vt) + \mathcal{S}_-\mathcal{R}_{-1} \exp -i(2\omega_0 x - Vt)] + \cdots] \quad (169)$$

The second and third brackets are clearly small, since the space average is equivalent to evaluating the spatial Fourier transform at $\omega = 0$, i.e.,

$$\int dx\, f(x) = \int dx\, f(x)\, e^{-i\omega x}\Big|_{\omega = 0} \qquad (170)$$

The first bracketed term in Eq. (169) is the only one whose spectrum is centered on $\omega = 0$.

Since $s(x - Vt)$ and $r(x)$ are assumed to be real,

$$\mathscr{S}_-(x - Vt) = \mathscr{S}_+{}^*(x - Vt), \qquad \mathscr{R}_{-l}(x) = \mathscr{R}_{+l}^*(x) \qquad (171)$$

and we find

$$R(-Vt) = 2\mathrm{Re} \int_{-\infty}^{\infty} dx\ \mathscr{S}_+(x - Vt)(\exp -i\omega_0 Vt)\mathscr{R}_{-1}(x) \qquad (172)$$

If $r(x)$ were imaginary, $R_{-l}(x) = -R_{+l}^*(x)$ and $R(-Vt) = 0$. It will be useful to have the correlation integral expressed in these forms to facilitate comparison of correlator outputs.

Moreover, from Eq. (172), we may make a physical argument about the operation of low-frequency correlators. Under the assumptions we have made concerning the band-limited nature of the driving signal, the weakness of the interaction, and the Raman–Nath limit of operation, the light emerging from the ULM consists of three groups of plane waves. The zeroth order consists of the undiffracted light which continues along the $+z$ axis. Symmetrically disposed first orders comprise the modulation imposed by the ULM. The $+1$ set of plane waves is Doppler-shifted upward in frequency by the acoustic frequency; the -1 set is down-shifted. In passing through a reference modulator, these three spatial components are rediffracted. When the reference modulation is exactly matched to the signal modulation, the rediffraction will result in recollimating, or realigning, some of these plane waves. Thus, the $+1$ order from the signal modulator will be partially rediffracted by the -1 action of the reference and will emerge parallel to the light which was undiffracted by either modulator. Another case occurs when light undiffracted by the signal modulator is diffracted into first order by the reference modulator and is thereby aligned with the $+1$ order from the signal modulator.

This realignment or recollimation is perfect when the signal and reference modulations are exactly matched. It allows for heterodyne detection without the introduction of an external reference—one of the beams serves as local oscillator. This automatic alignment of the light components is referred to as "collinear heterodyning" (Carleton et al., 1969). The frequency shift introduced by the ULM (and not by a passive mask) allows the beat frequency of the heterodyning process to equal the modulator frequency so that undesired contributions to the detection process can be easily removed by band-pass filtering.

We will consider only coherent Raman–Nath processors. They are easily rendered incoherent by eliminating that component of light which serves as local oscillator (L.O.) and removing the band-pass filter from the output. In some processors, the intrinsic L.O. beam is suppressed and a second external L.O. is injected at the photodetector. Such a beam must be collinear with the signal beam within a fraction of an optical wavelength if efficient linear detection is to be obtained. This degree of precision is difficult to obtain and maintain. Since it is always possible to obtain a collinear L.O. directly from the processing aperture, we will not devote space to a discussion of the more involved and unstable case of the externally injected L.O.

b. Coherent Processors Employing Postreference Fourier-Plane Filtering. One means of optically calculating the correlation integral uses the system of Fig. 26. A slit located in the frequency plane (4) provides the optical

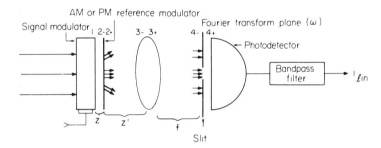

Slit location	Ref.type	Output $I_{\ell in}$
$\omega = 0$	AM PM	NO YES
$\omega = \pm \omega_0$	AM PM	YES YES

FIG. 26. A coherent, Raman–Nath processor employing postreference, Fourier plane filtering. A slit in the ω-plane selects one or another group of collinear wave-constituting a particular composite diffraction order.

filtering required to obtain a useful correlator output. This approach has its origin in a letter by Slobodin (1963) where he used a knife edge to block all but one set of off-axis orders (Schlieren filtering). The off-axis slit was introduced by Arm *et al.* (1964), who at first disputed Slobodin (1964). See also Lambert *et al.* (1965) and Felstead (1967). A modified frequency-offset method is described by Felstead (1968), but this technique results in an incoherent output. (Note that Felstead uses the words incoherent and coherent to refer to the illumination system, whereas we use them to describe the video or heterodyne detection process.)

The light front emerging from the signal modulator in Fig. 26 is

$$\exp[-ivt + i\alpha s(x - Vt)] \approx e^{-ivt}[1 + i\alpha s(x - Vt)] \tag{173}$$

for small α. Thus,

$$\mathscr{E}(x, t, 1) = 1 + i\alpha s(x - Vt) \tag{174}$$

We initially locate the signal modulator immediately adjacent $(z = 0)$ to a reference transparency. The reference function $r(x)$ is expressed by the *amplitude* transmittance $w_0 + w(x)$, where the bias term w_0 accounts for the average dc amplitude transmittance of the transparency. Then,

$$\mathscr{E}(x, t, 2_+) = [1 + i\alpha s(x - Vt)] [w_0 + w(x)] \tag{175}$$

We will assume as in Eq. (138) that the phase modulation is weak, so that $w(x)$ is purely imaginary for a "phase reference" and purely real for an "amplitude reference." Further, in what follows, only the Fourier components of $w(x)$ near $\omega \approx \pm \omega_0$ will enter. We will neglect all higher-frequency components $(\pm 2\omega_0, \ldots)$ since the higher orders contribute nothing to the understanding of the processors described. Because of the phase modulation, a photodetector placed in plane 2_+ yields, for $w(x)$ real, $i_{1\text{lin}} = 0$ [Eq. (140)]. A perfect cylinder lens added in plane 3 yields, for $z' = f$, apart from unimportant factors,

$$\mathscr{E}(x, t, 4_-) = E(\omega, t, 2_+), \qquad \omega = kx/f \tag{176}$$

No magic is accomplished by the transform process, so that a detector in plane 4_- again yields $i_{1\text{lin}} = 0$. If, however, a slit is placed in this plane, a nonzero output current can be obtained. Moreover, the current will be of the form $i_{1\text{lin}} \propto \mathsf{R}$ [Eqs. (165)–(172)]. To show this, we write, in the spirit of Eqs. (166), (167), and (171), with r taken to be amplitude reference expressed by the transmissivity $w(x) \equiv \mathscr{W}_+(x) \exp(i\omega_0 x) + \mathscr{W}_-(x) \exp(-i\omega_0 x) + \cdots$,

$$\mathscr{E}(x, t, 2_+)$$

$$= [1 + i\alpha\mathscr{S}_+(x - Vt) \exp i\omega_0(x - Vt) + i\alpha\mathscr{S}_-(x - Vt) \exp -i\omega_0(x - Vt)]$$

$$\times [w_0 + \mathscr{W}_+(x) \exp i\omega_0 x + \mathscr{W}_-(x) \exp -i\omega_0 x] \tag{177}$$

$$= w_0 + i\alpha[\mathscr{S}_+(x - Vt)\mathscr{W}_-(x) \exp -i\omega_0 Vt + \mathscr{S}_-(x - Vt)\mathscr{W}_+(x) \exp i\omega_0 Vt]$$

$$+ [i\alpha w_0 \mathscr{S}_+(x - Vt)(\exp -i\omega_0 Vt) + \mathscr{W}_+(x)](\exp i\omega_0 Vt)$$

$$+ [i\alpha w_0 \mathscr{S}_-(x - Vt)(\exp i\omega_0 Vt) + \mathscr{W}_-(x)](\exp -i\omega_0 x)$$

$$+ i\alpha\{\mathscr{S}_+(x - Vt)\mathscr{W}_+(x) \exp(i2\omega_0 x - i\omega_0 Vt)$$

$$+ \mathscr{S}_-(x - Vt)\mathscr{W}_-(x) \exp(-i2\omega_0 x + i\omega_0 Vt)\} \tag{178}$$

From Eq. (176) and the definitions

$$\mathscr{S}_\pm(x) \leftrightarrow S_\pm(\omega)$$

$$\mathscr{W}_\pm(x) \leftrightarrow W_\pm(\omega) \tag{179}$$

$$\mathscr{S}_\pm(x - Vt)\mathscr{W}_\mp(x) \leftrightarrow P_{\pm\mp}(\omega, t)$$

we have for the amplitude distribution in the back focal plane

$$\mathscr{E}(x, t, 4_-) = w_0\, \delta(\omega) + i\alpha[P_{+-}(\omega, t)\, \exp -i\omega_0\, Vt + P_{-+}(\omega, t)\, \exp i\omega_0\, Vt]$$
$$+ \{i\alpha w_0 S_+(\omega - \omega_0)[\exp -i(\omega + \omega_0)Vt] + W_+(\omega - \omega_0)\}$$
$$+ \{i\alpha w_0 S_-(\omega + \omega_0)[\exp i(\omega + \omega_0)Vt] + W_-(\omega + \omega_0)\}$$
$$+ i\alpha S_+(\omega - 2\omega_0)W_+(\omega - 2\omega_0)[\exp -i(\omega + \omega_0)Vt)]$$
$$+ i\alpha S_-(\omega + 2\omega_0)W_-(\omega + 2\omega_0)\exp[i(\omega + \omega_0)Vt] \tag{180}$$

consisting of terms centered at $\omega = 0$, $\pm\omega_0$, $\pm 2\omega_0$. If the slit passes the spatial frequency band about $\omega = \omega_0$ (first-order detection), a correlation output is achieved. We find from Eq. (180)

$$\mathscr{E}(x, t, 4_+) = i\alpha w_0 S_+(\omega - \omega_0)[\exp -i(\omega + \omega_0)Vt] + W_+(\omega - \omega_0) \tag{181}$$

and the average linear photocurrent will be

$$i_{1\mathrm{in}}(t) = \frac{1}{2}\,\mathrm{Re}\int dx\,\mathscr{E}\mathscr{E}^* = \frac{1}{2\pi}\frac{1}{2}\,\mathrm{Re}\int d\omega\, EE^*$$

$$= \frac{1}{4\pi}\,\mathrm{Re}\int d\omega[2\,\mathrm{Re}\, i\alpha w_0 S_+(\omega - \omega_0)W_+{}^*(\omega - \omega_0)\,\exp -i(\omega + \omega_0)Vt]$$
$$\tag{182}$$

$$= \frac{\alpha w_0}{2\pi}\,\mathrm{Re}\left[i\int d\omega\, S_+(\omega - \omega_0)e^{-i\omega Vt}W_+{}^*(\omega - \omega_0)\,\exp -i\omega_0\, Vt\right] \tag{183}$$

Using Parseval's theorem,

$$\int_{-\infty}^{\infty} dx\, f(x)g^*(x) = (1/2\pi)\int_{-\infty}^{\infty} d\omega\, F(\omega)G^*(\omega) \tag{184}$$

we may write from Eq. (183)

$$i_{1\mathrm{in}}(t) = \alpha w_0\,\mathrm{Re}\int_{-\infty}^{\infty} dx\,\mathscr{S}_+(x - Vt)\mathscr{W}_+{}^*(x)\exp(-i\omega_0\, Vt + i\pi/2) \tag{185}$$

The result agrees with Eq. (172) apart from an unimportant factor of $\frac{1}{2}e^{i\pi/2}$. If r is chosen to be a pure phase reference, $r(x)$ is replaced by $iw(x)$ and the resulting current is again as in Eq. (185) except for an additional factor of $e^{i\pi/2}$. Thus, a slit which passes only a band of spatial frequencies about ω_0 produces an output current proportional to the correlation of $s(x)$ with $w(x)$ for real or imaginary reference functions. Moreover, this output depends *only* on the first-order diffraction from the signal and from the reference modulators.

The output from two slits at $\omega = \pm\omega_0$ could be combined in the photodetector providing the optical path lengths were properly chosen and held within $\lambda/4$. This conclusion is easily reached from a study of Eq. (185).

Central field detection is also of interest. The slit may be moved to $x = 0$ to pass a band centered on $\omega = 0$. For an amplitude reference [$w(x)$ real], Eq. (180) reveals immediately that the appropriate field contributions at

$\omega \approx 0$ are

$$\mathcal{E}(x, t, 4_+) = w_0 \delta(\omega) + i\alpha P_{+-}(\omega, t) \exp(-i\omega_0 Vt)$$
$$+ i\alpha P_{-+}(\omega, t) \exp(+i\omega_0 Vt) \tag{186}$$

For $w(x)$ real, $W_+(\omega) = W_-*(-\omega)$, $P_{+-}(\omega) = P^*_{-+}(-\omega)$, and

$$\mathcal{E}(x, t, 4_+) = w_0 \delta(\omega) + i2\alpha \operatorname{Re}[P_{+-}(\omega, t) \exp -i\omega_0 Vt] \tag{187}$$

It is clear that only the cross term will contribute to $i_{1\mathrm{in}}$ in the band of interest and that this cross term is imaginary. Thus, $i_{1\mathrm{in}} = 0$. For a central slit filter with amplitude reference, no correlation is possible.

Correlation is possible, however, for central field detection and a phase reference [$w(x)$ imaginary]. By Eq. (140), a wide-aperture photodetector in plane 2_+ yields an output $i_{1\mathrm{in}} = 0$. For a slit at $\omega = 0$, Eq. (186) still holds, but now $W_+(\omega) = -W_-*(-\omega)$, $P_{+-}(\omega) = -P^*_{-+}(-\omega)$, so that

$$i = \tfrac{1}{2} \operatorname{Re} \int dx \, \mathcal{E}\mathcal{E}*$$

$$i_{1\mathrm{in}} = \tfrac{1}{2} \operatorname{Re} \int d\omega \, w_0 \delta(\omega) \, 4 \operatorname{Re}[i\alpha P_{+-}(\omega, t) \exp -i\omega_0 Vt] \tag{188}$$

$$= 2w_0 \alpha \operatorname{Re}[iP_{+-}(0, t) \exp -i\omega_0 Vt]$$

By Eq. (170),

$$i_{1\mathrm{in}} = 2w_0 \alpha \operatorname{Re} \int dx \, \mathcal{S}_+(x - Vt) i \mathcal{W}_-(x) \exp -i\omega_0 Vt \tag{189}$$

which is just twice the off-axis slit value of Eq. (185). We conclude that central plane slit detection can be employed with a phase reference, and first-order slit detection with phase or amplitude reference. The workable combinations of reference types and slit positions are summarized in Fig. 26.

It is worth pointing out that, if the phase reference is supplied by a second ULM, then the motion of the reference wave must be considered. This motion is easily taken into account by changing V to $V - V'$, where V' is the velocity of the reference wave along the $+x$ axis. In this case [see Eq. (189)],

$$i_{1\mathrm{in}} = -2w_0 \alpha \operatorname{Re} \int dx \, \mathcal{S}_+[x - (V - V')t]$$
$$\times \mathcal{W}_+*(x) \exp[-i\omega_0(V - V')t + i\pi/2] \tag{190}$$

For identical materials in signal and reference ULM's, V' will be $-V$ and $i_{1\mathrm{in}}$ will be centered on a carrier frequency $2\Omega = 2\omega_0 V$. The center frequency of the band-pass filter must be appropriately changed.

The influence on correlation of a spatial gap between signal and reference modulators will now be examined. Unless a liquid line is used with the reference modulator submerged in it, there will always be some space between the modulators. When $z \neq 0$, the phase modulation in plane 1 (Fig. 26)

will be partially converted to amplitude modulation, via Fresnel diffraction. If $z = (2n + 1)Z$, this conversion is complete [Eq. (161)]. In the amplitude reference case, $\mathscr{E}(x, t, 2_+)$ then becomes a product of two amplitude modulation signals. The lens and slit may be dispensed with, and the wide-aperture processor results. We will take up this special case later in Section B,3c.

Since the output of the correlator with a slit at $\omega \approx \omega_0$ depends only on $\mathscr{S}_+$ and $\mathscr{W}_-$, we may easily take into account the effect of $z \neq 0$. The "exact" approach is to replace $S_\pm$ in Eq. (181) by a new function (see Section B,2a)

$$S_\pm'(\omega) = S_\pm(\omega)\Phi(\omega, z) \tag{191}$$

Then

$$\begin{aligned} S_+'(\omega)e^{-i\omega V t} &= S_+(\omega)e^{-i\omega V t}\exp[(-z/2k)\omega^2] \\ &\approx S_+(\omega)e^{-i\omega V t}\exp\{-i(z/2k)[\omega_0^2 + 2\omega_0(\omega - \omega_0) + (\omega - \omega_0)^2]\} \end{aligned} \tag{192}$$

The ω_0^2 term is a fixed phase shift and the linear term a time shift of the instant of correlation. The $(\omega - \omega_0)^2$ term introduces a quadratic phase delay which, as we have seen in the "free-space filter," is negligible for $(2n + 1)[(\omega_{max} - \omega_0)/\omega_0]^2 < \frac{1}{4}$. We will accordingly ignore it. As in Section B,2f, the other terms will change Eq. (185) to

$$\begin{aligned} i_{lin}''(t) = \alpha w_0 \, \text{Re} \int_{-\infty}^{\infty} dx \, \mathscr{S}_+ \left(x - Vt - \frac{2n + 1}{2}\Lambda_0\right) \\ \times \mathscr{W}_+^*(x) \exp[-i\omega_0 Vt + i(n + 1)\pi] \end{aligned} \tag{193}$$

To deduce the effect of $z \neq 0$ on central field ($\omega \approx 0$) detection, we must return to Eq. (186) replacing

$$\begin{aligned} \mathscr{E} &\to \mathscr{E}' \\ \mathscr{S} &\to \mathscr{S}' = \mathscr{S} * \phi \\ P &\to P' \end{aligned} \tag{194}$$

Then, we proceed to calculate

$$\begin{aligned} i_{lin}'' = \tfrac{1}{2} \, \text{Re} \int d\omega \, w_0 \, \delta(\omega)\alpha[2 \, \text{Re} \, iP'_{+-}(\omega, t)(\exp -i\omega_0 Vt) \\ + 2 \, \text{Re} \, iP'_{-+}(\omega, t) \exp i\omega_0 Vt] \\ = \alpha w_0 \, \text{Re}[iP_{+-}'(0, t)(\exp -i\omega_0 Vt) + iP_{-+}'(0, t) \exp i\omega_0 Vt] \end{aligned} \tag{195}$$

where now $P'_{+-} \neq P'^*_{+-}$. For $z = (2n + 1)Z$, we find, by Eqs. (162), (178), and (191),

$$\ddot{i}_{1\text{in}} = \alpha w_0 \,\text{Re} \int_{-\infty}^{\infty} dx [\mathscr{S}_+'(x - Vt) i \mathscr{W}_-(x)(\exp -i\omega_0 Vt)$$
$$+ \mathscr{S}_-'(x - Vt) i \mathscr{W}_+(x) \exp i\omega_0 Vt] \qquad (196)$$

$$\approx \alpha w_0 \,\text{Re} \int dx\, \mathscr{S}_+\!\left(x - Vt - \frac{2n+1}{2}\Lambda_0\right) \mathscr{W}_-(x) \exp[-i\omega_0 Vt + i\pi(n+1)]$$
$$+ \alpha w_0 \,\text{Re} \int dx\, \mathscr{S}_-\!\left(x - Vt + \frac{2n+1}{2}\Lambda_0\right) \mathscr{W}_+(x) \exp[i\omega_0 Vt + i\pi(n+1)]$$
$$\qquad (197)$$

$$i_{1\text{in}} \approx \alpha w_0\, e^{i(n+1)\pi}\,\text{Re} \int dx\, \left\{ \mathscr{S}_+\!\left(x - Vt - \frac{2n+1}{2}\Lambda_0\right) \right.$$
$$\left. + \mathscr{S}_-^*\!\left(x - Vt + \frac{2n+1}{2}\Lambda_0\right) \right\} \mathscr{W}_-(x) \exp -i\Omega_0 t \qquad (198)$$

Here, we have used $\omega_0 V = \Omega_0$ and the fact that $w(x)$ is real ($\mathscr{W}_+ = \mathscr{W}_-^*$) and we have made the usual assumptions of Section B.2f. The output is clearly the sum of two terms—paired responses—shifted symmetrically in time by $\frac{1}{2}(2n+1)\Lambda_0/V$. Fresnel diffraction has converted the phase modulation from the ULM into an amplitude modulation. On the other hand, the product of this amplitude image with a phase reference ($\mathscr{W}_+ = -\mathscr{W}_-^*$) yields a subtraction of the paired responses and almost complete cancellation. The subtraction is incomplete because the shifted responses no longer overlap completely in time. This "null" will repeat for even integer multiples of Z. We could, if we wished, correlate a phase image ($z = 2nZ$) against a phase mask ($\mathscr{W}_+ = -\mathscr{W}_-^*$) and achieve the correlation given by Eq. (198). We conclude that for central field detection it is necessary that a phase image of the ULM aperture occur in the plane of a phase reference or an amplitude image in the plane of an amplitude reference.

We next consider the effect on correlator output of locating the reference modulator away from the lens input plane 3. If the reference modulator in Fig. 26 is not exactly in the front focal plane of the lens ($z' \neq f$), Eq. (176) becomes [cf. Eq. (136)]

$$\mathscr{E}(x, t, 4_-) = \{\exp[(if/2k)(1 - z'/f)\omega^2]\} E(\omega, t, 2_+) \qquad (199)$$

There is clearly no effect of this diffraction on $i_{1\text{in}}$ for either slit position since, in calculating $i = \text{Re} \int dx\, \mathscr{E}\mathscr{E}^*$, the Fresnel term in Eq. (199) disappears. Physically, the diffraction orders which will end up at a given slit position have already been aligned essentially parallel by the reference modulator—hence, no differential phase shift can be introduced by subsequent optics. This insensitivity of "collinear heterodyning" systems to postreference optical-path perturbations is a highly desirable feature in a practical system.

We may say a few words about the effect of transverse fluctuations in the optical path between signal and reference modulators. These may be taken into account schematically if we assume $\mathscr{S}_+$ to contain a factor $e^{i\gamma(x)}$, where γ is assumed to vary slowly enough in x that Fresnel diffraction due to it may be neglected. This phase factor will then occur in Eqs. (185) or (189) and can, if the fluctuation is serious, destroy the correlation, e.g., $\gamma_{max} \approx \lambda/4$ is very severe. Thus, optical tolerances must be held on all surfaces. As discussed before, optical-path perturbations subsequent to the reference are unimportant unless they result in loss of light through the detecting aperture. The equations show that the output term consists of $\mathscr{S}_+\mathscr{W}_+{}^*$. One may think of postreference perturbations as affecting $\mathscr{S}_+$ and $\mathscr{W}_+$ equally. The conjugation of $\mathscr{W}_+$ causes these perturbations to cancel out.

  c. Coherent Wide-Aperture Processors. We will now take up the wide-aperture processor introduced above and shown in Fig. 27. It has been

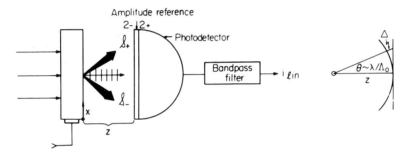

FIG. 27. Diagram of a coherent, Raman–Nath processor employing Fresnel diffraction to effect phase-to-amplitude modulation conversion. The phase-modulated light carrier emerging from the modulator at plane 1 is converted to an amplitude-modulated carrier in propagating to plane 2. Here, $(z + \Delta) \cos \theta = z$; $(z + \Delta) \times [1 - \frac{1}{2}(\lambda/\lambda_0)^2] = z$; $\Delta = \lambda/2 \Rightarrow z = \Lambda_0{}^2/2\lambda = Z$.

treated theoretically and experimentally by Meltz and Maloney (1968), Carleton *et al.* (1969), Atzeni and Pantani (1969a), Maloney and Meltz (1969), and Atzeni and Pantani (1969b).

  In this configuration, no lenses or slits are employed. All the light from the reference aperture is collected onto the photodetector. As before,

$$\mathscr{E}(x, t, 1) = 1 + i\alpha s(x - Vt) \tag{200}$$

If $z = 0$,

$$\mathscr{E}(x, t, 2_+) = [1 + i\alpha s(x - Vt)][r_0 + r(x)] \tag{201}$$

where the reference function will be taken to be an amplitude reference. The photodetector receives all the light in plane 2_+, calculates the intensity at each x and sums over all x. The output current linear in s is

$$i_{lin} = \tfrac{1}{2} \operatorname{Re} \int dx\, \mathscr{E}\mathscr{E}^* = \alpha \operatorname{Re} \int dx\, is(x - Vt)|r_0 + r(x)|^2 = 0 \tag{202}$$

As we have seen before, a phase image of the ULM aperture does not correlate with an amplitude reference. (Note that, in what follows, one may choose to have $|r(x)|^2$ convey the desired reference information, i.e., an intensity mask is permissible. We will therefore express $|r_0 + r(x)|^2$ as $w_0 + w(x)$, where the w's now represent intensity transmission.) The phase image in plane 1 may, however, be changed to an amplitude image by allowing Fresnel diffraction to perform the conversion. Thus, if, as in Eq. (161), we take $z = (2n + 1)Z$, we find

$$\mathcal{E}(x, t, 2_-) = 1 + i\alpha s'[x - Vt, (2n + 1)Z] = 1 + i\alpha s(x - Vt) * \phi[x, (2n + 1)Z]$$

$$= 1 + (-1)^{n+1}\alpha\left\{\mathcal{S}_+\left(x - Vt - \frac{2n + 1}{2}\Lambda_0\right)\right.$$

$$\left. + \mathcal{S}_-\left(x - Vt + \frac{2n + 1}{2}\Lambda_0\right)\right\} \tag{203}$$

using the usual narrow-band assumptions. These paired images $\mathcal{S}_+$ and $\mathcal{S}_-$ are really the signals formed by the plus and minus spectra, respectively, i.e., $\mathcal{E}(x, t, 2_-)$ is the sum of two signals—each spatially shifted from its original position by $|\Delta x| = \frac{1}{2}(2n + 1)\Lambda_0$. In an amplitude plane (for small n), they add to produce a close approximation to $s(x - Vt)$. In a phase plane, $z = 2nZ$, the zeroth-order contribution ($\omega = 0$) approximately cancels the first-order contribution ($\omega \approx \pm\omega_0$). For a carrier frequency of 30 MHz in a fused silica ULM operating in the shear mode, $Z \approx 1.2$ cm, which is a very noncritical adjustment.

When the reference is placed in an amplitude plane, $z = (2n + 1)Z$, $\mathcal{E}(x, t, 2_-)$ correlates with $w_0 + w(x)$ to produce

$$i_{\text{lin}} = \alpha \operatorname{Re} \int dx \, i(i)^{2n+1}s'[x - Vt, (2n + 1)Z]w(x)$$

$$= (-1)^n\alpha \int dx \, s'[x - Vt, (2n + 1)Z]w(x) \tag{204}$$

which is very nearly equal to the correlation of $s(x)$ with $w(x)$.

There is no problem with Fresnel diffraction following the reference because the detector sees only the average, "slowly varying" light power emerging from the reference aperture and because this light power is constant in any plane to the right of 2_+. Optical tolerances are relaxed as in all collinear heterodyning schemes. An auxiliary lens may be inserted to collect light onto a small photocathode without fear that it can affect the correlation. No optical tolerances need be imposed on it. As before, QPE is negligible for modest n. Using a plastic Fresnel lens as a collector, Meltz and Maloney (1968) have compressed a 20-μsec fm pulse to less than 1 μsec at a center frequency of 30.4 MHz.

d. Coherent Processors Employing Prereference Fourier-Plane Filtering.
It is clear from the preceding treatment that an amplitude image of the signal modulator can be correlated against an amplitude reference (or preferably an intensity reference). Fresnel diffraction, properly utilized, can

effect this modulation conversion. Various other types of F_1 filters can be employed as well. Figure 28 illustrates a system featuring this prereference filtering. We will not analyze these systems in detail, but the operation

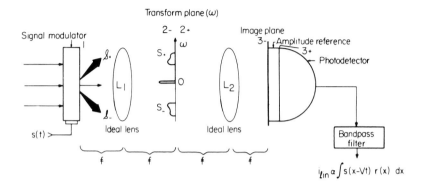

FIG. 28. A coherent, Raman–Nath processor using prereference Fourier-plane filtering. Modulation conversion is accomplished by optical filtering operations performed in the Fourier-plane 2.

should be obvious from the earlier discussion of Section B,2. The signal modulator output is as usual a phase object and

$$
\begin{aligned}
\mathcal{E}(x, t, 1) &= 1 + i\alpha s(x - Vt) \\
&= 1 + i\alpha \mathcal{S}_+(x - Vt)[\exp i\omega_0(x - Vt)] \\
&\quad + i\alpha \mathcal{S}_-(x - Vt) \exp -i\omega_0(x - Vt)
\end{aligned}
\tag{205}
$$

In the focal plane of L_1, we have

$$
\begin{aligned}
\mathcal{E}(x, t, 2_-) &= E(\omega, t, 1) \\
&= \delta(\omega) + i\alpha S_+(\omega - \omega_0)e^{-i\omega Vt} + i\alpha S_-(\omega + \omega_0)e^{i\omega Vt}
\end{aligned}
\tag{206}
$$

By inserting the proper form of spatial filter in plane 2, we can, as described in the first part of this chapter, convert the modulation at plane 3 to an amplitude form. This amplitude spectrum $E'(\omega, t, 1)$ is retransformed by L_2 to

$$
\mathcal{E}(x, t, 3_-) = -\mathcal{F}^{-1}[E'(\omega, t, 1)]
\tag{207}
$$

where $\mathcal{F}^{-1}$ indicates the inverse transform process and the minus arises from the fact that lenses take only direct and not inverse transforms, thereby reversing the coordinates. The amplitude image now existing in plane 3 is correlated against the amplitude reference in plane 3 to produce an $i_{1\text{lin}}$ proportional to the correlation integral of s with r.

The most common forms of this processor involve the Schlieren filter, which consists merely of blocking both $\omega = 0$ and $\omega > 0$ components in plane 2. This results in a loss of the "second beam" (L.O.) needed for heterodyning. Accordingly, in Schlieren processors, it is necessary to reinsert a reference light beam at the detector to achieve linear detection. Typical of this approach are processors described by Lambert (1965), King *et al.* (1967), and Izzo (1965). The difficulties of reinserting a reference would make field operation of such a device difficult. The other filtering possibilities, such as the half-plane phase delay filter U_1 of Lowenthal and Balvaux (1967), the half-plane stop U_2, or the Zernicke phase plate, all effect modulation conversion while preserving the reference beam inherent in the processor.

In general, systems involving double optical transforms require high-quality lenses and careful alignment. They offer no real advantages over the simpler systems, such as those described above or those based on polarization discrimination to be considered next.

e. Coherent Processors Employing Prereference Polarization Discrimination. If the laser light is incident on an isotropic ULM polarized parallel to or at right angles to the propagation direction of a transverse ultrasonic wave (one whose material displacement is at right angles to the propagation direction of both the ultrasonic and light waves), then odd diffraction orders from the ULM will be polarized at right angles to the even orders. (Carleton and Maloney, 1967; Mueller, 1938). For weak modulation, $\alpha^2 \ll 1$, this, in effect, means that the diffracted light orders $(\nu \pm \Omega)$ will be polarized at right angles to the undiffracted (ν) light. Similarly, if the incident light is polarized at $\pm 45°$ to the propagation direction of a longitudinal acoustic wave, the diffracted light will again be cross-polarized. If the emergent light is repolarized, for example, by an analyzer inserted after the modulator, the resulting modulation still constitutes a phase image. Now, however, the method of Zernicke (Born and Wolf, 1965) may be employed to convert the modulation. A quarter-wave plate inserted after the ULM with principal axes aligned along the two polarization directions will introduce a 90° relative phase shift between the two components (Carleton *et al.* 1969). The modulation emerging from a repolarizing analyzer will now be an amplitude image. Thus, the two lenses and Fourier-plane filter of Fig. 28 can be replaced by a quarter-wave plate sandwiched between the signal and the reference modulators (Fig. 29). Note that this distance z between modulators must be kept small $(z^2 \ll Z^2)$ to avoid further undesired conversion by Fresnel diffraction.

Polarization discrimination is particularly useful in the case of a phase reference, such as a second ULM. If both ULM's are operated to produce cross-polarization of the diffracted light, then an analyzer used *after* the reference modulator can replace the focal-plane filter of Fig. 26. Figure 30 shows such a configuration. Consider for the moment that the polarizer in plane 2 of Fig. 30 has been removed and there is no interval between the modulators (for example, both acoustic beams are in the same "block of glass").

Let the signal and reference modulation be identical and let $\hat{a}$ and $\hat{b}$ be two perpendicular transverse directions such that, for incident light polarized

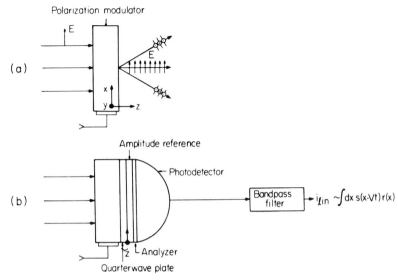

FIG. 29. Coherent Raman–Nath processor featuring modulation conversion by polarization discrimination. Cross-polarization of carrier and modulation allows a 90° relative phase shift to be introduced by a quarter-wave plate. The analyzer realigns polarizations so that heterodyning can take place.

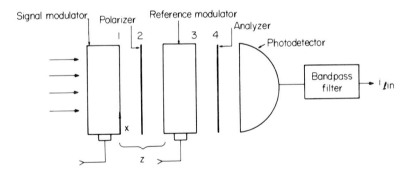

FIG. 30. Coherent Raman–Nath processor featuring modulation conversion by polarization discrimination in conjunction with polarization changing signal and reference modulators.

along $\hat{a}$, diffracted light is polarized along $\hat{b}$. Then,

$$\mathscr{E}(x, t, 1) = \hat{a} + i\alpha s(x - Vt)\hat{b} \tag{208}$$

and

$$\mathscr{E}(x, t, 3) = [1 + \alpha\beta s(x - Vt)r(x - V't)]\hat{a} + i[\alpha s(x - Vt) + \beta r(x - V't)]\hat{b} \tag{209}$$

Here, we have defined β as the modulation index for the reference modulation. An analyzer in plane 4 can select either the $\hat{a}$ or $\hat{b}$ signal or a linear combination of both. If $\hat{a}$ is selected,

$$i_{\text{lin}} = \alpha\beta \ \text{Re} \int dx \ s(x - Vt)r(x - V't) \tag{210}$$

which is the desired output. This correlator is completely equivalent to the (postreference) central field processor, but no ideal transforming lens is required. If the $\hat{b}$ output polarization is chosen,

$$i_{\text{lin}} = \alpha\beta \ \text{Ré} \int dx \ s(x - Vt)r(x - V't) \tag{211}$$

and this correlator is directly analogous to one employing both first-order slits ($\omega \approx \pm\omega_0$). If no polarization change were provided by the ULM's, the output of the correlator would be zero as in Eq. (202). In both Eqs. (210) and (211), the output current will be at the second harmonic of the input signal under the usual situation where $V' = -V$. A polarizer in plane 2 can be used to block the undiffracted light. If the analyzer in plane 4 is crossed with respect to that in plane 2, $i_{\text{lin}} = 0$, but the device will function as an incoherent correlator (Wilmotte, 1963). If the modulators can be located close enough together, no problem from Fresnel diffraction need arise. No tolerances beyond flatness are imposed; rigidity is obtained. All these polarization-discrimination approaches have the advantages of simplicity, ruggedness, and ease of adjustment.

f. Coherent Processors As Signal Generators. Any of the coherent processors described above, except those employing active reference modulators, may be operated as pulse generators. Consider that a correlator whose output in a band about ω_0 is given by

$$i_{\text{lin}} = \int s(x - Vt)r(x) \ dx \tag{212}$$

is driven by a short rf pulse. When this pulse has a constant-amplitude spectrum across the band of interest, we may approximate it as $\delta(t)$ resulting in an acoustic signal $\delta(x - Vt)$. The ensuing output current

$$i_{\text{lin}} = \int dx \ \delta(x - Vt)r(x) = r(-Vt) \tag{213}$$

is a generated signal having a time variation specified by the spatial reference function.

The signal to noise ratio (S/N) of such a pulse generator is generally poor, however, as the following considerations will show. In pulse compression, the energy input to the modulator over a relatively long period of time is compressed to yield a large-amplitude current pulse for a short interval. An enhancement of S/N by a factor of the time–bandwidth product

(TW) squared results. On the other hand, in the generation mode, a short, high-amplitude pulse is spread out to make a long pulse of much lower amplitude. The reference light (the local oscillator which permits linear operation) floods the entire aperture of the signal modulator. Shot noise is generated by photoemission due to light from this entire aperture, while the only useful L.O. power is that which passes through the region of the ULM actually occupied by the pulse. If a coherent flying-spot scanner could be built which would illuminate only the acoustic pulse as it traveled across the aperture, an enhancement in S/N by a factor of $(TW)^2$ would be obtained. If synchronization and sweep-linearity problems prevented the use of this type of scanner, an ultrasonic shutter (Maloney, 1969) can yield a more modest improvement of a factor of approximately TW. Such a shutter consists of a ULM identical with the signal modulator, operating in the same acoustic mode and placed in the light beam ahead of the signal ULM. It is sandwiched between crossed polars oriented so that only the diffracted light is passed. When it carries a pulse synchronized with the modulator pulse, its light output will illuminate the modulator pulse and travel across the ULM aperture in synchronism with it.

Another form of complex signal generator (Whitman *et al.*, 1967) employs frequency-plane heterodyning of the diffracted light from a Bragg modulator and a spherically diverging local oscillator wavefront to generate a linear fm output pulse. If the spherical wave is replaced by an oblique plane wave, the input pulse is reproduced but delayed by an amount determined by the angle of incidence of the plane wave. More complex signals can be generated by appropriate design of the local oscillator wavefront.

C. High-Frequency Signal Processors (The Bragg Limit)

1. General Comments

The desire for more rapid processing and higher range resolution leads to increased bandwidths and higher frequencies. We now turn our attention to the UHF region, where processors are characterized by wide bandwidth and shorter processing times. Because such devices operate in the Bragg limit ($Q > 1$), we will refer to them as Bragg processors. The requirement that light rays must strike the acoustic wavefronts at the Bragg angle Θ leads to difficult illumination problems where large fractional bandwidth is sought. It is necessary to restrict either the signal bandwidth or the choice of signal waveform. At these higher frequencies, the increasing acoustical losses in amorphous materials such as liquids and glasses require substitution of crystalline materials such as $LiNbO_3$ or sapphire. Thin-plate cemented transducers give way to evaporated CdS or ZnO films or to directly excited piezoelectric modulators. Limited dimensions of available single crystals restrict processing time to one or two tens of microseconds. The possibility of gigahertz bandwidth can, however, permit time–bandwidth products in excess of 10,000.

2. *Divergent Illumination*

The prototype Bragg processor shown in Fig. 31 is based on a configuration suggested by Gerig and Montague (1964). They noted that the diffraction angle $\theta \approx \lambda/\Lambda = K/k = \lambda f/V$ is linear in the acoustic frequency f, and that the

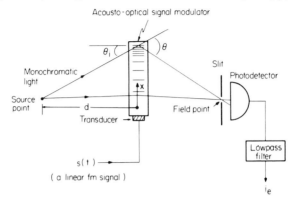

FIG. 31. Prototype Bragg pulse-compression system employing linear fm modulation and divergent illumination. The output is incoherent.

linear fm acoustic signal results in diffracted light rays whose diffraction angle θ varies linearly with x. Thus, for not-too-large bandwidths, light diffracted from a linear fm signal will focus naturally to a spot. As the acoustic signal propagates down the modulator, this spot will propagate along in a plane parallel to the modulator. If a narrow slit is placed in this plane, so that it is illuminated at the instant the signal fills the modulator aperture, pulse compression will be achieved as the spot traverses the slit. The spot will ideally have a spatial shape determined by the Fourier transform of the ULM aperture. Although Gerig and Montague first suggested this scheme for use with collimated illumination incident parallel to the wavefronts (Raman–Nath illumination), it is more usefully adapted to the Bragg processor (Cutrona, 1967), whose output is by nature incoherent ($i \propto |R|^2$).

The natural focusing principle is used twice in Fig. 31: once to satisfy the Bragg incidence condition (McMahon, 1967b) at each x and again to refocus the diffracted rays onto the detecting slit. The theory has been developed by McMahon and by Zahn (1968). Experimental compression of a 2-μsec pulse has been reported by Schulz *et al.* (1967), with bandwidths of 60 MHz and compression ratios of 111 at 1.16 GHz.

It is obvious that only a linear fm signal can be compressed in this manner. The bandwidth may be estimated by noting from Fig. 31 that the light rays obey the law $\tan \theta_1 = x/d$, while the Bragg condition requires $\sin \theta_1 = \lambda f/2V$. Thus, divergent point illumination can satisfy the Bragg condition [Eq. (75)] over the range of frequencies for which $\tan \theta_1 \approx \sin \theta_1 \approx \theta_1$ or $\theta_1 < 0.1$. One then finds by differentiating the Bragg equation that $df \approx (2V/\lambda) d\theta_i$ [Eq. (94)]. For 0.6-μm light and $V = 0.6 \times 10^6$ cm/sec, a bandwidth in excess of 1 GHz is theoretically possible. One cannot count on using the diffraction spreading of the acoustic beam to increase band width, since, over a wide processing aperture, significant spreading would lead to nonuniform interaction efficiencies. Larger bandwidths could, in principle, be achieved by modifying the illuminating wavefront with an aspheric compensating plate to convert the "tangent illumination" to "sine illumination" or by employing a nonlinear frequency sweep which "matches" the sine illumination.

3. Collimated Illumination, Isotropic Modulator

Collimated light incident at the average Bragg angle Θ_0 is also a suitable source of illumination where more limited bandwidths are acceptable. As discussed in Section IV,D,2a, however, efficient light–sound interaction can occur only between the light ray and that part of the sound beam which intersects the light rays at the correct Bragg angle. If acoustic-beam diameter is chosen so that the diffraction spread of the acoustic beam is appreciable, then some parts of the sound beam will satisfy the Bragg condition over the entire bandwidth. A working processor is then possible. Figure 20(a) illustrates this condition. Note that satisfaction of the Bragg requirement is made possible by fanning out the acoustic beam in space. The wider the required signal bandwidth, the greater the required diffraction spread, the weaker the interaction.

The performance factor $(\eta \, \Delta f)_0$ of the modulator in this configuration is given by Eq. (100) and the appropriate figure of merit is M_1. We note that $(\eta \, \Delta f)_0$ falls off as $1/f_0$. Thus, in this case, one does not gain bandwidth by moving to higher frequency.

4. Collimated Illumination, Birefringent Modulator

Another means of satisfying the Bragg condition over a wide bandwidth is suggested by Dixon (1967a) and by Lean *et al.* (1967). They recognized that, in an optically birefringent material operated so that the diffracted light is cross-polarized with respect to the incident light, it is possible to satisfy the Bragg condition approximately over a much wider bandwidth. This is clear from Fig. 8, which reveals that the (pseudo)momentum conservation is possible over a wide range of K if the magnitude of the birefringence is properly chosen. Figure 32 is a diagram of a Bragg processor devised by Collins *et al.* (1967) and based on this type of illumination. Collimated light is incident at an angle calculated to satisfy the modified Bragg condition (Eq. 33) at the average acoustic frequency f_0. A new bandwidth calculation yields (Eq. 114)

$$\Delta f \approx 2V(2n/L\lambda_0)^{1/2} \approx 2f_0(\lambda_0/LB)^{1/2}, \qquad f_0 = \Omega_0/2\pi \qquad (214)$$

where B is the birefringence $(n - n')$.

The performance factor $(\eta \, \Delta f)_a$ appropriate to anisotropic modulation is

$$(\eta \, \Delta f)_a = \frac{\pi^2 \sqrt{2}}{\lambda_0^{5/2}} \frac{P_{AC} \, L^{1/2}}{w} \frac{n^{13/2} p^2}{\rho V^2} \qquad (215)$$

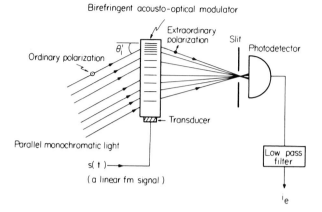

Birefringent acousto-optical modulator

FIG. 32. Modified Bragg compression system using a birefringent signal modulator to permit illumination by collimated light.

Of great significance is the fact that $(\eta\,\Delta f)_a$ is independent of f_0. If $(\eta\,\Delta f)_a$ is compared to $(\eta\,\Delta f)_0$, we find [Eq. (117)]

$$\frac{(\eta\,\Delta f)_a}{(\eta\,\Delta f)_0} = f_0(2\lambda_0\,L/nV^2)^{1/2} = 2(BL/\lambda_0)^{1/2} \qquad (216)$$

The anisotropic case thus is superior at higher frequencies, [but only because $(\eta\,\Delta f)_0$ is deteriorating!]. Further, it improves with increasing L. The choice of frequency is not arbitrary but is closely tied to the birefringence of available modulator materials. The choice which yields greatest bandwidth is $f_0 = f' = (V/\lambda_0)|n^2 - n'^2|^{1/2} \approx (V/\lambda_0)|2nB|^{1/2}$. Since suitable modulator materials ($LiNbO_3$, $LiTaO_3$, sapphire, etc.) have fixed values of B, the restriction imposed by this relationship is by no means unimportant. As an example of the improvement possible, let us consider a specific case: $n = 2$, $L = 0.001$ m, $\lambda_0 = 633$ nM, $V = 0.4 \times 10^4$ m/sec, $f_0 = 1$ GHz. Then $(\eta\,\Delta f)_a/(\eta\,\Delta f)_0 = 6.3$.

It is clear that a definite gain by as much as a factor of six in bandwidth at fixed modulation index is possible provided a material of the proper birefringence can be obtained. Too much birefringence results in unattainably large θ, while too little leads back to the isotropic case. A certain amount of adjustment can be obtained by rotating the sample about the acoustic axis until the desired amount of birefringence has been "stirred in" (Lean et al. 1967; Maloney and Gravel, 1969). For lithium niobate, at $f_0 = 500$ MHz, this rotation need be only a few degrees. This small rotation is negligible in its effect on all parameters except the index difference. Larger angles in other materials would demand a proper solution of the problem of Bragg scattering in an arbitrary crystalline direction and would lead to considerable complication.

An analysis of the birefringent Bragg processor has been carried out by

Barrett and Zahn (1968). Collins *et al.* (1967) have compressed a 240-nsec linear fm pulse to 15 nsec ($TW = 16$) in a sapphire modulator at 1560 MHz. Output signal-to-noise ratio was 12 dB. A serious drawback in all birefringent media is that, once the light and sound directions have been chosen to provide the desired refractive indices and birefringence, the relevant photoelastic constants for this configuration turn out to be small. Thus, while a gain of a factor of six in bandwidth may be indicated for equal modulation indices, in fact, much larger acoustic powers will be needed in the birefringent case to achieve the same modulation. All the above Bragg processors are designed to work solely with a linear fm signal. This code is frozen into the design.

5. *Bragg Processors with Phase Reference*

The processor shown in Fig. 33 is based on work of Squire and Alsup (1968). It is more flexible in that it allows for more general code selection. Except for the tilting of the modulators (both signal and reference) to satisfy

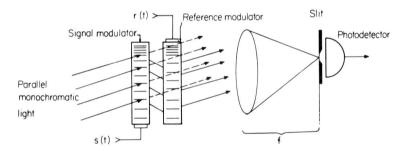

Fig. 33. Bragg processor with active reference modulator and coherent output. At the instant of correlation, the reference modulator rediffracts light rays back parallel to the undiffracted light to permit heterodyne detection. Rays are parallel only if spatial signals in modulators are matched.

the Bragg condition at Ω_0, the processor is similar to the Raman–Nath correlator with Schlieren preprocessing. A second ULM is provided as a reference. When both modulator apertures contain identical acoustic signals, the light rays from the first modulator are recollimated in passing through the second modulator. This parallel beam is then focused on a slit by a transforming lens. The rays will be recollimated and the slit will be illuminated only at the instant when the two acoustic signals are in register in the aperture and then only when the signals exactly match. Any acoustic waveform is allowable for which the Bragg condition can be approximately satisfied. Considerable coding flexibility is present. The output is coherent.

If the undiffracted light from the first modulator is blocked, correlator output will be incoherent. Squire and Alsup (1967) report experimental time–bandwidth products of 30 for a bandwidth of 300 MHz at 1 GHz.

A modification of the coherent processor is possible, similar to that employed in the Raman–Nath processors. The preprocessing may be accomplished by polarization discrimination if modulators are used whose diffracted and undiffracted light are cross-polarized, resulting in a simplification of the optical system.

Another version of this processor has been reported by Jernigan (1968). Here, the second modulator is replaced by a coded phase plate (actually a corrugated mirror). If the spatial modulation introduced by the phase plate exactly cancels that introduced by the signal modulator, light will be recollimated and correlation can be achieved.

TW products of 10,000 (20 μsec, 500 MHz) are theoretically possible. Actual products achieved with Bragg processors to date have not been outstanding, although bandwidths of 300 MHz (Squire and Alsup, 1967) and processing times of 2 μsec have been reached. The S/N is limited by light power and the rather weak interaction efficiencies, particularly in the birefringent modulator. It is doubtful that a S/N of 40 dB will ever be exceeded without a great deal of effort. Longer processing times await the development of crystal-growing technology and will hopefully not be long in coming. Typical system parameters for representative Raman–Nath and Bragg processors are presented in Table II.

TABLE II

Two Typical Correlators[a]

	Raman–Nath	Bragg
Acoustic mode	Shear	Longitudinal
Material	Fused silica	LiNbO$_3$
Processing time (μsec)	50	10
Processing aperture (cm)	19	5.5
Bandwidth (MHz)	40	300
TW product	2000	3000
Modulation index (α^2)	5%	5%
S/N (dB)	55	43
Acoustic power (W)	7	1.2
Signal power required	28	5

[a] 50-mW helium–neon laser, S-20 photocathode, 10 dB loss in optics, 6 dB transduction loss.

D. Spectrum Analyzers

1. Low-Frequency Spectrum Analyzers

As we mentioned above, once it is realized that the spatial frequencies emitted by a ULM aperture are directly related to the temporal frequencies applied to the transducer and that the back focal plane of an ideal lens exhibits these spatial frequencies, it is obvious that a spectrum analyzer can be devised. One type proposed by King *et al.* (1967) is shown in Fig. 34.

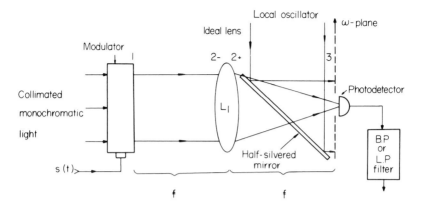

FIG. 34. Coherent acoustooptical spectrum analyzer.

A Raman–Nath modulator driven by an electrical signal $s(t)$ is illuminated by parallel light and followed by a transforming lens so that the signal appearing in plane 3 is

$$\mathscr{E}(x, t, z) = \delta(\omega) + i\alpha S(\omega)e^{-i\omega Vt}, \qquad \omega = kx/f \tag{217}$$

A photodetector located at a given ω-position in plane 3 measures

$$i = \tfrac{1}{2}\,\mathrm{Re}\,|\,\delta(0) + i\alpha S(0)|^2, \qquad \omega = 0 \tag{218}$$

$$= \tfrac{1}{2}\,\mathrm{Re}\,\alpha^2|S(\omega)|^2, \qquad \omega \neq 0 \tag{219}$$

Since ω is related to the frequency Ω by $\omega = \Omega/V$, we have an output current proportional to the intensity of each spectral component. The intensity spectrum can be plotted by moving the detector, or by use of a line of small detectors or a line of optical fibers each terminating in photodetectors.

A coherent phase-preserving analyzer can be made if a local oscillator reference beam is injected as shown in Fig. 34. Then

$$\mathscr{E}(x, t, z) = \delta(\omega) + i\alpha S(\omega)e^{-i\omega Vt} + 1 \tag{220}$$

and a linear output is possible from the detector, namely

$$i_{1\mathrm{in}} = \tfrac{1}{2}\,\mathrm{Re}\,2i\alpha S(\omega)e^{-i\omega Vt} \tag{221}$$

If we express the phase of $S(\omega)$ explicitly by writing $S(\omega) = |S(\omega)|\,e^{i\psi}$, we have

$$i_{1\mathrm{in}} = -\alpha|S(\omega)|\,\sin[\omega Vt + \psi] \tag{222}$$

$$= -\alpha|S(\Omega/V)|\,\sin[\Omega t + \psi] \tag{223}$$

which expresses an output current proportional to $|S(\omega)|$ and preserving the phase ψ as well.

2. High-Frequency Spectrum Analyzers

An incoherent Bragg spectrum analyzer can be designed along exactly the same lines as that of Fig. 34, since each direction of diffraction at the modulator will result in a spot at an appropriate ω-position in plane 3.

A Bragg-region spectrum analyzer is also derivable from the processor of Fig. 33. If the slit is removed and replaced by a wide-aperture photo-detector, we have the system of Fig. 35. Light rays, in passing through the signal modulator, are bent away from their incident direction through angles

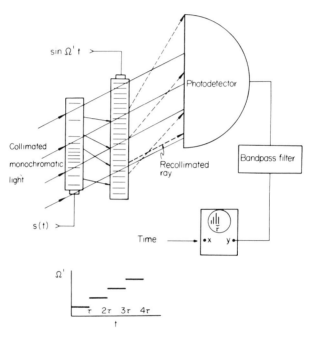

FIG. 35. Coherent Bragg-region spectrum analyzer based on the processor of Fig. 33. Here, $1/\tau$ is the repetition rate of $s(t)$.

proportional to the local acoustic frequency, $\theta = (K_0/k)\Omega/\Omega_0$. If a sinusoidal signal of frequency Ω' is applied to the second modulator, the rays will be rediffracted. For the frequencies such that $\Omega = \Omega'$, the twice-diffracted rays will be realigned with the incident direction, and heterodyne detection is possible. The band-pass filter will block all current components not centered on Ω_0. Thus, as the acoustic signal propagates down the first modulator, the output signal current will be proportional to the amplitude of the frequency component at Ω'. When the next signal pulse arrives, Ω' will have changed to Ω'' and the strength of the Ω'' component will be measured. As the reference modulator frequency is slowly swept over the signal band, a series of spikes of current will occur proportional to the spectrum of $s(t)$ at successive frequencies. Although Ω' is shown in Fig. 35 as changing in steps, a reference whose frequency changed slowly in time would also be suitable.

In this section, we have seen that a wide range of useful processing functions can be accomplished acoustooptically from frequencies of a few megahertz to frequencies in the gigahertz range. To make these approaches more attractive, advances will have to be made in developing acoustooptical materials with lower acoustical loss, increased interaction efficiency, and, in the case of crystalline media, larger physical size. Improved transducer technology is also desired. Uniformly illuminating large apertures with collimated, coherent light remains a difficult problem. Nevertheless, the ability of acoustooptical processors to operate in real time, over wide bandwidths, on complicated waveforms (often easily and rapidly changeable) makes them appealing candidates for many applications.

REFERENCES

Aas, H. G., and Erf, R. K. (1964). *J. Acoust. Soc. Am.* **36**, 1906–1913.

Adler, R. (1967). *IEEE Spectrum* **4**, 42–54.

Adrianova, I. I. (1962). *Opt. i Spektroskopiya* **12**, 99; and *Opt. Spectry.* (*USSR*) **12**, 48.

Adrianova, I. I. (1963). *Opt. Spectry* (*USSR*) **14**, 70–74.

Am. Inst. Phys. Handbook, 2nd Edition (1963). McGraw-Hill, New York.

Arm, M., Lambert, L., and Weissman, I. (1964). *Proc. IEEE* **52**, 842.

Atzeni, C., and Pantani, L. (1969a). *Proc. IEEE* **57**, 344–346.

Atzeni, C., and Pantani, L. (1969b). *Proc. IEEE* **57**, 1317–1319.

Auld, B. A., and Wilson, D. A. (1967a). *J. Appl. Phys.* **38**, 3331–3336.

Auld, B. A., and Wilson, D. A. (1967b). *Appl. Phys. Letters* **11**, 368–370.

Auth, D. C., and Mayer, W. G. (1967). *J. Appl. Phys.* **38**, 5138–5140.

Barrett, H. H., and Zahn, M. (1968). Raytheon Company, Research Division, Tech. Memo T-785. Waltham, Massachusetts.

Becker, H. E. R., Hanle, W., and Maercks, O. (1936). *Phys. Zeits.* **37**, 414.

Beiser, L. (1967). *IEEE J. Quantum Electronics* **QE-3**, 560–567.

Bergmann, L. (1938). "Ultrasonics and Their Scientific and Technical Applications." G. Bell and Sons, Ltd., London.

Bhatia, A. B., and Noble, W. J. (1953). *Proc. Roy. Soc.* (*London*) **A 220**: (I) 356–368; (II) 369–385.

Born, M., and Wolf, E. (1965). "Principles of Optics," 3rd edition. Pergamon Press, New York.

Brienza, M. J. (1968). *Appl. Phys. Letters* **12**, 181–184.

Brienza, M. J., and De Maria, A. J. (1966). *Appl. Phys. Letters* **9**, 312–314.

Brienza, M. J., and De Maria, A. J. (1967). *IEEE J. Quantum Electronics* **QE-3**, 567–575.

Brillouin, L. (1922). *Ann. Phys.* (*Paris*) **17**, 88.

Broneus, H. A., and Jenkins, W. H. (1960). *Proc. Nat. Electron. Conf.* **16**, 835–839.

Brugger, K. (1965). *J. Appl. Phys.* **36**, 759.

Caddes, D. E., and Wilkinson, C. D. W. (1966). *IEEE J. Quantum Electronics* **QE-2**, 330–331.

Caddes, D. E., Osternik, L. M., and Targ, R. (1968). *Appl. Phys. Letters* **12**, 74–76.

Cahen, O., Dieulesaint, E., and Torguet, R. (1968). *C. R. Acad. Sc. Paris* **266**, 1009–1011.

Carleton, H. R., and Maloney, W. T. (1967). *Proc. IEEE* **55**, 1077–1078.

Carleton, H. R., and Maloney, W. T. (1968). *Appl. Opt.* **7**, 1241–1243.

Carleton, H. R., and Soref, R. A. (1966). *Appl. Phys. Letters* **9**, 110–112.

Carleton, H. R., Maloney, W. T., and Meltz, G. (1969). *Proc. IEEE* **57**, 769–775.

Cohen, M. G., and Gordon, E. I. (1965). *Bell System Tech. J.* **44**, 693–721.

Collins, J. H., and Wilson, D. A. (1968). *Appl. Phys. Letters* **12**, 331–333.

Collins, J. H., Lean, E. G. H., and Shaw, H. J. (1967). *Appl. Phys. Letters* **11**, 240–242.

Cook, C. E., and Bernfeld, M. (1967). "Radar Signals." Academic Press, New York and London.

Cummins, H. Z., and Knable, N. (1963). *Proc. IEEE* **51**, 1246.

Cummins, H., Knable, N., Gampel, L., and Yeh, Y. (1963). *Appl. Phys. Letters* **2**, 62–64.

Cutrona, L. J. (1967). *IEEE International Conv. Record*, Part II, 99–105.

Cutrona, L. J., Leith, E. N., Palermo, L. J., and Procello, L. J. (1960). *IRE Trans. on Information Theory* **IT-6**, 386–400.

Damon, R. W. (1963). *In* "Magnetism" (G. T. Rado and H. Suhl, eds.), Chapter 11, Vol. I. Academic Press, New York and London.

David, E. (1937). *Phys. Zeits* **38**, 587–596.

Debye, P., and Sears, F. W. (1932). *Proc. Nat. Acad. Sci.* **18**, 409–414.

De Maria, A. J. (1963). *J. Appl. Phys.* **34**, 2984.

De Maria, A. J., and Gagosz, R. (1962). *Proc. IEEE* **50**, 1522.

De Maria, A. J., Gagosz, R., and Barnard, G. (1963). *J. Appl. Phys.* **34**, 453.

Distler, R. C. (1966). *J. Appl. Phys.* **37**, 3319–3320.

Dixon, R. W. (1967a). *IEEE J. Quantum Electronics* **QE-3**, 85–93.

Dixon, R. W. (1967b). *J. Appl. Phys.* **38**, 3634.

Dixon, R. W. (1967c). *J. Appl. Phys.* **38**, 5149–5153.

Dixon, R. W. (1967d). *Appl. Phys. Letters* **11**, 340–344.

Dixon, R. W., and Chester, A. N. (1966). *Appl. Phys. Letters* **9**, 190–192.

Dixon, R. W., and Cohen, M. G. (1966). *Appl. Phys. Letters* **8**, 205–207.

Dixon, R. W., and Gordon, E. I. (1967). *Bell System Tech. J.* **46**, 367–389.

Dixon, R. W., and Matthews, H. (1967). *Appl. Phys. Letters* **10**, 195–197.

Extermann, R., and Wannier, G. (1936). *Helv. Phys. Acta* **9**, 520.

Felstead, E. B. (1967). *IEEE Trans. on Aerospace and Electronic Systems* **AES-3**, 907–914.

Felstead, E. B. (1968). *Appl. Opt.* **7**, 105–108.

Foster, L. C., Ewy, M. D., and Crumly, C. B. (1965). *Appl. Phys. Letters* **6**, 6–8.

Gerig, J. S., and Montague, H. (1964). *Proc. IEEE* **52**, 1753.

Giarola, A. J., and Billeter, T. R. (1963). *Proc. IEEE* **51**, 1150–1151.

Goodman, J. W. (1968). "Introduction to Fourier Optics." McGraw-Hill, New York.

Goodwin, R. W., and Pedinoff, M. E. (1966). *Appl. Phys. Letters* **8**, 60–61.

Gordon, E. I. (1966a). *IEEE J. Quantum Electronics* **QE-2**, 104–105.

Gordon, E. I. (1966b). *Appl. Opt.* **5**, 1629–1639 and *Proc. IEEE* **54**, 1391–1401.

Gordon, E. I., and Cohen, M. G. (1965). *IEEE J. Quantum Electronics* **QE-1**, 191–198.

Hance, H. V., and Parks, J. K. (1965). *J. Acoust. Soc. Am.* **38**, 14–23.

Hance, H. V., Parks, J. K., and Tsai, C. S. (1967). *J. Appl. Phys.* **38**, 1981–1983.

Hargrove, L. E. (1962). *J. Acoust. Soc. Am.* **34**, 1547–1552.

Hargrove, L. E., and Achyuthan, K. (1965). *In* "Physical Acoustics" (W. P. Mason ed.) Chap. 12, Vol. II B. Academic Press, New York and London.

Hargrove, L. E., Fork, R. L., and Pollock, M. A. (1964). *Appl. Phys. Letters* **5**, 4–5.

Hiedemann, E. (1935). *Z. Phys.* **96**, 273.

Hiedemann, E. A., and Breazeale, M. A. (1959). *J. Opt. Soc. Am.* **49**, 372–375.

Hope, Lawrence L. (1968). *Phys. Rev.* **166**, 883–892.

Ingenito, F., Crandall, A. J., and Cook, B. D. (1967). *J. Acoust. Soc. Am.* **42**, 1178.

Ippen, E. P. (1967). *Proc. IEEE* **55**, 248–249.

Izzo, N. F. (1965). *Proc. IEEE* **53**, 1740–1741.

Jernigan, J. L. (1968). *Proc. IEEE* **56**, 374.

King, M., Bennett, W. R., Lambert, L. B., and Arm, M. (1967). *Appl. Opt.* **6**, 1367–1375.

Klein, W. R., and Cook, B. D. (1967). *IEEE Trans. Sonics Ultrasonics* **SU-14**, 123–134.

Klemens, P. G. (1965). *In* "Physical Acoustics" (W. P. Mason, ed.), Vol. III B, p. 201. Academic Press, New York and London.

Kolb, J., and Loeber, A. P. (1954). *J. Acoust. Soc. Am.* **26**, 249–251.

Korpel, A. (1966). *Appl. Phys. Letters* **9**, 425–427.

Korpel, A. (1968). *IEEE Trans. Sonics Ultrasonics* **SU-15**, 153–157.

Korpel, A., Adler, R., Desmares, P., and Smith, T. M. (1965). *IEEE J. Quantum Electronics* **QE-1**, 60–61.

Korpel, A., Adler, R., Desmares, P., and Watson, W. (1966). *Proc. IEEE* **54**, 1429–1437.

Korpel, A., Laub, L. J., and Sievering, H. C. (1967). *Appl. Phys. Letters* **10**, 295–297.

Krischer, C. (1968). *Appl. Phys. Letters* **13**, 310.

Krokstad, J., and Svaasand, L. O. (1967). *Appl. Phys. Letters* **11**, 155.

Lambert, L. B. (1965). "Optical Correlation" *in* "Modern Radar" (R. W. Berkowitz, ed.). Wiley, New York.

Lambert, L. B., Arm, M., and Aimette, A. (1965). *In* "Optical and Electro-Optical Information Processing," pp. 715–748. MIT Press, Cambridge, Massachusetts.

Landry, J., Powers, J., and Wade, G. (1969). *Appl. Phys. Letters* **15**, 186–188.

Lean, E. G. H. (1967). Microwave Labroatory Report No. 1543. W. W. Hansen Laboratories of Physics, Stanford University, Stanford, California.

Lean, E. G. H., and Tseng, C. C. (1969). *1969 IEEE Ultrasonics Symposium*, Paper A2, St. Louis, Missouri.

Lean, E. G. H., Quate, C. F., and Shaw, H. J. (1967). *Appl. Phys. Letters* **10**, 48–51.

Lee, H. W. (1938). *Nature* **192**, 59–62.

Liben, W., and Twigg, L. A., Jr. (1962). *J. Appl. Phys.* **33**, 249.

Lipnick, R., Reich, A., and Schoen, G. A. (1964). *Proc. IEEE* **52**, 853–854.

Lipnick, R., Reich, A., and Schoen, G. A. (1965). *Proc. IEEE* **53**, 321.

Lowenthal, S., and Belvaux, Y. (1967). *Appl. Phys. Letters* **11**, 49–51.

Lucas, R., and Biquard, P. (1932). *J. Phys. Radium* **3**, 464–477.

Maloney, W. T. (1969). *Appl. Opt.* **8**, 443–446.

Maloney, W. T., and Carleton, H. R. (1967). *IEEE Trans. Sonics Ultrasonics* **SU-14**, 135–139.

Maloney, W. T., and Gravel, R. L. (1969). *Proc. IEEE* **57**, 1332–1333.

Maloney, W. T., and Meltz, G. (1969). *Proc. IEEE* **57**, 1316–1317.

Maloney, W. T., Meltz, G., and Gravel, R. L. (1968). *IEEE Trans. Sonics Ultrasonics* **SU-15**, 167–172.

McMahon, D. H. (1967a). *IEEE Trans. Sonics Ultrasonics* **SU-14**, 103–108.

McMahon, D. H. (1967b). *Proc. IEEE* **55**, 1602–1612.

McMahon, D. H. (1968). *J. Acoust. Soc. Am.* **44**, 1007.

Mason, W. P. (1965). *In* "Physical Acoustics," Chapter 6, Vol. III B. Academic Press, New York and London.

Mayer, W. G., Lamers, G. B., and Auth, D. (1967). *J. Acoust. Soc. Am.* **42**, 1255–1257.

Melngailis, J., Maradudin, A. A., and Seeger, A. (1963). *Phys. Rev.* **131**, 1972–1975.

Meltz, G., and Maloney, W. T. (1968). *Appl. Opt.* **7**, 2091–2099.

Minkoff, J., Bennett, W. R., Lambert, L. B., Arm, M., and Bernstein, S. (1967). *Proc. Symp. Modern Optics*, pp. 703–715. Polytechnic Press, New York.

Mittra, R., and Ransom, P. L. (1967). *Proc. Symp. Modern Optics*, pp. 619–647. Polytechnic Press, New York.

Mueller, H. (1938). *Z. Krist.* **A99**, 122–141.

Okolicsanyi, F. (1937). *Wireless Engineer* **14**, 527–536.

Papoulis, A. (1968). "Systems and Transforms with Applications in Optics." McGraw-Hill, New York.

Parker, J. H., Jr., Kelly, E. F., and Bolef, D. I. (1964). *Appl. Phys. Letters* **5**, 7–9.

Pinnow, D. A., and Dixon, R. W. (1968). *Appl. Phys. Letters* **13**, 156–158.

Pinnow, D. A., Van Uitert, L. G., Warner, A. W., and Bonner, W. A. (1969). *Appl. Phys. Letters* **15**, 83–86.

Quate, C. F., Wilkinson, C. D. W., and Winslow, D. K. (1965). *Proc. IEEE* **53**, 1604–1623.

Raman, C. F., and Nath, N. S. N. (1935 a, b; 1936). *Proc. Indian Acad. Sci.*: I, **2**, 406–412; II, **2**, 413–430; III, **3**, 75–84; IV, **3**, 119–125.

Reintjes, J., and Schulz, M. B. (1968). *J. Appl. Phys.* **39**, 5254–5258.

Richardson, B. A., Thompson, R. B., and Wilkinson, D. C. W. (1967). *J. Acoust. Soc. Am.* **44**, 1608.

Robinson, D. M. (1939). *Proc. IRE* **27**, 483–486.

Salzmann, E., Plieninger, T., and Dransfeld, K. (1968). *Appl. Phys. Letters* **13**, 14.

Schaefer, A., and Bergmann, L. (1934). *Naturwiss.* **22**, 685.

Schulz, M. B., Holland, M. G., and Davis, L., Jr. (1967). *Appl. Phys. Letters* **11**, 237–240.

Siegman, A. E., Quate, C. F., Bjorkholm, J., and Francois, G. (1964). *Appl. Phys. Letters* **5**, 1–2.

Slobodin, L. (1963). *Proc. IEEE* **51**, 1782.

Slobodin, L. (1964). *Proc. IEEE* **52**, 842.

Slobodnik, A. J., Jr. (1969a). *Appl. Phys. Letters* **14**, 94.

Slobodnik, A. J., Jr. (1969b). IEEE Ultrasonics Symposium, Paper A4. St. Louis, Missouri.

Smith, A. W. (1967). *Appl. Phys. Letters* **11**, 1967.

Smith, A. W. (1968a). *Phys. Rev. Letters* **20**, 334–337.

Smith, A. W. (1968b). *IEEE Trans. Sonics Ultrasonics* **SU-15**, 161–167.

Smith, T. M., and Korpel, A. (1965). *IEEE J. Quantum Electronics* **QE-1**, 283–284.

Spencer, E. G., Lenzo, P. V., and Nassau, K. (1966). *IEEE J. Quantum Electronics* **QE-2**, 69–70.

Spencer, E. G., Lenzo, P. V., and Ballman, A. A. (1967). *Proc. IEEE* **55**, 2074–2108.

Squire, W. D., and Alsup, J. M. (1967). IEEE Ultrasonics Symposium, Paper C10. Vancouver, Canada.

Strauss, W. (1968). *In* "Physical Acoustics" (W. P. Mason, ed.), Vol. IV B. Academic Press, New York and London.

Thurston, R. N. (1964). *In* "Physical Acoustics" (W. P. Mason, ed.), Vol. I A, pp. 1–110. Academic Press, New York.

Thurston, R. N. (1965). *Proc. IEEE* **53**, 1320.

Tsai, C. S., and Hance, H. V. (1967). *J. Acoust. Soc. Am.* **42**, 1345–1347.

Turin, G. L. (1960). *IRE Trans. Information Theory* **IT-6**, 311–329.

Vander Lugt, A. (1966). *Proc. IEEE* **54**, 1055–1063.

Whitman, R., Korpel, A., and Lotsoff, S. (1967). *Proc. Symp. Modern Optics*, pp. 243–256. Polytechnic Press, New York.

Wilkinson, C. D. W., and Caddes, D. E. (1966). *J. Acoust. Soc. Am.* **40**, 498–499.

Wilmotte, R. M. (1963). U. S. Pat. 3,111,666.

Zahn, M. (1968). MIT Microwave and Quantum Magnetics Group, Dept. of Elec. Eng., Technical Report 20, Cambridge, Massachusetts.

Zworykin, V. K., and Morton, G. A. (1964). "Television." Wiley, New York.

Author Index

Numbers in italics refer to the pages on which the complete references are listed.

367

Subject Index